Advances in Intelligent Systems and Computing

Volume 256

Series Editor

Janusz Kacprzyk, Polish Academy of Sciences, Warsaw, Poland
e-mail: kacprzyk@ibspan.waw.pl

For further volumes:
http://www.springer.com/series/11156

About this Series

The series "Advances in Intelligent Systems and Computing" contains publications on theory, applications, and design methods of Intelligent Systems and Intelligent Computing. Virtually all disciplines such as engineering, natural sciences, computer and information science, ICT, economics, business, e-commerce, environment, healthcare, life science are covered. The list of topics spans all the areas of modern intelligent systems and computing.

The publications within "Advances in Intelligent Systems and Computing" are primarily textbooks and proceedings of important conferences, symposia and congresses. They cover significant recent developments in the field, both of a foundational and applicable character. An important characteristic feature of the series is the short publication time and world-wide distribution. This permits a rapid and broad dissemination of research results.

Mohammad S. Obaidat · Joaquim Filipe
Janusz Kacprzyk · Nuno Pina
Editors

Simulation and Modeling Methodologies, Technologies and Applications

International Conference, SIMULTECH 2012
Rome, Italy, July 28–31, 2012
Revised Selected Papers

Springer

Editors
Mohammad S. Obaidat
Monmouth University
New Jersey
USA

Joaquim Filipe
Polytechnic Institute of Setúbal/INSTICC
Setúbal
Portugal

Janusz Kacprzyk
Polish Academy of Sciences
Systems Research Institute
Warsaw
Poland

Nuno Pina
Superior School of Technology of Setúbal/IPS
Setúbal
Portugal

ISSN 2194-5357 ISSN 2194-5365 (electronic)
ISBN 978-3-319-03580-2 ISBN 978-3-319-03581-9 (eBook)
DOI 10.1007/978-3-319-03581-9
Springer Cham Heidelberg New York Dordrecht London

Library of Congress Control Number: 2013954678

Printed on acid-free paper

Springer is part of Springer Science+Business Media (www.springer.com)

Preface

This book includes extended and revised versions of a set of selected papers from the 2012 International Conference on Simulation and Modeling Methodologies, Technologies and Applications (SIMULTECH 2012) which was sponsored by the Institute for Systems and Technologies of Information, Control and Communication (INSTICC) and held in Rome, Italy. SIMULTECH 2012 was technically co-sponsored by the Society for Modeling & Simulation International (SCS), GDR I3, Lionphant Simulation, Simulation Team and IFIP and held in cooperation with AIS Special Interest Group of Modeling and Simulation (AIS SIGMAS) and the Movimento Italiano Modellazione e Simulazione (MIMOS).

This conference brings together researchers, engineers, applied mathematicians and practitioners interested in the advances and applications in the field of system simulation. We believe the papers here published, demonstrate new and innovative solutions, and highlight technical problems that are challenging and worthwhile.

SIMULTECH 2012 received 125 paper submissions from 38 countries in all continents. A double blind paper review was performed by the Program Committee members, all of them internationally recognized in one of the main conference topic areas. After reviewing, only 38 papers were selected to be published and presented as full papers, i.e. completed work (10 pages in proceedings/30' oral presentations) and 26 papers, describing work-in-progress, were selected as short papers for 20' oral presentation. Furthermore there were also 10 papers presented as posters. The full-paper acceptance ratio was thus 30%, and the total oral paper acceptance ratio was less than 44%.

The papers included in this book were selected from those with the best reviews taking also into account the quality of their presentation at the conference, assessed by the session chairs. Therefore, we hope that you find the papers included in this book interesting, and we trust they may represent a helpful reference.

We wish to thank all those who supported and helped to organize the conference. On behalf of the conference Organizing Committee, we would like to thank the authors, whose work mostly contributed to a very successful conference and the members of the Program Committee, whose expertise and diligence were instrumental to ensure the quality of final contributions. We also wish to thank all the members of the Organizing

Committee whose work and commitment were invaluable. Thanks also are due to the organizations that technically co-sponsored the conference. Last but not least, we would like to thank INSTICC for sponsoring and organizing the conference.

September 2013

Mohammad S. Obaidat
Joaquim Filipe
Janusz Kacprzyk
Nuno Pina

Organization

Conference Chair

Mohammad S. Obaidat	Monmouth University, USA

Program Co-chairs

Nuno Pina	EST-Setúbal/IPS, Portugal
Janusz Kacprzyk	Systems Research Institute - Polish Academy of Sciences, Poland

Organizing Committee

Helder Coelhas	INSTICC, Portugal
Andreia Costa	INSTICC, Portugal
Bruno Encarnação	INSTICC, Portugal
Carla Mota	INSTICC, Portugal
Raquel Pedrosa	INSTICC, Portugal
Vitor Pedrosa	INSTICC, Portugal
Cláudia Pinto	INSTICC, Portugal
Susana Ribeiro	INSTICC, Portugal
José Varela	INSTICC, Portugal
Pedro Varela	INSTICC, Portugal

Program Committee

Erika Ábrahám, Germany
Manuel Alfonseca, Spain
Giulia Andrighetto, Italy
Jan Awrejcewicz, Poland
Gianfranco Balbo, Italy
Simonetta Balsamo, Italy
Isaac Barjis, USA
Ana Isabel Barros, The Netherlands
Fernando Barros, Portugal
Ildar Batyrshin, Mexico
Lucian Bentea, Norway
Marenglen Biba, Albania
Louis Birta, Canada
Ipek Bozkurt, USA

Christiano Braga, Brazil
Srinivas Chakravarthy, USA
Naoufel Cheikhrouhou, Switzerland
E. Jack Chen, USA
Jean-Yves Choley, France
Ricardo Choren, Brazil
Lawrence Chung, USA
Bruno Ciciani, Italy
Claudio Cioffi-Revilla, USA
Kendra Cooper, USA
Andrea D'Ambrogio, Italy
Gabriella Dellino, Italy
Atakan Dogan, Turkey
Werner Dubitzky, UK
Stephan Eidenbenz, USA
Andreas Ernst, Germany
Roland Ewald, Germany
Denis Filatov, Mexico
Paul Fishwick, USA
Ian Flood, USA
José Manuel Galán, Spain
Nicolas R. Gauger, Germany
Nikolaos Geroliminis, Switzerland
Charlotte Gerritsen, The Netherlands
Daniele Gianni, The Netherlands
Brian Goldiez, USA
Alexandra Grancharova, Bulgaria
Zhi Han, USA
Monika Heiner, Germany
Brian Hollocks, UK
Polly Huang, Taiwan
Eric S. Imsand, USA
Mhamed Itmi, France
Mura Ivan, Italy
Luis Izquierdo, Spain
Segismundo Samuel Izquierdo, Spain
András Jávor, Hungary
Hans Jense, The Netherlands
Cara Kahl, Germany
Korina Katsaliaki, Greece
Peter Kemper, USA
Juš Kocijan, Slovenia
Petia Koprinkova-Hristova, Bulgaria
Samuel Kounev, Germany
Raymond Kristiansen, Norway
Jirí Kunovský, Czech Republic
Stephanie Jane Lackey, USA
Béla Lakatos, Hungary
Kennard Laviers, USA
Loo Hay Lee, Singapore
Sanghyun Lee, USA
Johannes Lüthi, Austria
Emilio Jiménez Macías, Spain
Carla Martin-Villalba, Spain
Radek Matušu, Czech Republic
Yuri Merkuryev, Latvia
Adel Mhamdi, Germany
Qi Mi, USA
Federico Milani, Italy
Gabriele Milani, Italy
Michael Möhring, Germany
Jairo Montoya-Torres, Colombia
Jairo R. Montoya-Torres, Colombia
Spyridon Mouroutsos, Greece
Navonil Mustafee, UK
Ines Mynttinen, Germany
Àngela Nebot, Spain
Manuel Noguera, Spain
Michael J. North, USA
James J. Nutaro, USA
Peter Csaba Ölveczky, Norway
Stephan Onggo, UK
Ioannis Paraskevopoulos, UK
James Parker, Canada
Ana Peleteiro, Spain
Petr Peringer, Czech Republic
L. Felipe Perrone, USA
H. Pierreval, France
Katalin Popovici, USA
Francesco Quaglia, Italy
Jacinto A. Dávila Quintero, Venezuela
Stanislav Racek, Czech Republic
Manuel Resinas, Spain
M. R. Riazi, Kuwait
José Risco-Martín, Spain
Stewart Robinson, UK
Theresa Roeder, USA
Paolo Romano, Portugal
Maarouf Saad, Canada
Jean-François Santucci, France

Rune Schlanbusch, Norway
Avraham Shtub, Israel
Yuri Skiba, Mexico
Jaroslav Sklenar, Malta
Yuri Sotskov, Belarus
James C. Spall, USA
Flaminio Squazzoni, Italy
Giovanni Stea, Italy
Steffen Straßburger, Germany
Nary Subramanian, USA
Claudia Szabo, Australia
Antuela A. Tako, UK
Elena Tànfani, Italy
Pietro Terna, Italy
Emmanuel Tsekleves, UK
Bruno Tuffin, France
Alfonso Urquia, Spain
Mayerlin Uzcategui, Venezuela
Timo Vepsäläinen, Finland
Anil Vullikanti, USA
Natalie van der Wal, The Netherlands
Frank Werner, Germany
Philip A.Wilsey, USA
Muzhou Xiong, China
Nong Ye, USA
Levent Yilmaz, USA
Gregory Zacharewicz, France
František Zboril, Czech Republic
Durk Jouke van der Zee, The Netherlands
Yabing Zha, China
Lin Zhang, China
Laurent Zimmer, France
Armin Zimmermann, Germany
Konstantinos Zografos, Greece

Auxiliary Reviewers

Rogerio Batista, Brazil
Xin Chen, Germany
Florian Corzilius, Germany
Nils Jansen, Germany
Ely Miranda, Brazil

Invited Speakers

David M. Nicol	University of Illinois, Urbana-Champaign, USA
Tuncer Ören	University of Ottawa, Canada
Simon Taylor	Brunel University, UK
Anthony John Jakeman	Australian National University, Australia

Contents

Part I

Invited Papers

The Richness of Modeling and Simulation and an Index of Its Body of Knowledge

Tuncer Ören

University of Ottawa, School of Electrical Engineering and Computer Science, Ottawa, Ontario, K1N 6N5 Canada
oren@eecs.uottawa.ca

Abstract. The richness of modeling and simulation (M&S) and its increasing importance are emphasized. The three aspects of professionalism as well as stakeholders of M&S are documented. Work being done by the author on M&S body of knowledge (BoK) is outlined. Several other BoK and M&S BoK studies are referred to. The conclusions section includes the emphases that wide-spread application and ever increasing importance of modelling and simulation necessitate an agreed on body of knowledge index and its elaboration as well as the preservation of the integrity of the M&S discipline.

Keywords: Richness of Modeling and Simulation, Importance of Modeling and Simulation, Simulation Body of Knowledge Index, Simulation Terminology, Exascale Simulation, Ethics in Simulation.

1 Introduction

Modeling and simulation (M&S) discipline provides a powerful and vital infrastructure for many disciplines as well as for large number of application areas [1]. M&S –like mathematics– has its own core knowledge which benefits from appropriate research and development and is also essential for many disciplines.

Several aspects of the richness and comprehensiveness of M&S are elaborated in section 2. In section 3, clarification of some aspects of the importance of M&S is offered. In section 4, the importance of the synergy of M&S and education is highlighted. Aspects of professionalism in as well as stakeholders of M&S are reviewed in sections 5 and 6. The importance of M&S necessitates development of a proper body of knowledge (BoK) index. In section 7, highlights of an Index of an M&S body of knowledge study is given. Section 8 consists of the conclusions and future work. Two indices complete the chapter: A list of over 500 terms denoting several types of simulation as a testimony of the richness of the M&S discipline and an M&S BoK index being developed by the author.

2 Richness of M&S

Compared to some traditional disciplines such as mathematics, physics, and astronomy, modeling and simulation is a young discipline. However, it has been

M.S. Obaidat et al. (eds.), *Simulation and Modeling Methodologies, Technologies and Applications*, Advances in Intelligent Systems and Computing 256,
DOI: 10.1007/978-3-319-03581-9_1, © Springer International Publishing Switzerland 2014

maturing steadily [2]. Several articles and book chapters depict comprehensive views of many aspects of M&S [3-5]. A recent article provides a systematic collection of about 100 definitions of M&S and emphasizes some of the many aspects of M&S [6]. Another article offers a critical review of the definitions and shows that some of the legacy definitions are not appropriate anymore [7].

Two aspects of simulation are particularly important: experimentation and experience. From the point of view of *experimentation*, "Simulation is performing goal directed experiments with models of dynamic systems" [7]. A taxonomy of experiments as well as some additional clarifications about experiments are also given by Ören [7]. From the point of view of *experience*, simulation is providing experience under controlled conditions for training or for entertainment. For *training* purposes, simulation is providing experience under controlled conditions for gaining / enhancing competence in one of the three types of skills: (1) motor skills (by virtual simulations), (2) decision and/or communication skills (by constructive simulations; serious games), and (3) operational skills (by live simulations)" Ören [7]. For *amusement* purposes: "Simulation is providing experience for entertainment purpose (gaming simulations). Some aspects of gaming simulation make it a source of inspiration for education as well as for serious games used for training purposes. These aspects include advanced visualization techniques and specification of environments and scenarios. Gaming simulation can also be combined to explore experimentation for scientific research. An example is eyewire project of MIT which is gamified for crowdsourcing to have large cooperation of simulation game players to explore how connectomes of retina work [8-9].

M&S has many other aspects, each of which covering a wealth of concepts [10]. However, in this chapter, only the aspects pertinent to experimentation and experience for training purposes are elaborated on. As a testimony of the richness of simulation, we can cite a large number of types of simulation as well as M&S terms. In Appendix 1, one can see over 500 terms which denote mostly application area-independent types of simulation. An ontology-based dictionary of these terms is planned to be developed. An ontology-based dictionary is a relational dictionary built on top of a taxonomy of the terms. An example is an ontology-based dictionary of terms of machine understanding where over 60 terms are covered [11].

Another evidence of the richness of the M&S discipline is the number of terms it uses. An early trilingual (English-French-Turkish) modeling and simulation dictionary that the author was involved included about 4000 terms [12]. A bilingual (Chinese-English, English-Chinese) version prepared with 30 Chinese contributors has over 9000 terms [13].

3 Importance of M&S

M&S provides very important infrastructure for hundreds of application areas in science, engineering, social as well as health sciences. In this chapter, the importance of M&S is elaborated in three sets of disciplines, i.e., simulation-based science and engineering, simulation-based social sciences, and computational neuroscience. Several other fields, such as simulation-based (civilian as well as military) training and simulation-based learning benefit from the contribution of simulation.

Possibilities that extreme scale simulation offers, provide another dimension to the importance of M&S.

3.1 The Impact of Extreme Scale Computing on Simulation

Extreme scale computers are high-speed computers such as teraflop, petaflop, or exaflop computers. They perform, respectively, 10^{12} (i.e., one thousand times one billion), 10^{15} (i.e., one million times one billion), or 10^{18} (i.e., one billion times one billion) floating point operations per second. Simulations performed on these types of computers are called, extreme-scale simulation, terascale simulation, petascale simulation, or exascale simulation, respectively.

Simulation on exascale computer –or exascale simulation in short– is anticipated to contribute mainly in the following application areas: aerospace, airframes and jet turbines, astrophysics (including cosmology and the nature of dark energy and black holes); biological and medical systems, climate and weather, energy (including combustion, nuclear fusion, solar energy, and nuclear fission), materials science, national security, socioeconomic modelling; and in the following technical areas: mathematics and algorithms, software, hardware, and cyber infrastructure [14].

The ASCAC report also examines what applications may be transformed by going to the exascale simulations [14, p. 25]:

> "*From biology to nuclear engineering, computing at the exascale promises dramatic advances in our capabilities to model and simulate complex phenomena, at levels of fidelity that have the potential to dramatically change both our understanding and our ability to comprehend. Thus, there are almost certain to be great benefits to going to the exascale*" [14].

While exaflop computers are in active research and development, petaflop supercomputers already exist. As of November 2011, "*Japan's K Computer maintained its position atop the newest edition of the TOP500 List of the world's most powerful supercomputers, thanks to a full build-out that makes it four times as powerful as its nearest competitor. Installed at the RIKEN Advanced Institute for Computational Science (AICS) in Kobe, Japan, the K Computer achieved an impressive 10.51 petaflop/s . . . in second place is the Chinese Tianhe-1A system with 2.57 petaflop/s performance*" [15]. As a practical importance of petascale simulation, one can point out that, if one billion entities are represented in a simulation model, every second, over a million operations can be performed for each object represented.

3.2 Simulation-Based Science and Engineering

Simulation-based engineering science (SBES) is a well established and important concept [16]. The major findings and principal recommendations show the crucial importance of simulation in all branches of engineering and engineering applications:

> "*SBES is a discipline indispensable to the nation's continued leadership in science and engineering. It is central to advances in biomedicine, nanomanufacturing, homeland security, microelectronics, energy and environmental sciences, advanced materials, and product development. There is ample evidence that*

developments in these new disciplines could significantly impact virtually every aspect of human experience." [16, p. xvi]. "*Meaningful advances in SBES will require dramatic changes in science and engineering education*" [16, p. 56].

Interested readers may benefit reading the report. Here, some points from the conclusions part of the report are highlighted:

"*First, computer modeling and simulation will allow us to explore natural events and engineered systems that have long defied analysis, measurement, and experimental methodologies, . . .*

Second, modeling and simulation will have applications across technologies—from microprocessors to the infrastructure of cities.

. . .

Fifth, modeling and simulation will expand our ability to cope with problems that have been too complex for traditional methods. . . .

Sixth, modeling and simulation will introduce tools and methods that apply across all engineering disciplines—electrical, computer, mechanical, civil, chemical, aerospace, nuclear, biomedical, and materials science. . . .

3.3 Simulation-Based Social Sciences

Simulation-based social sciences include anthropology, archaeology, economics, geography, government, linguistics, management, political science, and sociology. The interest areas of simulation-based social sciences of the Centre for Research on Simulation in the Social Sciences (CRESS) [17] are specified as follows:

"*Simulation is a novel research method in most parts of the social sciences, including sociology, political science, economics, anthropology, geography, archaeology and linguistics. It can also be the inspiration for new, process-oriented theories of society.*"

A selection from the aims of World Congress on Social Simulation (WCSS 2012) [18] reads as follows:

"*Social sciences are moving in a direction in which their various constituent parts are sharing a common set of foundations, languages and platforms, which makes the social sciences be unprecedentedly behavioral, algorithmic and computational. At the turn of the 21st century, a group of computer scientists and social scientists worked together to initiate new series of conferences and to establish new academic organizations to give momentum to this emerging integration now known as computational social sciences. . . . WCSS is sponsored by the three regional scientific associations on social simulations*:" The European Social Simulation Association (ESSA) [19], Pacific Asian Association for Agent-based Approach in Social Systems Sciences (PAAA) [20] and Computational Social Science Society of the Americas (CSSSA) [21].

3.4 Computational Neuroscience

Computational neuroscience is a subfield of neuroscience that uses mathematical methods to simulate and understand the function of the nervous system [22]. "A connectome is a comprehensive map of neural connections in the brain" [23]. "The Human Connectome Project aims to provide an unparalleled compilation of neural data, an interface to graphically navigate this data and the opportunity to achieve never before realized conclusions about the living human brain" [24]. Advanced simulation is an integral part of the connectome project. A recent article about the Neuropolis project, to be centered in Switzerland and preselected by European Union, presents it as the future capital of the virtual brain [25].

4 M&S and Education

An important corollary of the importance of M&S is proper education and training in modeling and simulation at every level, starting at primary and secondary education to be followed by education at colleges and universities at undergraduate, graduate, and post graduate levels. Vocational training in modeling and simulation is also of particular importance. There are already several graduate M&S degree programs to educate future simulationists. However, for future professionals such as all types of engineers, scientists, including social scientists, proper M&S education will definitely be an asset [26-27]. It is expected that the education of future policy makers and public administrators may also benefit from several possibilities simulation offer. Lack of proper simulation-based professional training may even be invitation to disasters. A recent contribution to university-level education in M&S with several current references is done by Mielke et al. [28]. Another slightly dated report describes the contents of a Microsoft Access database developed in support of the Workforce Modeling and Simulation Education and Training for Lifelong Learning project. The catalogue contains searchable information about 253 courses from 23 U.S. academic institutions" [29].

5 Aspects of Professionalism in M&S

Three aspects of professionalism in M&S are activities, knowledge, and conduct and monitoring [30].

M&S Activities. Three groups of activities are involved in professionalism of M&S. (1) Generation of products, services, and/or tools to solve problems. This is normally done by industry. M&S activities also include generation of simulation-based solutions. (2) Generation & dissemination of knowledge. This is normally done by academia and R&D establishments. (3) Funding which is done by owner(s) of the project, governmental agencies, users, or interest groups.

Knowledge. Five types of knowledge are needed for professional M&S activities. (1) The core knowledge of the M&S discipline, i.e., M&S Body of Knowledge (M&S BoK). It is elaborated in section 7 and an Index being developed by the author is given as Appendix 2. (2) Knowledge of relevant Science (including system science), Engineering (including systems engineering), and Technology. (3) Knowledge of business and industry. (4) Knowledge of application area(s). (5) Knowledge of how to behave, i.e., code of professional ethics (Simulation_ethics).

Monitoring and Certification. Two types of monitoring are needed. (1) Professional and ethical conduct (both voluntary (responsibility) and required (accountability)). (2) Certification of professionalism of: (2.1) Individuals, as M&S professionals, (2.2) Companies for their maturity levels (yet to be specified). (3) Educational programs and institutions.

6 Stakeholders of M&S

M&S stakeholders include individuals, institutions, discipline and market, and countries [30].

Individuals include: Researchers/educators, practitioners, experienced as well as novice learners, customers and users of products/services, people (to be) affected by simulation projects (done or not yet done), industrialists, and technicians.

Institutions include: Customer or user organizations (that may be government organizations as well as profit and non-profit organizations), agencies for licensing or certification of individuals and organizations, funding agencies, professional societies, standardization organizations, educational institutions, industrial or professional groups and centers, and commercial organizations. The fact that there are over 100 associations and well over 40 groups and centers is also a good indicator of the variety of stakeholders of M&S [31].

Discipline and Market

Countries. A higher level stakeholder is "countries" which may have or may acquire leading edge superiority over other countries by exploring possibilities offered by extreme scale simulation, especially by exascale simulation. But not limited to this possibility alone.

7 M&S Body of Knowledge

7.1 M&S Bok: Preliminary

A body of knowledge (BoK) of a discipline is “structured knowledge that is used by members of a discipline to guide their practice or work” [32]. "While the term *body of knowledge* is also used to describe the document that defines that knowledge – the *body of knowledge* itself is more than simply a collection of terms; a professional reading list; a library; a website or a collection of websites; a description of professional functions; or even a collection of information. It is the accepted ontology for a specific domain" [33]. It is worthwhile underlining the fact that a BoK of a specific domain or discipline is its ontology. This fact necessitates that the development of a BoK should be systematic. A BoK Index is a set of systematically organized pointers to the content of a BoK. Desired BoK Index features include: "*(1) Supporting a variety of users within the M&S Community of Practice (CoP). (2) Identifying and providing access to BoK topics/content. (3) Providing configuration managed views to content that changes over time*" [34].

7.2 Bok of Other Disciplines

It would be useful to have an idea about the body of knowledge studies of other disciplines. Over 30 such BoK and their URLs are listed at the M&S BoK website being developed and maintained by Ören [35]. These BoK studies are grouped under the following categories: Business/Management, Civil Engineering, Database, Family and Consumer Sciences, Geography, Mechanical Engineering, Medicine, Project Management, Quality, Research Administration, Safety, Software Engineering / Computer Science, Systems Engineering, Usability, and Utility Infrastructure Regulation. Given that many of these disciplines/fields have more than one BoK is indicative that M&S may also have more than one BoK. This requirement may stem especially from the domain of application. However, at a higher level, the identification of a comprehensive view of M&S and elaboration of several of its aspects may be useful to avoid the problems of having narrow vision.

7.3 M&S Bok: Previous and Ongoing Studies/Activities

Previous M&S BoK studies, developed by domain experts cast light on several aspects of M&S [36-40]. A recent M&S BoK is developed for DoD [41]. As a distinguishing feature from a large number of BoK studies of other disciplines, the M&S BoK developed for DoD combines Bloom's taxonomy of learning with the BoK. However, the combination of the M&S BoK with Bloom's taxonomy detracts the attention from what is the core knowledge of M&S and confuses them with the levels of knowledge needed by different categories of users.

7.4 M&S Bok Being Developed by the Author

A website has been developed and maintained by Ören for an M&S BoK Index [35]. Some of the work done by Ören –alone or with colleagues– include the following [30, 36, 42-45]. The M&S BoK being developed by Ören consists of six parts: Background, core areas, supporting domains, business/industry, education/training, and relevant references [35] and is given as Appendix 2. Especially the Business/Industry section benefited from the previous work [46].

8 Conclusions and Future Work

M&S provides a vital infrastructure for large number of disciplines and application areas. With the advancements in high-speed computers, new vistas are being opened for M&S to tackle problems which would be unimaginable a few decades ago. The synergy of systems engineering, software agent and M&S opens new possibilities for advanced simulation [47]. The richness of the M&S field is well documented by its many types, by its own rich and discriminating terminology, and by the content of the M&S BoK studies.

Theoretical basis of M&S was laid down by Zeigler [48]. The Discrete Event System Specification (DEVS) formalism that he created is a well established theoretical basis for discrete event system simulation and has already many varieties [49]. There are many other publications about theories of simulation models, e.g., Barros [50]. A recent article elaborates on some aspects of axiomatic system theoretical foundations of M&S [51].

A multilingual (English-French-Italian-Spanish-Turkish) M&S dictionary is being developed by international co-operation of over 70 simulationists [52]. (M&S dictionary project). The author is committed to enhance the M&S BoK study that he started; and already an invitation for the final phases of its preparation is open [45]. Wide-spread application and ever increasing importance of modelling and simulation necessitate the preservation of its integrity. The word integrity is used as defined in Merriam-Webster: "an uncompromising adherence to a code of moral, artistic, or other values: utter sincerity, honesty, and candor: avoidance of deception, expediency, artificiality, or shallowness of any kind." Such code of ethics already exists for simulation and is adopted by many concerned groups [53]. It is expected that the list of adherent groups will grow.

References

1. Ören, T.I.: Uses of Simulation. In: Sokolowski, J.A., Banks, C.M. (eds.) Principles of Modeling and Simulation: A Multidisciplinary Approach, All Chapters by Invited Contributors, ch. 7, pp. 153–179. John Wiley and Sons, Inc., New Jersey (2009a)
2. Ören, T.I.: Keynote Article. Maturing Phase of the Modeling and Simulation Discipline. In: Proceedings of ASC - Asian Simulation Conference 2005 (The Sixth International Conference on System Simulation and Scientific Computing, ICSC 2005 (2005a)
3. Ören, T.I.: The Importance of a Comprehensive and Integrative View of Modeling and Simulation. In: Proceedings of the 2007 Summer Computer Simulation Conference, San Diego, CA, July 15-18 (2007)
4. Ören, T.I.: Modeling and Simulation: A Comprehensive and Integrative View. In: Yilmaz, L., Ören, T.I. (eds.) Agent-Directed Simulation and Systems Engineering. Wiley Series in Systems Engineering and Management, pp. 3–36. Wiley, Berlin (2009b)
5. Ören, T.I.: Simulation and Reality: The Big Picture (Invited paper for the inaugural issue) International Journal of Modeling, Simulation, and Scientific Computing (of the Chinese Association for System Simulation - CASS) 1(1), 1–25 (2010), `http://dx.doi.org/10.1142/S1793962310000079`
6. Ören, T.I.: The Many Facets of Simulation through a Collection of about 100 Definitions. SCS M&S Magazine 2(2), 82–92 (2011b)
7. Ören, T.I.: A Critical Review of Definitions and About 400 Types of Modeling and Simulation. SCS M&S Magazine 2(3), 142–151 (2011c)
8. Anthony, S.: MIT crowdsources and gamifies brain analysis, `http://www.extremetech.com/extreme/117325-mit-crowdsources-and-gamifies-brain-analysis`
9. Eyewire, `http://eyewire.org/`

10. Ören, T., Yilmaz, L.: Philosophical Aspects of Modeling and Simulation. In: Tolk, A. (ed.) Ontology, Epistemology, & Teleology for Model. & Simulation. ISRL, vol. 44, pp. 157–172. Springer, Heidelberg (2013)
11. Ören, T.I., Ghasem-Aghaee, N., Yilmaz, L.: An Ontology-Based Dictionary of Understanding as a Basis for Software Agents with Understanding Abilities. In: Proceedings of the Spring Simulation Multiconference (SpringSim 2007), Norfolk, VA, March 25-29, pp. 19–27 (2007) (ISBN: 1-56555-313-6)
12. Ören, T.I., et al.: Modeling and Simulation Dictionary: English-French-Turkish, Marseille, France, p. 300 (2006) ISBN: 2-9524747-0-2
13. Li, B., Ören, T.I., Zhao, Q., Xiao, T., Chen, Z., Gong, G., et al.: Modeling and Simulation Dictionary: Chinese-English, English Chinese - (about 9000 terms). Chinese Science Press, Beijing (2012) ISBN 978-7-03-034617-9
14. ASCAC. The Opportunities and Challenges of Exascale Computing. Summary Report of the Advanced Scientific Computing Advisory Committee (ASCAC) Subcommitte. U.S. Department of Energy, Office of Science (2010), http://science.energy.gov/~/media/ascr/as-cac/pdf/reports/Exascale_subcommittee_report.pdf
15. Top 500 Supercomputer Sites, http://www.top500.org/
16. Oden, J.T., et al.: Simulation-based Engineering Science –Revolutionizing Engineering Science through Simulation. Report of the National Science Blue Ribbon Panel on Simulation-based Engineering Science, NSF, USA (2006), http://www.nsf.gov/pubs/reports/sbes_final_report.pdf
17. CRESS – Centre for Research on Simulation in the Social Sciences, http://cress.soc.surrey.ac.uk/web/home
18. WCSS 2012 - World Congress on Social Simulation, September 4-7, Taipei, Taiwan (2012), http://www.aiecon.org/conference/wcss2012/index.htm
19. ESSA – the European Social Simulation Association, http://www.essa.eu.org/9.Eyewire, http://eyewire.org/
20. AAA – Pacific Asian Association for Agent-based Approach in Social Systems Sciences, http://www.paaa.asia/
21. CSSSA – Computational Social Science Society of the Americas, http://computationalsocialscience.org/
22. Scholarpedia. Encyclopedia of Computational Neuroscience, http://www.scholarpedia.org/article/Encyclopedia_of_computational_neuroscience
23. Wiki-connectome. Connectome, http://en.wikipedia.org/wiki/Connectome
24. Human Connectome Project, http://www.humanconnectomeproject.org/
25. Spencer, E.: Neuropolis, future capitale du cerveau virtuel. Sciences et Avenir 790, 8–12 (2012)
26. Kincaid, J.P., Westerlund, K.K.: Simulation in Education and Training. In: Proceedings of the 2009 Winter Simulation Conference, pp. 273–280 (2009), http://www.informs-simulation.org/wsc09papers/024.pdf
27. Sokolowski, J.A., Banks, C.M.: The Geometric Growth of M&S Education: Pushing Forward, Pushing Outward. SCS M&S Magazine 1(4) (2010)
28. Mielke, R.R., Leathrum Jr., J.F., McKenzie, F.D.: A Model for University-Level Education in Modeling and Simulation. MSIAC M&S Journal, Winter edition, 14–23 (2011), http://www.dod-msiac.org/pdfs/journal/MS%20Journal%202011%20winter.pdf

29. Catalano, J., Didoszak, J.M.: Workforce Modeling & Simulation Education and Training for Lifelong Learning: Modeling & Simulation Education Catalog; NPS-SE-07-M01. Naval postgraduate School, Monterey (2007), http://edocs.nps.edu/npspubs/scholarly/TR/2007/NPS-SE-07-M01.pdf
30. Ören, T.I.: A Basis for a Modeling and Simulation Body of Knowledge Index: Professionalism, Stakeholders, Big Picture, and Other BoKs. SCS M&S Magazine 2(1), 40–48 (2011a)
31. M&S Associations, http://www.site.uottawa.ca/~oren/links-MS-AG.htm
32. Ören, T.I.: Body of Knowledge of Modeling and Simulation (M&SBOK): Pragmatic Aspects. In: Proc. EMSS 2006 - 2nd European Modeling and Simulation Symposium, Barcelona, Spain, October 4-6, pp. 2006–2002 (2006)
33. Wiki-BoK (updated March 18, 2012), http://en.wikipedia.org/wiki/Body_of_Knowledge
34. Lacy, L.W., Waite, B.: Modeling and Simulation Body of Knowledge (BoK) Index Prototyping Effort Status Report. Presentation at SimSummit (2011), http://www.simsummit.org/Simsummit/SimSummit11/BoK%20F11%20SIW%20082411.pdf (December 1, 2011)
35. Modeling and Simulation Body of Knowledge (M&S BoK) – Index (by T. I. Ören) v.11, http://www.site.uottawa.ca/~oren/MSBOK/MSBOK-index.htm
36. Birta, L.G.: The Quest for the Modelling and Simulation Body of Knowledge. In: Keynote presentation at the Sixth Conference on Computer Simulation and Industry Applications, February 19-21. Instituto Tecnologico de Tijuana, Mexico (2003), http://www.site.uottawa.ca/~lbirta/pub2003-02-Mex.htm
37. Elzas, M.S.: The BoK Stops Here! Modeling & Simulation 2(3) (July/September 2003)
38. Loftin, B.R., et al.: Modeling and Simulation Body of Knowledge (BOK) and Course Overview. Presentation at DMSO Internal Program Review (2004)
39. Petty, M., Loftin, B.R.: 2004. Modeling and Simulation "Body of Knowledge" Version 5b (April 17, 2004)
40. Waite, W., Skinner, J.: Body of Knowledge Workshop, 2003 Summer Computer Simulation Conference (2003)
41. M&S BoK_DoD. USA Department of Defense, Modeling and Simulation Body of Knowledge (BOK) (2008), http://www.msco.mil/documents/_25_M&S%20BOK%20-%2020101022%20Dist%20A.pdf
42. Lacy, L.W., Gross, D.C., Ören, T.I., Waite, B.: A Realistic Roadmap for Developing a Modeling and Simulation Body of Knowledge Index. In: Proceedings of SISO (Simulation Interoperability Standards Organization) Fall SIW (Simulation Interoperability Workshop) Conference, Orlando, FL, September 20-24 (2010)
43. Ören, T.I.: Invited Tutorial. Toward the Body of Knowledge of Modeling and Simulation (M&SBOK). In: Proc. of I/ITSEC (Interservice/Industry Training, Simulation Conference), Orlando, Florida, November 28-December 1, pp. 1–19 (2005); paper 2025
44. Ören, T.I., Waite, B.: Need for and Structure of an M&S Body of Knowledge. In: Tutorial at the I/ITSEC(Interservice/Industry Training, Simulation Conference), Orlando, Florida, November 26-29 (2007)
45. Ören, T.I., Waite, B.: Modeling and Simulation Body of Knowledge Index: An Invitation for the Final Phases of its Preparation. SCS M&S Magazine 1(4) (October 2010)

46. Waite, W., Ören, T.I.: A Guide to the Modeling and Simulation Body of Knowledge Index (Draft Edition – version 1.5). Prepared as an activity of the SimSummit Round Table (2008), http://www.sim-summit.org
47. Yilmaz, L., Ören, T.I. (eds.): All Chapters by Invited Contributors. Agent-Directed Simulation and Systems Engineering. Wiley Series in Systems Engineering and Management, p. 520. Wiley, Berlin (2009)
48. Zeigler, B.P.: Theory of Modelling and Simulation. Wiley, New York (1976)
49. Wiki_DEVS, http://en.wikipedia.org/wiki/DEVS
50. Barros, F.J.: Modeling Formalisms for Dynamic Structure Systems. ACM Transactions on Modeling and Computer Simulation 7(4), 501–515 (1997)
51. Ören, T.I., Zeigler, B.P.: System Theoretic Foundations of Modeling and Simulation: A Historic Perspective and the Legacy of A. Wayne Wymore. Special Issue of Simulation – The Transactions of SCS 88(9), 1033–1046 (2012), doi:10.1177/0037549712450360
52. M&S dictionary project, http://www.site.uottawa.ca/~oren/SCS_MSNet/simDic.htm
53. Simulation_ethics, http://www.scs.org/ethics

Appendix 1

Over 500 terms denoting several types of simulation

3d simulation
A--
ab initio simulation
abstract simulation
academic simulation
accurate simulation
activity-based simulation
ad hoc distributed simulation
adaptive simulation
adaptive system simulation
adiabatic system simulation
advanced simulation
advanced distributed simulation
advanced numerical simulation
agent simulation
agent-based simulation
agent-based participatory simulation
agent-controlled simulation
agent-coordinated simulation
agent-directed simulation
agent-initiated simulation
agent-monitored simulation
agent-supported simulation
aggregate level simulation
autotelic system simulation
B--
backward simulation
base case simulation
baseline simulation
bio-inspired simulation
biologically-inspired simulation
bio-nano simulation
block-oriented simulation
bond-graph simulation
branched simulation
built-in simulation
C--
case-based simulation
cellular automaton simulation
classical simulation
closed-form simulation
closed-loop simulation
cloud simulation
cloud-based simulation
cluster simulation
coercible simulation
cognitive simulation
cokriging simulation
computer-aided simulation
computer-based simulation
computerized simulation
computer-mediated simulation
concurrent simulation
condensed-time simulation
conditional simulation
conjoint simulation
conservative simulation
constrained simulation
constructive simulation
constructive training simulation
continuous simulation
continuous-change simulation
continuous-system simulation
continuous-time simulation
conventional simulation
convergent simulation
cooperative hunters simulation
cooperative simulation
coopetition simulation
co-simulation
coupled simulation
credible simulation

AI-controlled simulation
AI-directed simulation
all software simulation
all-digital simulation
all-digital analog simulation
allotelic system simulation
analog simulation
analog computer simulation
analytic simulation
anticipatory perceptual simulation
appropriate simulation
approximate simulation
approximate zero-variance simulation
as-fast-as-possible simulation
asymmetric simulation
asynchronous simulation
audio simulation
augmented live simulation
augmented reality simulation
direct numerical simulation
direct simulation
disconnected simulation
discrete arithmetic-based simulation
discrete event line simulation
discrete event simulation
discrete simulation
discrete-change simulation
discrete-system simulation
discrete-time simulation
distributed agent simulation
distributed DEVS simulation
distributed interactive simulation
distributed real-time simulation
distributed simulation
distributed web-based simulation
distributed-parameter system simulation
DNA-based simulation
dynamic simulation
dynamic system simulation
dynamically composable simulation
E--
electronic gaming and simulation
collaborative component-based simulation
collaborative distributed simulation
collaborative simulation
collaborative virtual simulation
collocated cokriging simulation
collocated simulation
combined continuous-discrete simulation
combined simulation
combined system simulation
competition simulation
component simulation
component-based collaborated simulation
component-based distributed simulation
composable simulation
composite simulation
compressed-time simulation
computational simulation
computer network simulation
computer simulation
exascale simulation
expanded-time simulation
experience-aimed simulation
experiment-aimed simulation
experimental simulation
explanatory simulation
exploration simulation
exploratory simulation
ex post simulation
extensible simulation
extreme scale simulation
F--
fast simulation
fault simulation
fault tolerant simulation
faulty simulation
federated simulation
first degree simulation
full system simulation
fully coupled simulation
functional simulation
fuzzy simulation
fuzzy system simulation
critical event simulation
customizable simulation
customized simulation
D--
data-driven simulation
data-intensive simulation
decision simulation
degree 1 simulation
degree 2 simulation
degree 3 simulation
demon-controlled simulation
descriptive simulation
detached eddy simulation
deterministic simulation
DEVS simulation
digital analog simulation
digital computer simulation
digital quantum simulation
digital simulation
hands-on simulation
hardware-in-the-loop simulation
heterogeneous simulation
hierarchical simulation
high-fidelity simulation
high-level simulation
high-resolution simulation
historical simulation
HLA-based simulation
HLA-compliant simulation
holographic simulation
holonic simulation
holonic system simulation
HPC simulation
human-centered simulation
human-in simulation
human-in-the-loop simulation
human-machine simulation
hybrid computer simulation
hybrid gaming simulation
hybrid simulation
I--
identity simulation

live system-supporting simulation
logic simulation
logical simulation
low level simulation
Lindenmayer system simulation
L-system simulation

M--

machine simulation
machine-centered simulation
man-centered simulation
man-in-the-loop simulation
man-machine simulation
man-machine system simulation
manual simulation
Markov simulation
massively multi-player simulation
mathematical simulation
mental simulation
mesh-based simulation
mesoscale simulation
microanalytic simulation
microcomputer simulation
process simulation
process-based discrete event simulation
process-oriented simulation
proof-of concept simulation
proxy simulation
pseudosimulation
public domain simulation
pure software simulation
purpose of simulation

Q--

qualitative simulation
quantitative simulation
quantum simulation
quasi-analytic simulation
quasi-Monte Carlo simulation

R--

random simulation
rare-event simulation
real-system enriching simulation
real-system support simulation
real-time continuous simulation
multiphysics simulation
multi-player simulation
multiple-fidelity simulation
multi-processor simulation
multirate simulation
multiresolution simulation
multiscale simulation
multi-simulation
multistage simulation
mutual simulation

N--

nano simulation
nano-scale simulation
narrative simulation
nested simulation
net-centric simulation
networked simulation
network-oriented simulation
non-convergent simulation
non-deterministic simulation
non-equation-oriented simulation
non-linear system simulation
self-organizing system simulation
self-replicating system simulation
self-stabilizing system simulation
semiotic simulation
sequential Gaussian simulation
sequential simulation
serial simulation
serious simulation
service-based simulation
shape simulation
simulation
simultaneous simulation
single-aspect simulation
single-component simulation
single-processor simulation
skeleton-driven simulation
smart phone activated simulation
smoothness simulation
spatial simulation
spreadsheet simulation
stand-alone simulation
ordinary kriging simulation
outcome-driven simulation
outcome-oriented simulation

P--

parallax simulation
parallel discrete-event simulation
parallel simulation
parallelized simulation
partial equilibrium simulation
participative simulation
participatory agent simulation
participatory simulation
peace simulation
peer-to-peer simulation
perceptual simulation
petascale simulation
Petri net simulation
physical simulation
physical system simulation
portable simulation
predictive simulation
prescriptive simulation
teleonomic system simulation
terminating simulation
texture simulation
third degree simulation
thought controlled simulation
thought experiment simulation
thought simulation
throttled time-warp simulation
time-driven simulation
time-slicing simulation
time-stepping simulation
time-varying system simulation
time-warp simulation
trace-driven simulation
tractable simulation
training simulation
trajectory simulation
transfer function simulation
transparent reality simulation
trial simulation
trustworthy simulation

real-time data-driven simulation
real-time decision making simulation
real-time simulation
reasonable simulation
reasoning simulation
recursive simulation
regenerative simulation
regenerative simulation
related simulation
reliable simulation
remote simulation
replicative simulation
retro-simulation
retrospective simulation
reverse simulation
risk simulation
role playing simulation
role-play simulation
rule-based simulation
rule-based system embedded simulation

S--

scalable simulation
scaled real-time simulation
scientific simulation
second degree simulation
self-organizing simulation
wearable computer-based simulation
wearable simulation
Web service-based simulation
Web-based simulation
static simulation
steady-state simulation
stochastic simulation
strategic decision simulation
strategic simulation
strategy simulation
strong simulation
structural simulation
structure simulation
successor simulation
suitable simulation
swarm simulation
symbiotic simulation
symbolic simulation
symmetric simulation
system simulation
systematic simulation
system-of-systems simulation
systems-theory-based simulation

T--

tactical decision simulation
tactical simulation
tandem simulation
technical simulation
teleogenetic system simulation
teleological system simulation
Web-centric simulation
Web-enabled simulation
yoked simulation

U--

ultrascale simulation
uncertainty simulation
unconstrained simulation
uncoupled simulation
unified discrete and continuous simulation
unsuitable simulation
utilitarian simulation

V--

variable fidelity simulation
variable resolution simulation
very large eddy simulation
very large simulation
video game simulation
virtual simulation
virtual time simulation
virtual training simulation
virtualization simulation
visual interactive simulation
visual simulation

W--

war simulation
warfare simulation
weak classical simulation
weak simulation

Z--

zero-sum simulation
zero-variance simulation

Appendix 2

Modeling and Simulation Body of Knowledge (M&S BoK) Index

(*adopted* from M&S BoK-Ören which is being updated)

- Background
- Core Areas
- Supporting Domains
- Business/Industry
- Education/Training
- References

Background

1. Preliminary

1.1 **Approach to the preparation of the Index**

1.2 **Version history**

1.3 **Members of the review committee**

1.4 **Recommendations and how they have been taken into account**

2. Introduction

2.1 Aspects and basic definitions of simulation

2.2 Scope of the study

2.3 Stakeholders of M&S

2.4 M&S associations, groups, and centers

2.5 Why an M&S BoK? Rationale and possible usages

2.6 Importance of M&S

2.7 High level recognition of M&S

3. Terminology

3.1 **Some definitions of M&S (and their critical reviews)**

3.2 Dictionaries of M&S

3.3 **Ontology-based dictionaries (what? examples, benefits)**

4. Comprehensive View

4.1 **Challenges and benefits of comprehensive view (Big picture) of M&S**

Core Areas

1. Science/Methodology of M&S

1.1 Data

- *Issues (types of data and related terms)*
- *Variables* **(types of variables and related terms)**
    - **-Input variables (types of input values and related terms)**
- *Values* **(types of values and related terms)**

1.2 Models

- *Dichotomy of reality and its representation (model)*
- *Models and modeling*
    - *-Conceptual models, conceptual modeling*
    - *-Model attributes: fidelity, (multi-)resolution, scaling*
    - *-Modeling formalisms: Discrete, cellular, continuous*
    - *-Model composability, dynamic model composability*
    - *-Modelling physical systems*
    - *-Visual models*
    - *-Modeling qualitative systems*
    - *-Types of models and related terms*
    - *-Taxonomies of simulation formalisms and simulation models*
- *Model building (model synthesis)*
    - *-Modeling issues: reusability*

-Model composition (and dynamic model composition)

- *Model-base management*

  -Model search (including semantic model-search)

- *Model parameters and parameter-base management*

  -Parameters, **auxiliary parameters (deterministic/stochastic)**

- *Model characterization (Descriptive model analysis)*

  -For model comprehensibility

  --Model documentation (static/dynamic)

  --Model ventilation
  (to examine assumptions, deficiencies, limitations)

  -For model usability (model referability)

- *Model evaluation (Evaluative model analysis) with respect to*

  -a modelling formalism

  --Model consistency with respect to a modeling formalism

  --Model robustness

  -another model

  --Structural model comparison
  (model verification, checking, equivalencing)

  -real system (for analysis problems)

  --Model validity

  -technical system specification (for design and control problems)

  --Model qualification
  (model realism, adequacy, correctness analysis)

  -goal of the study -Model relevance

  --Model relevance

- *Model transformation*

  -Types of model transformation

  (copying, elaboration, reduction (simplification, pruning), isomorphism, homomorphism, endomorphism)

1.3 Experiments

- *Issues related with experiments*

  -Types of experiments

  -Terms related with experiments and experimentation

  -Experimental conditions (experimental frames)

  --Applicability of an experimental frame to a model

  -Experimentation parameters

- *Statistical design of experiments (issues and types)*

  -Computer-aided systems for design of experiments

  -Computer-aided systems for execution of experiments

1.4 Model behavior

- *Main issues related with model behavior*

  -Types of and terms related with model behavior

- *Generation of model behavior (for each modeling formalisms)*

-For discrete event models

-For discrete-time systems
(time-slicing simulation, finite state automata)

-For Petri nets

-For (and piece-wise) continuous-change systems
--Integration, processing discontinuity, stiffness

-For combined discrete/continuous-change systems

• Symbolic behaviour-generation

-Mixed numerical/symbolic behaviour generation

• *Analysis of model behavior*

-Processing and compression of model behavior (analytical, statistical processing)

-Visualization of model behavior (3-D, animation)

1.5 M&S interoperability

• *DIS*

• *ALPS*

• *HLA*

• *TENA*

• *other approaches*

1.6 **Computational intelligence and M&S**

• Agent-directed simulation

-Agent-based models & agent-based modeling

-Agent simulation, Agent-supported simulation

-Agent-monitored simulation, Agent-initiated simulation

• **Soft computing and M&S**

-Neural networks and simulation

-Fuzzy logic and simulation (fuzzy simulation)

-Evolutionary computation and M&S

--Genetic computation and simulation (genetic simulation)

--Swarm intelligence and simulation (swarm simulation)

1.7 Types of simulation

• **Types and terms**

• **Taxonomy of simulations**

1.8 Reliability & Quality Assurance of M&S

• Errors

-Types, terms, taxonomy

• **Validation**

-Types, terms, taxonomy

-Validation techniques and tools

• **Verification**

-Types, terms, taxonomy

-Verification techniques and tools

• **Quality assurance (QA)**

-Basic concepts, Types, terms, taxonomy

-Built in quality assurance

• **Failure avoidance**

-In traditional simulation, in inference engines used in simulation, in cognitive and affective simulations

1.9 Life cycles of M&S

- **For experimentation**
  - -For decision support
    - --Types of simulation-based decision support
  - -For understanding
  - -For education
- **To gain experience for training to enhance**
  - *-Motor skills*
    for civilian/military applications
    (virtual simulation; simulators, virtual simulators)
  - *-Decision making and communication skills*
    for civilian/military applications (constructive simulation – serious games; business gaming, peace gaming, war gaming)
  - *-Operational skills*
    for civilian/military applications (live simulation)
- **(To gain experience) for entertainment (simulation games)**
- *M&S for augmented reality*
- **Synergies of simulation games with other types of simulations**
- **Synergies of M&S with other disciplines**
- *Integrated LVB simulation*
- *Simulation integrated with decision support such as C4I*

2. Technology of M&S

2.1 M&S languages

- Programming languages
- Graphic languages
- Problem specification languages

2.2 M&S tools and environments

- *Human/system interfaces*
  - *-Types of interfaces (including immersion)*
  - *-Characteristics and design principles for human/system interfaces*
    - *--For front end interfaces*
    - *--For back-end interfaces*

2.3 Simulation-based problem solving environments for

- **Science**
- **Engineering**
- **Social sciences**
- **Health sciences**
- Defence

2.4 Infrastructure for M&S

- **M&S standards (By source, by topic)**
- **Lessons learned, codes of best practice**
- **Resource repositories**
  - **-Data, models, algorithms, heuristics, software, hardware**
  - **-Educational material**

3. History of M&S

3.1 Early simulations

3.2 Early simulators

3.3 Evolution of the M&S industry

4. Trends, challenges, and desirable features
- **4.1 Trends**
- **4.2 Challenges**
- **4.3 Desirable features**

Supporting Domains

1. Supporting science areas
- **1.1 System science (as bases for modeling & algorithmic model processing)**
  - **Emergence, emergent behavior**
  - **Complex adaptive systems**
- **1.2 Physics**
- **1.3 Mathematics (numerical analysis, probability, statistics)**
- **1.4 Queueing theory**

2. Supporting engineering areas
- **2.1 Systems engineering (for simulation systems engineering)**
- **2.2 Software engineering**
- **2.3 Visualization (including advanced display techniques)**
- **2.4 Sonorization** (including speech synthesis)

3. Computers
- **3.1 Digital, hybrid, analog; mobile, cloud**
  - Mobile simulation, cloud simulation
- **3.2 Extreme-scale computers (Petascale simulation, Exascale simulation)**
- **3.3 Soft computing and M&S**
  - **Neural networks and simulation**
  - **Fuzzy logic and simulation**
  - **Evolutionary computation and M&S**
    - **-Genetic computation and simulation**
    - **-Swarm intelligence and simulation**

Business/Industry

1. Management of M&S
- **1.1 Levels of M&S management**
  - **Enterprise level (Enterprise management)**
  - **Project level (Project management)**
  - **Product level (Product management)**
- **1.2 Life-cycle span of M&S management**
- **1.3 Resource management**
  - **Personnel management**
  - **Infrastructure management**
  - **Configuration management**
- **1.4 Quality management**
  - **Documentation management (Requirements, versions)**
  - **Uncertainty management**
  - **VV&A management**
  - Failure avoidance management
- **1.5 Cost-schedule management**
- **1.6 Success metrics for M&S management**

2. Types of employment in M&S
- **2.1 M&S for experimentation**
  - **For decision support (including simulation-based engineering)**
    - **-simulation-based acquisition**
  - ***F*or understanding (including simulation-based science)**

- ***F*****or education**

2.2 M&S for Training

- ***F*****or motor skills (for simulators, virtual simulators)**
- ***F*****or decision making and communication skills (for constructive simulation)**
- ***F*****or operational skills (for live simulation)**

2.3 M&S for entertainment (simulation games)

2.4 Visualisation for M&S

2.5 Sonorization for M&S

3. Domains of employment of M&S

3.1 Subject areas of applications for M&S

4. M&S business practice and economics

4.1 Economics of M&S

4.2 M&S market description

4.3 M&S-based enterprise

4.4 M&S investment and ROI

4.5 Reuse of M&S assets

4.6 Cost of M&S asset development, use, etc.

4.7 Value of M&S

4.8 Economic impact

5. M&S workforce

5.1 Community of practice

5.2 Workforce development needs / requirements

5.3 Curricular management

5.4 Professional certification for practitioners

6. Maturity levels and certification of:

6.1 Individuals

6.2 Organizations

6.3 Curricula

6.4 Educational programs

7. Ethics for M&S

7.1 Code of professional ethics for simulationists

7.2 Rationale

Education/Training

1. Education for M&S

1.1 Academic programs and curricula (for M&S professionals)

- Undergraduate programs and curricula
- Graduate programs and curricula
- Doctoral programs and curricula

1.2 Non-academic programs

- Executive seminars
- Professional appreciation and development seminars/courses

2. M&S for education

2.1 Academic programs and curricula at:

- *Universities*
    - For service courses in:
        - --Science

--Engineering

--Humanities

• *High schools*

• *Elementary schools*

2.2 General education

• *Continuing education*

• *To inform public*

References

1. M&S resources

1.1 Lists of M&S resources

1.2 M&S bibliographies

1.3 M&S dictionaries, encyclopedia, M&S master plans, standards

1.4 Repositories of educational material

1.5 M&S Archives

• M&S Books, Journals, Proceedings, Reports

1.6 News resources on M&S

• Google (news on simulation)

1.7 M&S Social networks

1.8 M&S blogs

2. References

2.1 by authors

2.2 by topics

• **Body of knowledge (BoK)**

-BoK of other areas

-M&S BoK – early studies & other contibutions

• **Composability, Reusability, Interoperability**

• **Conceptual Models and Conceptual Modeling**

• **Economics of M&S**

M&S Epistemology, Ontologies, Taxonomies

M&S Lessons Learned - Best Practices

Simulation and Systems Engineering

Simulation Professionals & Needed Qualifications

Validation/Verification

2.3 by application areas

Modelling for Managing the Complex Issue of Catchment-Scale Surface and Groundwater Allocation

Anthony Jakeman[1], Rebecca Kelly (nee Letcher)[2], Jenifer Ticehurst[1], Rachel Blakers[1], Barry Croke[1], Allan Curtis[3], Baihua Fu[1], Sondoss El Sawah[1], Alex Gardner[4], Joseph Guillaume[1], Madeleine Hartley[4], Cameron Holley[1], Patrick Hutchings[1], David Pannell[5], Andrew Ross[1], Emily Sharp[3], Darren Sinclair[1], and Alison Wilson[5]

[1] Integrated Catchment Assessment and Management Centre,
The Fenner School of Environment and Society, Australian National University, Canberra, ACT, Australia and National Centre for Groundwater Research and Training (NCGRT)
[2] IsNRM Pty Ltd, PO Box 8017, Trevallyn, Tasmania, Australia
[3] Institute for Land, Water and Society, Charles Sturt University, Albury, Australia and NCGRT
[4] Law School, University of Western Australia, Crawley, Australia and NCGRT
[5] School of Agricultural and Resource Economics, University of Western Australia, Crawley, Australia and NCGRT
tony.jakeman@anu.edu.au

Abstract. The management of surface and groundwater can be regarded as presenting resource dilemmas. These are situations where multiple users share a common resource pool, and make contested claims about their rights to access the resource, and the best use and distribution of the resource among competing needs. Overshadowed by uncertainties caused by limited data and lack of scientific knowledge, resource dilemmas are challenging to manage, often leading to controversies and disputes about policy issues and outcomes. In the case of surface and groundwater management, the design of collective policies needs to be informed by a holistic understanding of different water uses and outcomes under different water availability and sharing scenarios. In this paper, we present an integrated modelling framework for assessing the combined impacts of changes in climate conditions and water allocation policies on surface and groundwater-dependent economic and ecological systems. We are implementing the framework in the Namoi catchment, Australia. However, the framework can be transferred and adapted for uses, including water planning, in other agricultural catchments.

Keywords: Integrated Modelling, Surface and Groundwater Management, Resource Dilemmas, Water Allocation.

1 Introduction

Water resource management issues are often described as: wicked [1] persistent [2] and dilemmas [3]. The main features of such issues include:

M.S. Obaidat et al. (eds.), *Simulation and Modeling Methodologies, Technologies and Applications,* Advances in Intelligent Systems and Computing 256,
DOI: 10.1007/978-3-319-03581-9_2, © Springer International Publishing Switzerland 2014

- ill-defined, multiple and conflicting goals leading to disputes and controversies
- interdependency among stakeholder activities, and their effects on the resource
- highly complex and interconnected social, technological, and biophysical processes
- uncertainty about system processes, and how they respond to change

Resource dilemmas cannot be solved in a sense of finding a final and risk-free solution that satisfies all preferences. However, they need to be managed by continually developing and adapting resource sharing policies that can best accommodate various present and future needs under different water availability scenarios. The design for collective and adaptive policies calls for *integration* among:

- policy issues to develop systemic and long term policies rather than short term and piecemeal decisions
- scientific disciplines in order to develop a multi-perspective stance about coupled socio-ecological processes that cannot be derived from isolated mono-disciplinary stances, and
- science-policy-stakeholders throughout the policy making lifecycle.

Models and modelling can play a key role in establishing and supporting these integration dimensions. They can be used as tools for synthesising and communicating our understanding of complex social-ecological systems. Modellers can help integrate methods and findings from different scientific fields (e.g. ecology, hydrology, economics and other social sciences) to present relevant policy and decision-making information. Participatory modes of modelling provide support for framing the issues of concern from multiple viewpoints, clarifying decision options, identifying and engaging with stakeholder groups, and sharing the knowledge generated.

This paper presents a project where modelling has been designed to help deal with over-allocation of surface and groundwater, a key policy issue in Australia and worldwide. The modelling project brings together a collaborative multi-disciplinary research team (i.e. social, economic, ecological, hydrological, legal and institutional disciplines) with the aim of developing an integrated modelling framework to identify the social, economic and environmental trade-offs under various water policy decisions and climate variations. The model allows the exploration of adaptation mechanisms, identified by our social science team, that water users are likely to accept in order to minimise the impacts of climate change and reductions in their water allocation. The modelling framework is implemented in the Namoi catchment, Australia. However, the framework can be transferred and adapted for use in other agricultural catchments.

The paper is structured as follows: in Section 2, we discuss the challenge of over-allocation in water planning in Australia, and briefly introduce the concept of integrated modelling. Section 3 presents water allocation in the Namoi catchment as the case study. The modelling framework is described in Section 4. We wrap up with the discussion and conclusion in Sections 5 and 6.

2 Background

2.1 Over-Allocation and Trade-Offs in Australia

In Australia, returning over-allocated surface and groundwater systems to sustainable levels is a key challenge for water planning. According to a recent national assessment [4] major catchments and aquifers are at or approaching the risk of being ecologically stressed as a result of flow regulation and/or consumptive water use. This may be exacerbated by predicted long term declines in rainfall, increases in temperature and in evapotranspiration, and variability in stream flows and aquifer recharge rates. [5] The National Water Initiative (NWI), the principal blueprint for water reform, stresses the need for water planners to make "trade-offs", or decisions that balance water requirements for the environment with the water demands of consumptive users. The requirements as well as challenges for designing trade-offs include [6]:

1. robustness: taking into consideration the possible impacts of climate variations (including climate change and variability) on environmental outcomes and consumptive use,
2. transparency: all information used to set up priorities and assess outcomes are clear and publicly accessible
3. risk-based: assessing consequences, associated risks and benefits under different water sharing scenarios
4. science-informed: assessments to be based on best available scientific knowledge and data, including socio-economic and ecological analysis

To meet these requirements, the design and implementation of trade-offs need to be informed by a holistic assessment of different water uses and how they may change under different climate and policy scenarios. This requires mechanisms for integrating knowledge, methods and tools from different scientific disciplines in order to analyse system elements and synthesise information that is useful and relevant to planners and managers. One of these mechanisms is integrated modelling.

2.2 Integrated Modelling

Integrated assessment and modelling is becomingly increasingly accepted as a way forward to address complex policy issues [7]. Many of the earlier concepts drew upon the integration of different types of models, or different types of data sources [8]. Models have been developed to integrate across more than one discipline. For example Pulido-Velazquez et al. [9] integrated across two disciplines (surface water and groundwater hydrology and economics), and Kragt et al. [10] integrated the hydrologic, economic and ecologic aspects of the management of a river basin in Tasmania. Barth et al. [11] included some social aspects of people's choices and responses to various water-related scenarios. However, only a few examples exist where more than three disciplines have been included, particularly social science disciplines considering behaviour, social impacts, law and institutions. One such example is Mongruel et al. [12] who accounted for governance by implementing

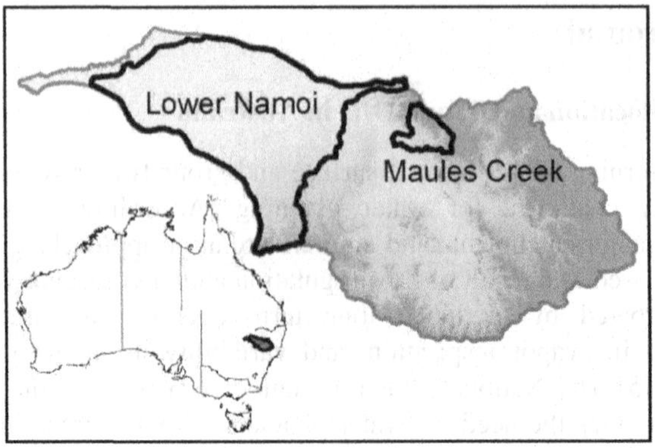

Fig. 1. The Namoi catchment and study locations

various policies, rules, laws and agreements, and social implications (e.g. recreational fishing and oyster growing), linked by hydrological and ecological consequences. To account for all major aspects in the management of surface and groundwater, an integrative approach would need to account for governance, economic, ecological, hydrological and social components of the system, and their relevant linkages.

3 Case Study

The Namoi Catchment is located in northern New South Wales in Australia (Fig. 1). The catchment is about 42,000 sq. km in size and extends from the Great Dividing Range in the east, with elevations of 1000m down to the flat plans in the west at only 250m. The average annual rainfall over the catchment varies accordingly, from over 1100mm to 470mm, and falls mainly in the summer in high intensity events. [13] This makes the catchment hydrologically complex, with many of the streams and drainage channels being ephemeral. The population in the catchment is over 100,000 people residing both in towns (mainly Tamworth and Narrabri) and rural settlements. The regional output is over AUS$1 billion, half of which comes from agriculture. Agricultural land uses include grazing on the steeper elevations, and cropping, both dryland and irrigated, on the flatter country. The cotton industry is a highly lucrative irrigated industry in the area. The latest national assessment by NWC [4] has rated the catchment as "highly" stressed, and over-allocated.

The Namoi has been well studied in the past. Kelly et al. [14] present an extensive list of studies of groundwater alone, including both modelling and data analysis in the Namoi catchment, yet still concluded that more than 15 other projects could be conducted to further our scientific understanding of the groundwater in that catchment. In addition there have been social studies, [15] economic studies, [16] hydrological studies [17] and ecological studies[18]; and a recent report by Hartley et al. [19] demonstrates the complexity of the governance issues surrounding groundwater across the Namoi catchment, with variation in the scale and governing bodies at different locations.

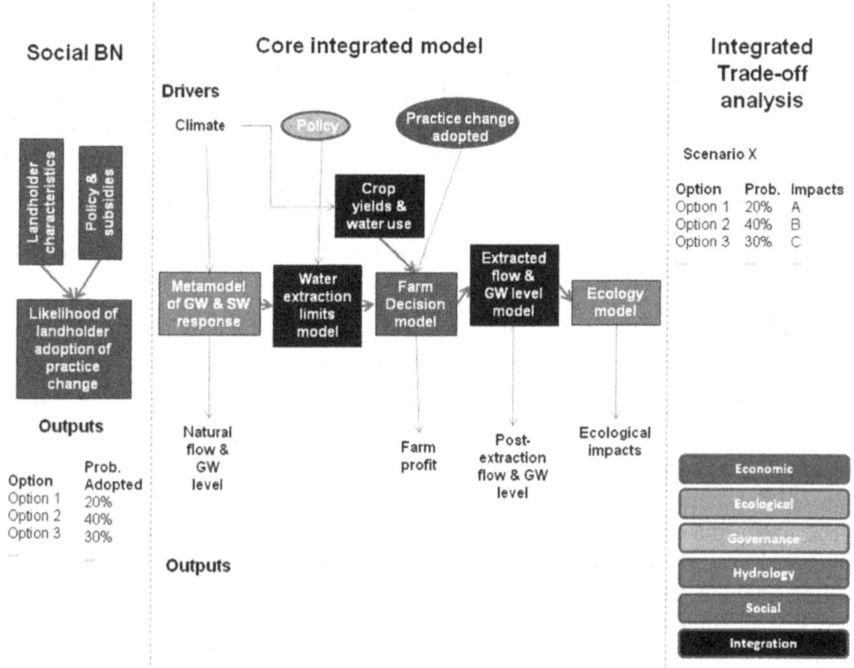

Fig. 2. Conceptual diagram of the integrated model

The work presented here builds upon that of Letcher et al. [20] by utilising the expertise of disciplinary research scientists to: add ecological and social models; improve the hydrological model to include a component for surface-groundwater interactions; update the data and information in the social, economic and crop models; and develop more informed policy (and adaptation and climate) scenarios using the expertise of governance and law researchers.

Within the Namoi catchment we are focusing upon two specific groundwater areas - the Lower Namoi which includes access to regulated surface water, and the smaller Maules Creek catchment which has unregulated surface water access. These case study areas capture the complexity of the social, hydrological and ecological components crucial to the wider area for the management of groundwater.

4 Model Description

The model has several components which are integrated into a single working model shown in Fig. 2. The various model components have been designed to run at various spatial scales (see below), with a temporal horizon of 20 years to allow for irrigation infrastructure investments.

The integrated model:

- Uses prediction of the natural surface and groundwater flow, and the policy scenarios to estimate the water extraction limits

- Determines the water use and crop yields given the climate and various crop types
- Uses the output from the likely behaviours and adoption of various actions by landholders from the social model, the water allocation levels and the crop yields and water use, to input into the farm decision model and determine the farm profit
- Calculates the extracted flow and groundwater levels remaining following farmer decisions, and
- Estimates the ecological impacts of the available surface and groundwater flows on the ecology.

Descriptions of each component of the integrated model now follow.

4.1 Model Components

Hydrological Model

At the core of the integrated model is a hydrological model that predicts the effects of surface and groundwater extraction regimes on surface flows, aquifer storage and discharge. These hydrological impacts are used to assess water availability and the resulting social, economic and ecological outcomes.

A key challenge in developing the hydrological model was identifying an appropriate level of spatial aggregation and parameterisation that provides satisfactory prediction of water storages and fluxes necessary for evaluating options for managing the water resources, given the uncertainties in modelled output and observed data. The social and economic components of the project, as is common with such models, divide the catchment into a number of large regions that are considered homogenous with respect to land use and farm management practices [20]. Hence the performance of the hydrological model only needs to be assessed at this large spatial scale. A second consideration was that the run times of the integrated model needed to be minimised to facilitate assessment of model performance and uncertainty analysis.

The selected hydrological model is a spatially lumped model that employs a catchment-scale conceptualisation of the key hydrological processes, and includes two groundwater layers: a shallow system that contributes baseflow to the river, and a deeper groundwater system that is used as a water resource. The model consists of three components: a rainfall-streamflow model representing runoff and baseflow from the shallow aquifer system for each subcatchment; a groundwater mass balance model for the deeper aquifer system, and a lag-route routing model [21] capturing the flux of water between nodes. The hydrological model represents the stream network as a series of nodes. Each node represents a sub-catchment and is comprised of two modules. The first is a non-linear loss module that takes rainfall and temperature data and produces 'effective rainfall' (rainfall that becomes runoff or recharge), accounting for losses due to evapotranspiration. The second is a linear routing module that converts effective rainfall to stream flow via two parallel transfer functions representing a quick-flow pathway (equated to surface runoff) and a slow-flow pathway (representing discharge from the shallow aquifer system).

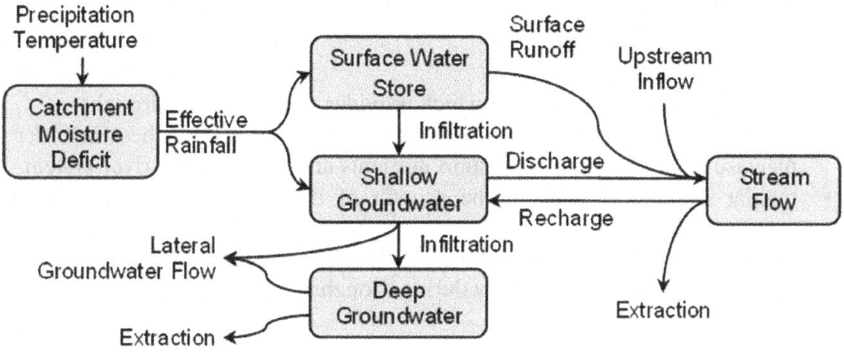

Fig. 3. Structure of the hydrological model for a single subcatchment and connected groundwater aquifers

A preliminary formulation of the model is detailed by Blakers *et al.* [22] and is represented in Fig. 3. The model is based on the IHACRES rainfall-runoff model [23] with the addition of groundwater and surface water interactions from Ivkovic *et al.* [24] and a two layer aquifer system by Herron and Croke [25]. The additional work by Blakers *et al.* [26] allows for better specification of the groundwater aquifers, which do not follow surface catchment boundaries, and improved representation of surface-groundwater interactions and groundwater flow.

The model takes rainfall, temperature and extraction data and other basic catchment information (e.g. catchment area, location of aquifers) to predict on a daily basis:

- surface water flows, including contributions from surface runoff and baseflow
- groundwater levels at selected locations in the catchment, for use in ecological modelling and water availability assessment.

Social Model

A Bayesian network (Bn) has been developed to represent the social model component. Bns capture the conceptual understanding of a system by causal links between variables with multiple states, and the strength of the links is represented by probability distributions. These models are becoming increasingly popular in natural resource management for their ability to incorporate quantitative and qualitative information, their implicit representation of uncertainty and their usefulness as communication tools. See Ticehurst et al. [27] for an example discussion on the advantages and disadvantages of using Bns in natural resource management. Bns have also been successfully used in the analysis of social data [28].

Here the Social Bn has been developed to map the likely behaviours of farmers in terms of compliance, changes in farming systems and water use efficiency depending upon various climate scenarios and policy options. It is based upon the findings of Sharp and Curtis [29].

Ecological Model

The ecological model for the Namoi is directed to healthy river function. This involves:

- a sustained level of base flow, which provides refuges during drought
- regular flushing at various levels of benches and anabranches, in order to increase habitat areas and transport nutrients and carbon to the river system
- regular flooding to sustain the growth of riverine vegetation and support regeneration
- suitable groundwater and salinity levels to allow the access of water by riverine vegetation, particularly during drought.

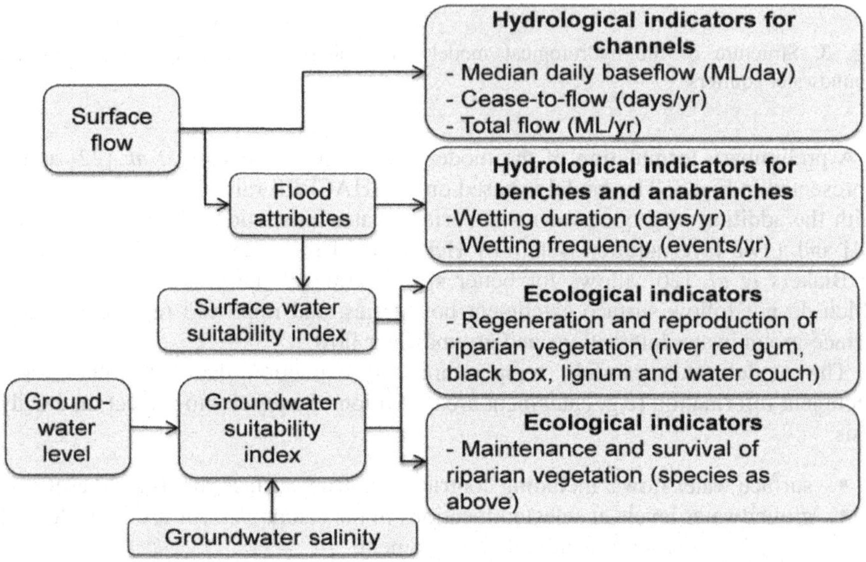

Fig. 4. Structure of the ecological model

The ecological component has been developed by Fu and Merritt [30]. The model, shown conceptually in Fig. 4, uses inputs of surface water flow and groundwater level and salinity to estimate hydrological and ecological indicators for niche ecological assets identified for the Namoi. The hydrological indicators include baseflow level, cease-to-flow days and total flow. Wetting duration and frequency for benches and anabranches are estimated at each asset. The ecological indicators report the water suitability index for the maintenance and regeneration of four riverine vegetation species: *Eucalyptus camaldulensis* (river red gum), *Eucalyptus largiflorens* (black box), *Muehlenbeckia florulenta* (lignum) and *Paspalum distichum* (water couch). The water suitability index is generated from both surface water and groundwater suitability indices. Variables considered for the surface water suitability index are flood duration, timing and interflood dry period. The groundwater suitability index is derived from the groundwater level index and adjusted by groundwater salinity. Groundwater salinity acts as a modifier: if the groundwater salinity level is greater than the salt tolerance threshold for a given species, the groundwater suitability index is reduced to 0; otherwise, the groundwater suitability index is equal to the groundwater level index. Finally, all ecological model outputs in annual time series are converted into exceedance probabilities for use in the integrated model.

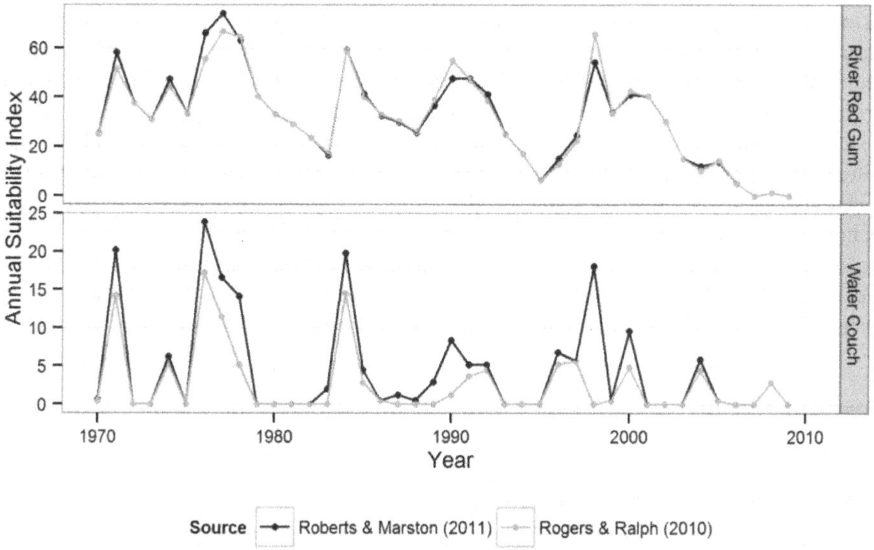

Fig. 5. Estimated annual water suitability index for river red gum and water couch over 1970-2010 at the river corridors between Mollee and Gunidgera (Asset 4), Namoi. Estimation was based on preference curves generated from Rogers and Ralph [34] and Roberts and Marston [35]. Note the overall declining trend for the river red gum, which reflects the decline in groundwater levels.

Preference curves are used to generate the surface water suitability index and groundwater level index. This approach was initially developed for the Murray Flow Assessment Tool [31] and then [32-33]. The key to this approach is to convert flood and groundwater attributes to suitability indices, based on data, literature and/or expert opinions. However, our knowledge of riverine ecosystems is imperfect, which contributes to the uncertainty in the generation of preference curves. Fu and Merritt [31] found that this uncertainty can have impacts on the estimated ecological outcomes. The level of impact varies depending on species and water regime. For example, water requirements of water couch are much less studied than river red gum, which is reflected in the lesser level of consistency in the preference curves and model outcomes (Fig. 5). In terms of water regime, the requirement for flood timing is most uncertain for most species and contributes to the variation in model outcomes. The implications of ecological uncertainty for the modelling will be further assessed through the comparison of model outcomes for different climate and policy scenarios. The integrated model will allow comparison of the impacts of hydrological uncertainties and ecological uncertainties on the same scales. Such analyses will provide valuable insights into the relative significance of ecological knowledge uncertainty in the integrated hydro-ecological model.

Farm Decision (economic) Model
The farm decision model builds on the approach previously developed by Letcher et al. [34] This model uses a multiple-period linear programming approach to capture farmer decisions relating to crop choice, area planted and irrigation water use. Farming system information, gross margin values and crop rotations options have been obtained through interviews with cotton growers, irrigation extension agents and irrigation engineers. Carryover rules and allocations in each of the three water systems (groundwater, and regulated and unregulated surface water) as well as the potential to carry water over in on-farm storages are also accounted for. Long term decisions, such as those relating to decisions to invest in changes in irrigation technology, develop water storages or to permanently sell water are simulated using the social Bn and input to the farm decision model. Representative farms of 940 ha, 4 000 ha and 11 000 ha with differing access to groundwater, regulated and unregulated surface water are used to represent the diversity of farmers in the case study areas.

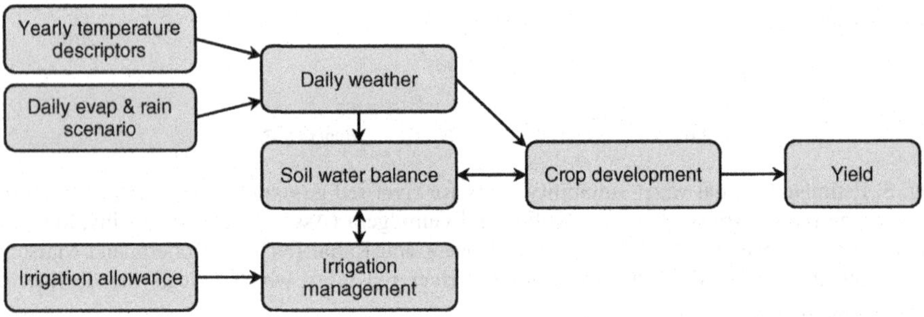

Fig. 6. Conceptual structure of the crop model

Crop Metamodel
A simplified crop metamodel is used to simulate the effects of applied water on crop yield for various commonly used crops in the region (Fig. 6). Cotton, wheat, chickpea and vetch components are used, based on simplified versions of industry standard models used in the APSIM package described by McCown *et al.* [35].

The metamodel is a collection of modules that rusn components of the model on a daily basis. The model inputs are yearly temperature descriptors, daily evaporation and rainfall, and seasonal irrigation water allowance.

The metamodel contains a soil water balance module based on CERES-maize, developed by Jones and Kiniry [36] that provides water input for the simulated crops. The module maintains a daily water balance using irrigation, weather and crop growth feedback as inputs. Water balance is outputted for crop development as a soil moisture index for a single depth layer that is defined for each crop.

The irrigation module provides input to the water balance module by supplying water from a yearly allocation pool. Timing and amount of irrigation is automated based on feedback from the water balance module to maintain a soil moisture index above a desired threshold during crop cycles.

The model takes daily evaporation and rainfall data from historic or generated climate scenarios as inputs. Temperature descriptors of average yearly and daily temperature amplitude and average yearly maximum temperature are input to generate daily temperature patterns to correlate with the evaporation and rainfall data. The daily weather and soil balance modules provide input to the crop development module to simulate daily crop development and generate yield estimates for each season.

To fit within the wide scope of the analysis, agronomic decisions are limited to seasonal irrigation quantities. Daily tasks such as sowing, irrigating and harvesting are automated by the metamodel using heuristics utilised by the crop models that it is based on.

4.2 Running the Model

The model can be used to explore the impact of landholder activities, climate and policy scenarios. Landholder scenarios include the maximum change in hydrological and ecological condition if all landholders were to adopt particular water efficiency practices. This information could be used to inform hydrological targets for the region.

The climate scenarios are predetermined predictions based upon the CSIRO predictions for the year 2030 [37]. The likely change in landholder actions and economic situation, and consequent hydrologic and ecological condition are predicted following changes in the climate.

The model can explore a number of governance and policy issues that pertain to the achievement of water extraction limits. These include existing experiences such as collaborative governance, participatory democracy, adaptive management, compliance and enforcement, Sustainable Diversion Limits, Water sharing rules, cease to pump rules, and trading rules as well as potential future/alternative policy approaches, such as self-management and co-regulation. More specifically, three scenarios that the integrated model will be used to test relate to the economic loss incurred as a result of reductions in access entitlements. Two of the scenarios model the loss incurred as a result of government actions, with the third relating to loss incurred from the actions of private water users.

These scenarios calculate the economic loss arising from:

i. climate change induced reductions in water availability leading to reductions in access entitlements;
ii. investments in water use efficiency when facing reductions in entitlements because of over-allocation; and
iii. the cost of water theft, where some licensees have their water access reduced as a result of other users taking water in breach of their licence entitlements.

The primary output of the integrated model is an integrated trade-off matrix for a selected set of scenario options, presented through reports, workshops and other presentations (e.g. Fig. 7). The matrix comprises the likelihood of the adoption of various practices under each scenario, as well as the impacts simulated from each of the integrated model components, which are:

- Natural flow and groundwater level
- Farm profit
- Post extraction flow and groundwater level, and
- Ecological impacts.

The model outputs for each of these are not necessarily numerical. They could also be presented as graphs, pictures, or a qualitative measure of impact compared to a baseline, or base case condition. It is likely that the trade-off matrix will rely on summary indicators from each of the components to explore the multi-disciplinary trade-offs associated with scenario options.

Scenario	Short-term profit (Base case)	Capital costs (Base case)	Government subsidies (Base case)	Groundwater impacts (Full line= Base case Dashed line = Scenario)	Ecological impacts (Full line = Base case Dashed line= Scenario)
1	(350,000) 300,000	(30,000) 25,000	(5,000) 20,000	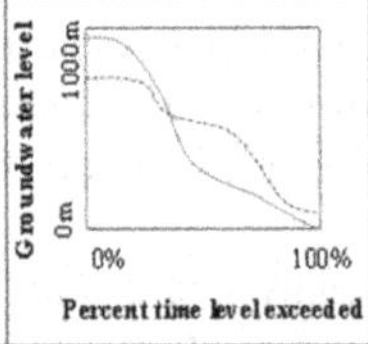	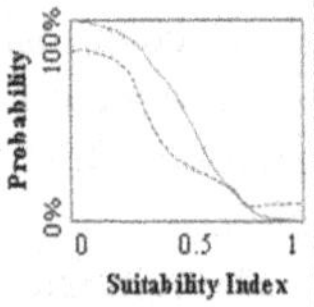
2	(350,000) 400,000	(30,000) 35,000	(5,000) 10,000	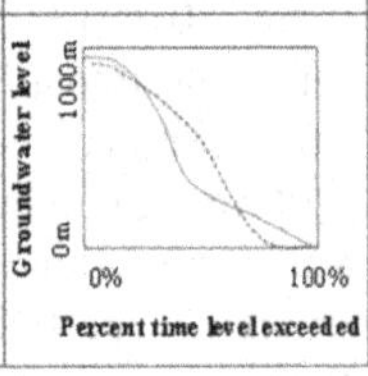	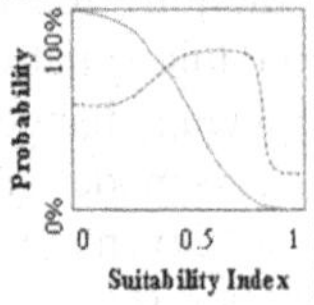

Fig. 7. Prototype of example model output showing trade-offs of various outputs

The model outputs from the specific scenarios discussed above will be utilised by the legal team to conceive a viable framework for introducing compensation measures in circumstances when entitlement reductions occur consequential to factors outside a licensee's control.

4.3 Testing Model Uncertainty

It is assumed that decision makers will naturally draw conclusions from the trade-off matrix and model outputs. However, model development involves a number of assumptions and modelling decisions about how uncertain data and knowledge are used, including the choice of parameters and model structures. An investigation will involve working with experts on each model component to identify alternate assumptions and decisions that would also be considered plausible. It will then search for a set of these decisions for which the conclusions drawn by decision makers would be shown to be wrong. This provides an audit of the model. If it fails, sensitivity analysis will help prioritise future improvements to ensure decision makers can draw robust conclusions.

5 Discussion

The novelty of the integrated model described here is its holistic capture of the issues in managing groundwater from five different disciplinary perspectives (i.e. ecology, hydrology, economics, governance and social). Importantly, the model has been developed by integrationists working with research scientists from each of these disciplines, all working concurrently in researching and collecting data to support the development and running of the integrated model. It is envisaged that, although results from scenario runs and the uncertainty testing are not yet available for publication, the outcomes from the scenario results will provide insight and a discussion focus for the local policy staff, government water managers and irrigators.

One trade-off in developing such an inclusive integrated model with a group of research scientists is that it takes a significantly longer period to consolidate the model and its components than if it had been developed in isolation by the integration team. However the additional benefits gained from being able to integrate the collective expertise of such a diverse team are substantial.

Another trade-off is that the model itself is quite complex and is not suitable to be distributed as a decision support tool without intensive training. Consequently scenarios of interest will be identified by the steering committee of the project, and the results of these will be analysed and then delivered in a facilitated workshop in the Namoi catchment during 2013.

Previous work with FARMSCAPE [38] suggests that, despite having spent a large amount of time (several years) running and testing models for local conditions, the end-users still benefit more from running, analysing and discussing the model results with the local researchers and advisors. Our experience also suggests that significant value can be gained in the discussion around the model development and analysis of the results, as opposed to focussing just on the direct model output.

6 Conclusions

This paper describes the development of an integrated model for use in the management of groundwater-surface water systems that are particularly under pressure due to past over-allocation and potential climate change. The model provides the opportunity to explore the socioeconomic, hydrologic and ecological trade-offs of various policy, adaptation and climate scenarios. The model is implemented for the Namoi catchment, but its components can be transferred to other agricultural catchments.

The model developed is the result of a collaborative research project by a team of disciplinary research scientists from ecological, economic, social, hydrological, governance and integrated modelling backgrounds. The team has been working together, with the project steering committee of local catchment water managers, irrigators and advisors since the outset of the project, to generate and share knowledge, research new ideas and collect data to inform the integrated model.

As a consequence of developing such an inclusive model, with such a large team, a considerable amount of time has been spent in the model development phase. Unsurprisingly, the resultant model is quite complex despite its identified components being kept as effectively simple as possible. The model scenario results will be presented as part of a facilitated workshop with the local water managers and irrigators throughout 2013. It is hoped that the findings from such an inclusive integrated model will be well-received by the local water managers as a tool to assist in unravelling the complexities and clarifying their options in groundwater and surface water management.

If identified as advantageous, further work may be completed to develop and modify a meta-model of the integrated model, to be used as a decision support tool at the farm scale. As with any wicked problem there will be no stopping point, the work contributing more to on-going problem resolution rather than the specification of black and white solutions. In this connection the model can be updated as new issues and information become available in order to shed light on appropriate tradeoffs.

Acknowledgements. This work was funded by the (Australian) National Centre for Groundwater Research and Training (NCGRT), and the Cotton Catchment Communities Cooperative Research Centre (CRC). Thanks go especially to the Project Steering Committee made up of representatives from the Cotton CRC, Namoi Catchment Management Authority, NSW Office of Water and Namoi Water.

References

1. Rittel, H.: On the planning crisis: Systems analysis of the first and second generations. Bedrifts Okonomen 8, 390–396 (1972)
2. Rotmans, J., Kemp, R., van Asselt, M.B.A.: More evolution than revolution: transition management in public policy. Foresight 3(1), 15–31 (2001)
3. Ison, R.L., Rolling, N., Watson, D.: Challenges to science and society in the sustainable management and use of water: investigating the role of social learning. Environmental. Science and Policy 10(6), 499–511 (2007)
4. NWC, Assessing water stress in Australian catchments and aquifers, National Water Commission, Canberra (2012)
5. Climate Commission: The Critical Decade: Climate Science, Risks and Responses. Department of Climate Change and Energy Efficiency, Canberra (2011)
6. NWI, Policy guidelines for water planning and management, Canberra (2010)
7. Jakeman, A.J., Letcher, R.A.: Integrated assessment and modelling: features, principles and examples for catchment management. Environmental Modelling and Software 18, 491–501 (2003)
8. Giupponi, C., Jakeman, A.J., Karssenberg, G., Hare, M.P. (eds.): Sustainable Management of Water Resources: an Integrated Approach. Edward Elgar Publishing, Cheltenham (2006)

9. Pulido-Velazquez, M., Andreu, J., Sahuquillo, A., Pulido-Velazquez, D.: Hydro-economic river basin modelling: the application of a holistic surface-groundwater model to assess opportunity costs of water use in Spain. Ecological Economics 66, 51–65 (2008)
10. Kragt, M.E., Newham, L.T.H., Bennett, J., Jakeman, A.J.: An integrated approach to linking economic valuation and catchment modelling. Environmental Modelling and Software 26, 92–102 (2011)
11. Barth, M., Hennicker, R., Kraus, A., Ludwig, M.: DANUBIA: An integrative simulation system for global change research in the upper Danube Basin. Cybernetics and Systems 35, 639–666 (2004)
12. Mongruel, R., Prou, J., Balle-Beganton, J., Lample, M., Vanhoutte-Brunier, A., Rethoret, H., Aguindez, J.A.P., Vernier, F., Bordenave, P., Bacher, C.: Modeling Soft Institutional Change and the Improvement of Freshwater Governance in the Coastal Zone. Ecology and Society 16 (2011), http://dx.doi.org/10.5751/ES-04294-160415
13. Donaldson, S., Heath, T.: Namoi River Catchment: Report on Land Degradation and Proposals for Integrated Management for its Treatment and Prevention. NSW Department of Land and Water Conservation (1996)
14. Kelly, B., Merrick, N., Dent, B., Milne-Home, W., Yates, D.: Groundwater knowledge and gaps in the Namoi Catchment Management Area, University of Technology, Sydney, Report 2.2.05 (March 2007), http://www.cottoncrc.org.au/catchments/Publications/Groundwater/Groundwater_Knowledge_Gaps
15. Mazur, K., Bennett, J.: Location differences in communities' preferences for environmental improvements in selected NSW catchments: A choice modelling approach. In: 54th Annual Conference of the Australian Agricultural and Resource Economics Society, Cairns, Australia, February 10-13 (2009)
16. Dudley, N.J.: A single decision-maker approach to farm and irrigation reservoir management decision making. Water Resources Research 24, 633–640 (1988a); Dudley, N.J.: Volume sharing of reservoir water. Water Resources Research 24, 641–648 (1988b); Greiner, R.: Catchment management for dryland salinity control: model analysis for the Liverpool Plains in New South Wales. Agricultural Systems 56, 225–251 (1998); Greiner, R.: An integrated modelling system for investigating the benefits of catchment management. Environmental International 25, 725–734 (1999)
17. McCallum, A.M., Anderson, M.S., Kelly, B.F.J., Giambastiani, B., Acworth, R.I.: Hydrological investigations of surface water-groundwater interactions in a sub-catchment in the Namoi Valley, NSW, Australia, Trends and Sustainability of Groundwater in Highly Stressed Aquifers. In: Proc. of Symposium JS.2 at the Joint IAHS & IAH Convention, Hyderabad, India, p. 157. IAHS Publ. 329 (September 2009)
18. Leonard, A.W., Hyne, R.V., Lim, R.P., Chapman, J.C.: Effect of Endosulfan Runoff from Cotton Fields on Macroinvertebrates in the Namoi River. Ecotoxicology and Environmental Safety 42, 125–134 (1999)
19. Hartley, M., Sinclair, D., Wilson, A., Gardner, A.: Governance Report: Namoi Catchment, A report on the governance of groundwater in the Namoi Catchment, National Centre for Groundwater Research and Training, National Water Commission and Australia Research Council (2011)
20. Letcher, R.A., Jakeman, A.J., Croke, B.F.W.: Model development for integrated assessment of water allocation options. Water Resources Research 40, W05502 (2004)
21. Croke, B.F.W., Letcher, R.A., Jakeman, A.J.: Development of a distributed flow model for underpinning assessment of water allocation options in the Namoi River Basin, Australia. Journal of Hydrology 39, 51–71 (2006)

22. Blakers, R.S., Croke, B.F.W., Jakeman, A.J.: The influence of model simplicity on uncertainty in the context of surface-groundwater modelling and integrated assessment. In: 19th International Conference of Modelling and Simulation, Perth, December 12-16 (2011), `http://mssanz.org.au/modsim2011`
23. Jakeman, A.J., Hornberger, G.M.: How much complexity is warranted in a rainfall-runoff model? Water Resources Research 29(8), 2637–2649 (1993); Croke, B.F.W., Jakeman, A.J.: A catchment moisture deficit module for the IHACRES rainfall-runoff model. Environmental Modelling & Software 19, 1–5 (2004)
24. Ivkovic, K.M., Letcher, R.A., Croke, B.F.W.: Use of a simple surface-groundwater interaction model to inform water management. Australian Journal of Earth Sciences 56, 71–80 (2009); Ivkovic, K.M., Letcher, R.A., Croke, B.F.W., Evans, W.R., Stauffacher, M.: A framework for characterising groundwater and surface water interactions: a case study for the Namoi catchment. In: 29th Hydrology and Water Resources Symposium, Water Capital, Engineers Australia, NSW, February 21-23 (2005a) (ISBN 085 825 8439); Ivkovic, K.M., Croke, B.F.W., Letcher, R.A., Evans, W.R.: The development of a simple model to investigate the impact of groundwater extraction on river flows in the Namoi Catchment. In: "Where Waters Meet" NSHS-IAH-NSSSS Conference, November 28-December 2, Auckland, New Zealand (2005b); Ivkovic, K.M., Letcher, R.A., Croke, B.F.W.: Use of a simple surface–groundwater interaction model to inform water management. Australian Journal of Earth Sciences 56, 71–80 (2009)
25. Herron, N.F., Croke, B.F.W.: Including the Influence of Groundwater Exchanges in a Lumped Rainfall-Runoff Model. Mathematics and Computers in Simulation 79, 2689–2700 (2009a), doi:10.1016/j.matcom.2008.08.007; Herron, N.F., Croke, B.F.W.: IHACRES-3S – A 3-store formulation for modelling groundwater-surface water interactions. In: Anderssen, R.S., Braddock, R.D., Newham, L.T.H. (eds) 18th World IMACS Congress and MODSIM09 International Congress on Modelling and Simulation. Modelling and Simulation Society of Australia and New Zealand and International Association for Mathematics and Computers in Simulation, 2009, pp. 3081–3087 (July 2009b), `http://www.mssanz.org.au/modsim09/I1/herron.pdf` ISBN: 978-0-9758400-7-8
26. Blakers, R.S., Croke, B.F.W., Jakeman, A.J.: The influence of model simplicity on uncertainty in the context of surface-groundwater modelling and integrated assessment. In: 19th International Conference of Modelling and Simulation, Perth, December 12-16 (2011), `http://mssanz.org.au/modsim2011`
27. Ticehurst, J.L., Letcher, R.A., Rissik, D.: Integration modelling and decision support: a case study of the Coastal Lake Assessment and Management (CLAM) tool. Mathematics and Computers in Simulation 78, 435–449 (2008)
28. Ticehurst, J.L., Curtis, A., Merritt, W.S.: Using Bayesian networks to complement conventional analysis to explore landholder management of native vegetation. Environmental Modelling and Software 26(1), 52–65 (2011)
29. Sharp, E., Curtis, A.: Namoi groundwater management survey 2011: A summary of findings. Institute for Land Water and Society Technical Report. ILWS: Albury, NSW (2012)
30. Fu, B., Merritt, W.: The impact of uncertain ecological knowledge on a water suitability model of riverine vegetation. In: Seppelt, R., Voinov, A.A., Lange, S., Bankamp, D. (eds.) International Conference on Environmental Modelling and Software (iEMSs 2012), Leipzig, Germany (2012), `http://www.iemss.org/society/index.php/iemss-2012-proceedings`

31. Young, W.J., Scott, A.C., Cuddy, S.M., Rennie, B.A.: Murray Flow Assessment Tool – a technical description. Client Report, CSIRO Land and Water, Canberra (2003)
32. Rogers, K., Ralph, T.: Floodplain Wetland Biota in the Murray-Darling Basin: Water and Habitat Requirements. CSIRO Publishing (2010)
33. Roberts, J., Marston, F.: Water Regime of Wetland and Floodplain Plants: A Source Book for the Murray-Darling Basin. National Water Commission, Canberra (2011)
34. Letcher, R.A., Jakeman, A.J., Croke, B.F.W.: Model development for integrated assessment of water allocation options. Water Resources Research 40, W05502 (2004)
35. McCown, R.L., Hammer, G.L., Hargreaves, J.N.G., Holzworth, D.P., Freebairn, D.M.: APSIM: a novel software system for model development, model testing and simulation in agricultural systems research. Agricultural Systems 50(3), 255–271 (1996)
36. Jones, C.A., Kiniry, J.R.: CERES-Maize: A simulation model of maize growth and development. Texas A&M University Press, College Station (1986)
37. Chiew, F., Teng, J., Kirono, D., Frost, A., Bathols, J., Vaze, J., Viney, N., Hennessy, K., Cai, W.: Climate data for hydrologic scenario modelling across the Murray-Darling Basin: A report to the Australian Government from the CSIRO Murray Darling Basin Sustainable Yields Porject, p. 35. CSIRO, Australia (2008)
38. Carberry, P.S., Hochman, Z., McCown, R.L., Dalgliesh, N.P., Foale, M.A., Poulton, P.L., Hargreaves, J.N.G., Hargreaves, D.M.G., Cawthray, S., Hillcoat, N., Robertson, M.J.: The FARMSCAPE approach to decision support: farmers', advisors', researchers' monitoring, simulation, communication and performance evaluation. Agricultural Systems 74, 141–177 (2002)

Part II
Papers

Kinetic Analysis of the Coke Calcination Processes in Rotary Kilns

E.M. Elkanzi, F.S. Marhoon, and M.J. Jasim

University of Bahrain, Isa Town, Kingdom of Bahrain
{elkanzi,fmarhoon}@uob.edu.bh

Abstract. Kinetic analysis of the green petroleum coke calcining processes using the simulation program HYSYS and actual industrial data is presented. The rates of physical and chemical phenomena of interest, such as the rate of moisture removal, rates of volatile matter release and combustion, rates of coke dust and sulphur combustion were all represented by their kinetic models. This paper gives a detailed description of the simulation of these processes using HYSYS "kinetic reactor" module. The results were compared with actual industrial rotary kiln data in order to validate the simulation and there was a reasonable agreement for the two different GPCs considered. The methodology of kinetics-based simulation described in this study may be used to predict coke calcining kilns performance regardless of the green coke composition.

Keywords: Rotary Kiln, Calcining Processes, Kinetics, Simulation.

1 Introduction

Calcined petroleum coke (CPC) suitable for aluminum industry anodes calls for a high quality standard. In this context, coke with no moisture, volatile matter and low sulfur contents and acquiring an appropriate crystalline structure are desired. Calcination is accomplished by the gradual heating of the green coke (GC) from ambient temperature to temperatures around 1390°C in a rotary kiln. The kiln is operated as a counter-flow heat exchanger and described elsewhere [4]. There are certain properties in green coke that a calciner must take into consideration such as: moisture, sulfur, volatile matter (VM), and metal content. Hence, when a calciner receives GPC with any of these properties failing to meet the CPC specifications, he must take action to resolve the problem. If the problem in the GPC is related to the metal content exceeding the specifications limit then blending of different green cokes is necessary to meet desired conditions because metals can not be removed by the calcining processes. Whereas in the case of exceeding moisture, sulfur, or VM contents it is possible to adjust certain operational parameters to reach the allowable limit.

Mathematical models have been developed in the past to describe petroleum coke calcinations processes in rotary kilns. Some reviews of these models and simulations were referenced elsewhere [4]. Some previous simulators were essentially written computer programs to solve the set of the simultaneous differential equations

M.S. Obaidat et al. (eds.), *Simulation and Modeling Methodologies, Technologies and Applications*, Advances in Intelligent Systems and Computing 256,
DOI: 10.1007/978-3-319-03581-9_3,

representing the material balances, heat balances and chemical reactions along the kiln [7, 13, 11, 2, 10]. In a previous publication [4], the rotary coke calcining kiln processes were simulated using a commercial simulator. The reactions were simulated as conversion reactions and the values of the conversions were obtained from real kiln data.

The objective of this study is to simulate the rotary kiln processes based on the kinetic expressions of the calcination reactions. The kinetic-based simulation would improve the prediction of the kiln operating conditions that control the contents of undesirable impurities in the calcined coke. The commercial software ASPEN HYSYS was used for this purpose and the reactions were assumed to take place in mixed reactors in series along the rotary kiln.

2 Rates of Calcination Processes

With reference to Figure 1, the calcining processes may be visualized to take place at three general zones inside the kiln. In the release zones water is driven off the coke at temperatures up to 400 °C, while the volatile matter (VOC) is driven off the coke between 400-600 °C. The *volatile* matter is combusted between 600-1000 °C in the VOC zone. In the fuel combustion and calcined coke zones, fuel is combusted, the coke delsulfurized and carbon oxidized at temperatures of 1250 to 1400 °C. All these zones were simulated as continuous stirred tank reactors

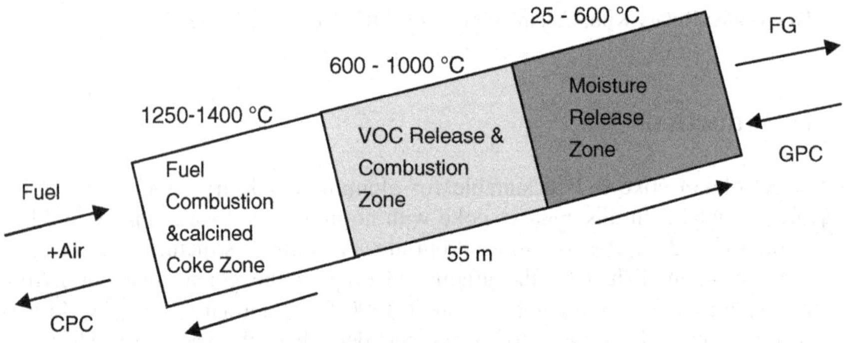

Fig. 1. Calcining Processes Zones

2.1 Moisture Release Rate

The water in the pores of the coke is heated by the counter flow of the flue gases in the kiln and is driven off the coke. The evaporation *"reaction"* may be represented by:

$$H_2O_{(L)} \longrightarrow H_2O_{(V)} \tag{1}$$

However, the use of kinetic reaction in HYSYS requires that either water liquid or water vapour be represented by a hypothetical compound having the same properties of water.

The water release rate from the bed to the vapour phase had been described as a first order reaction [9] and was given by:

$$R_w = k_w \, (X_c \, G_c \, / \, MW \, A \, u_b \,) \, X_w \, e^{-E/RT} \tag{2}$$

where k_w is a constant = $2.549 \times 10^7 \ s^{-1}$ and E is the release energy = 4.1942×10^4 J/mol

2.2 VOC Release Rates

Different GPC feed stocks contain different amounts and compositions of VOCs. The composition of one of the VOCs, GPC1, used in this study is 10.89 wt.% and its analysis showed that it contained 5.445 wt.% H_2, 4.062 wt.% CH_4 and 1.383 wt.% tar (ALBA,2005). To investigate further the type of the VOCs, FTIR spectroscopy analysis was carried out. Figure 2 shows the spectrum of the GPC1 that displays strong absorption band near 3440 cm^{-1} indicative of hydroxyl groups e.g. phenols. The peaks between 3000 and 2800 cm^{-1} correspond to saturated C-H of alkyl substituents and methylene groups in hydroaromaic compounds, the band between 1640-1680 cm^{-1} is indicative of carbon sulphides and that at 1437 corresponds to C-H bonding of methylene groups. The pattern of absorption bands between 874 and 740 cm^{-1} probably arise from vibration of aromatic C-H bonds. There are neither clear sulphur peaks at 3200-330 nor mercaptans at or near 2600. These results are similar to those of a number GPCs reported by a previous study [12].

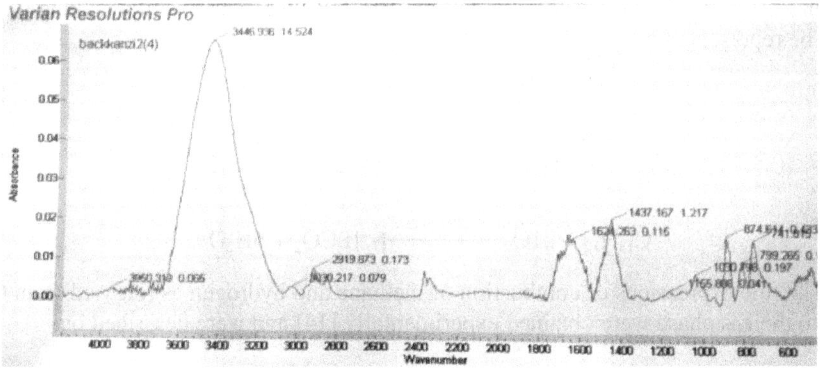

Fig. 2. FTIR Spectrum of GPC1

The VOCs entrained in the pores of the coke are heated by the counter flow of the flue gases in the kiln and evolve into the gas phase. The desorption and evaporation "*reactions*" may be represented by:

$$CH_{4(ad)} \longrightarrow CH_{4(g)} \tag{3}$$

$$H_{2(ad)} \longrightarrow H_{2(g)} \tag{4}$$

$$C_{18}H_{12(L)} \longrightarrow C_{18}H_{12\,(g)} \tag{5}$$

The same procedure was followed for the VOC release where hypothetical compounds were introduced with the same properties as the real ones.

The kinetic expressions for the release of VOCs from the bed to the vapour phase were obtained experimentally ([3] and reported by [10]) and are described by the empirical correlation:

$$R_v = k_V\,(X_c\, G_c / MW_v\, u_b\, A)\,(X_{v0} / [X_c\, X_{v0}]^n)\,(X'_v)^n\, e^{-E/RT} \quad (6)$$

The values of the constants in Eq. (6) are as shown in Table 1.

Table 1. Empirical constants for evaluating VOC release rates [3]

VM	$k_{VOC}(s^{-1})$	$E(J\ mol^{-1})$ (release)	n
H_2	$9.17*10^1$	$4.37*10^4$	1.5
CH_4	$1.49*10^6$	$1.75*10^5$	2.0
$C_{18}H_{12}$	$1.09*10^{-1}$	$1.25*10^5$	1.5

2.3 Combustion of VOCs

The combustion of VOCs was simulated by kinetic expressions in HYSYS. The VOCs released from the coke bed are oxidized by hot gases and would combust depending on the amount of tertiary air entering this zone. The reactions taking place may be represented by:

$$CH_4 + 2O_2 \longrightarrow CO_2 + 2H_2O \quad (7)$$

$$2H_2 + O_2 \longrightarrow 2H_2O \quad (8)$$

$$C_{18}H_{12} + 21O_2 \longrightarrow 18CO_2 + 6H_2O \quad (9)$$

The rate expressions of combustion of methane and hydrogen as released from the bed to the gas phase were obtained experimentally [14] and were described by:

$$R_{CH4} = 7 * 10^{11}\,(y_{CH4})^{-0.5}\,(y_{O2})^{1.5}\,e^{[-30196/Tg]}\,T_g \quad (10)$$

$$R_{H2} = 2.45 * 10^{11}\,[H_2]^{0.85}\,[O_2]^{1.42}\,e^{-20634/Tg} \quad (11)$$

The rate expression for the combustion of tar can be described [3,5] by the empirical equation:

$$R_{ch} = k_{ch}\,[C_{ch}]^1\,[C_{O2}]^{0.5}\,[C_{H2O}]^{0.5}\,e^{[-E/RT]} \quad (12)$$

The values of the constants were as shown in Table 2.

Table 2. Constants of the combustion of Tar rate expression

k_{ch} cm^3 / mole-s	E J/ mole	Reference
$1.8 * 10^{13}$	$1.32*10^5$	[3]
$1.3 * 10^{14}$	$1.25*10^5$	[5]

2.4 Desulfurization and Carbon Oxidation Rates

The desulfurization and carbon dust oxidation reactions were simulated as two reactors. Oxygen from tertiary air reacts with the sulphur and sulphides and the carbon dust to produce SO_2, CO_2 and CO according to the following reactions:

$$C + O_2 \longrightarrow CO_2 \quad (13)$$

$$C + CO_2 \longrightarrow 2CO \quad (14)$$

$$S + O_2 \longrightarrow SO_2 \quad (15)$$

$$COS + 3/2\ O_2 \longrightarrow CO_2 + SO_2 \quad (16)$$

$$CS_2 + 3O_2 \longrightarrow CO_2 + 2SO_2 \quad (17)$$

The rates of entrained coke fines burn-up were reported [7] and described by the following equations:

$$R_{CO2} = 6.84 * 10^{13}\ C_{O2}\ e^{[-30800/Tg]}\ (S_f\ \Delta L / V) \quad (18)$$

$$R_{CO} = [1.08 * 10^{12}\ e^{-62180/Tg}\ (P_{CO2}{}^{0.5})\ (S_f\ \Delta L / V)] / [12.01\ (1+ (P_{CO} / 2887 e^{-24624/Tg}))] \quad (19)$$

The rate of carbon monoxide can be simplified by neglecting the 1 in the dominator since the term $\{P_{CO} / 2887\ e^{-24624/Tg}\} \gg 1$ and hence equation (19) becomes:

$$R_{CO} = 2.62 * 10^{14}\ (P_{CO2})^{0.5}\ (P_{CO})^{-1}\ e^{[-86804/Tg]}\ (S_f\ \Delta L / V) \quad (20)$$

The rate of oxidation of sulphur was reported [8] and is described by:

$$R_S = 1.1 * 10^{-11} * e^{[-4360]/T} * C_{sv} \quad (21)$$

Since the units of C_{sv} are in molecule/ cm^3, multiplying by Avogadro's number yields:

$$R_S = 6.6 * 10^{12} * e^{[-4360]/T} * C_{sv} \quad (22)$$

2.5 Fuel Combustion Rate

The fuel used in the kiln was natural gas and hence the rate of the fuel combustion is as given by equation (10). The results of the kinetic equation (10) were compared with normal conversion reaction and showed great deal of agreement.

3 HYSYS Simulation

Figure 3 is the HYSYS simulation flow sheet of the coke calcinations processes using industrial input data [1] and showing the kiln calcinations zones. Process simulation assumes good mixing inside the kiln; which is accomplished by the kiln being rotating at an inclined position and by the presence of tumblers. Moreover, it was assumed that the reactions take place effectively in the gas phase. As a result of these assumptions, the calcinations processes were simulated to take place inside adiabatic continuous stirred-tank reactors. It is to be noted that there is no sharp demarcation

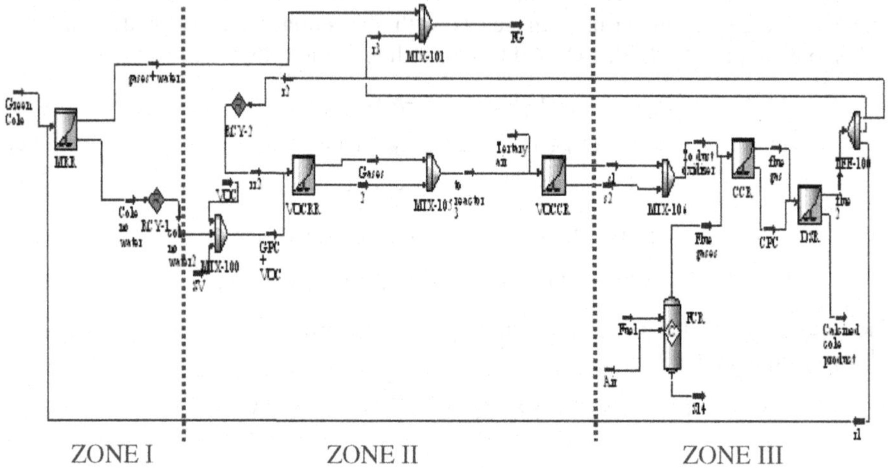

Fig. 3. HYSYS Simulation of Kiln Processes

between these reactors; in fact part of the VOCs is released in the moisture release "reactor" and a very little amount of VOCs is combusted in the VOC release "reactor" since the release zone is oxygen deficient. The simulation procedure is similar to the one used before [4] featuring the concept of using "recycle" to simulate counter-current mass flow that is not allowed by HYSYS. The combustion of fuel (CH_4) is simulated by fuel combustion reactor, FCR, which is a conversion reactor. The flue gas from the burner is mixed with the outputs from the VOC combustion reactor (VOCCR) and enters the carbon combustion reactor (CCR). The CCR is simulated using equations (16) and (18), for which the carbon dust was simulated by hypothetical "carbons" so as not to confuse them with the rest of the coke carbon that was assumed not to react in the solid phase. The output from CCR enters the desulphurization reactor (DSR) that was simulated using equation (20). The solid output from DSR is the calcined coke product. The gas output is the "*recycle stream*" which is split into three streams: the first is mixed with the fresh GPC and enters the moisture release reactor (MRR, simulated using equation2, for which the product water was simulated by a hypothetical water component), the second is sent to the VOC release reactor (VOCRR), and the third is mixed with the gas output from the MRR to form the flue gas that is sent the incinerator. It is to be noted here that the split ratio of the recycle stream is determined by trial and error in order for the temperature in MRR, VOCRR and VOCCR to fall within the operating range shown in Fig. 1. Since it is assumed that the VOCs and sulphur are not released in the MRR, they were mixed and sent to the VOC release reactor (VOCRR) that is simulated using equation (6) and the data of Table 1. The components in the release equations (3), (4) and (5) were simulated by hypothetical components. The output from VOCRR is mixed with tertiary air and sent to the VOCCR that was simulated using equations (10), (11), (12), and the data in Table 2.

Results and Discussion
Coke kiln industrial data [1] for two different types of green cokes are compared with the results of the simulated values using the same feed and other operating conditions for each feed. The compositions of the two GPCs are as shown in Table 3.

Table 3. GPCs Compositions

GPC1		GPC2	
Species	Wt.%	Species	Wt.%
Carbon	81.7	Carbon	80.0
VOC	10.89	VOC	6.34
H2O	5.26	H2O	6.57
Sulphur	2.17	Sulphur	6.66
Carbon Dust	0.635	Carbon Dust	0.43

At this initial stage of work, the comparison would be between compositions, flow rates, and temperatures of CPC, and flue gas streams.

The results of the simulations were as shown in Tables 4&5 together with the industrial values. The comparison reveals that the perentage of sulphur in the CPC, for GPC1, is zero and that the exiting CPC temperature is less by 0.9 % than the industrial value. Since the CPC real density is a function of temperature [6], the decrease in CPC predicted temperature would have very little effect on the real density. The zero sulphur is an over prediction since equation (21) is for the oxidation of sulphur in the vapour phase. The flue gas temperature was higher by about 11%. This can be explained by looking into the composition of the flue gas where almost all volatile components and carbon dust were oxidized adding more heat than that of the industrial case. Moreover, the assumption of adiabatic operation would add to the overall rise in temperature. The results of the simulation of GPC2 using the same conditions as the base case showed that an increase in the flow rate of GPC2 by1.4% from the base case is necessary to meet the same rate of production of CPC. The decrease of 16.4% and 19.2% in the CPC and flue gas temperatures , from the base case values, respectively, may be explained by the decrease of the VOCs in GPC2 resulting in less heat release. In this case the sulphur is considered in the form of carbon sulphides as predicted by the FTIR analysis of Fig.2 and equations (16&17).

Table 4. Simulation Results of GPC1

Stream / Condition		GPC 1	GPC 2	CPC 1	CPC 2	FG 1	FG 2	Tertiary 1	Tertiary 2	Fuel 1	Fuel 2	Fuel Air 1	Fuel Air 2
Temperature, K		293	293	1663	1650	1373	1499	293	293	293	293	293	293
Pressure, kPa		120		120		120		120		120		120	
Flow Rate, kg/s		9.76	9.76	7.44	7.27	20.60	20.62	14.0	14.0	0.276	0.276	4.77	4.77
Components Flow Rates, kg/s	C	7.925	7.925	7.27	7.27	0.24	-	-	-	-	-	-	
	C*	0.062	0.062	-		-	-			-	-	-	
	CH_4	0.394	0.394	-		0.242	-	-	-	0.276	0.276	-	-
	H_2	0.528	0.528	-		0.314	0.4	-		-		-	
	$C_{18}H_{12}$	0.134	0.134	-		0.072	-	-		-		-	
	H_2O	0.51	0.51	-		3.03	3.41	-		-		-	
	N_2	-		-		14.34	14.34	10.68	10.68	-		3.66	3.66
	CO_2	-		-		1.84	2.24			-		-	
	CO	-		-		0.43	0.077	-		-		-	
	SO_2	-		-		0.060	0.069	-		-		-	
	S	0.211	0.211	0.169	-	0.0055	-	-		-		-	
	O_2	-	-	-		-	0.057	3.24	3.24	-		11.53	11.53

1= Industrial data 2= this study C* = carbon dust

Table 5. Simulation Results of GPC2

Stream / Condition		GPC 1	GPC 3	CPC 1	CPC 3	FG 1	FG 3	Tertiary 1	Tertiary 3	Fuel 1	Fuel 3	Fuel Air 1	Fuel Air 3
Temperature, K		293	293	1663	1390	1373	1109	293	293	293	293	293	293
Pressure, kPa		120		120		120		120		120		120	
Flow Rate, kg/s		9.76	9.906	7.44	7.448	20.60	20.62	14.0	14.0	0.276	0.276	4.77	4.77
Components Flow Rates, kg/s	C	7.925	7.66	7.27	7.27	0.24	-	-	-	-	-	-	
	C*	0.062	0.043	-		-	-			-	-	-	
	CH_4	0.394	0.227	-		0.242	-	-	-	0.276	0.276	-	-
	H_2	0.528	0.304	-		0.314	0.2	-		-		-	
	$C_{18}H_{12}$	0.134	0.077	-		0.072		-		-		-	
	H_2O	0.51	0.629	-		3.03	2.48	-		-		-	
	N_2	-		-		14.34	14.34	10.68	10.68	-		3.66	3.66
	CO_2	-		-		1.84	1.24			-		-	
	CO	-		-		0.43	0.061	-		-		-	
	SO_2	-		-		0.060	0.92	-		-		-	
	S	0.211	0.637	0.169	0.0	0.0055	-	-		-		-	
	O_2	-	-	-		-	0.057	3.24	3.24	-		11.53	11.53

1= Industrial data 3= this study C* = carbon dust

4 Conclusions

The simulation of the processes that describe green petroleum coke calcinations was based on using the kinetic expressions of the calcination processes. The results were compared with actual industrial rotary kiln data and there was a reasonable agreement for the two different GPCs. The methodology of kinetics-based simulation described in this study may be used to predict coke calcining kilns performance regardless of the green coke composition.

5 Nomenclature

A = area per unit axial length, m.
C = species molar concentration, kgmol/m^3
E = activation energy, kj/kgmol
G = mass flow rate, kg/s
k = rate constant, s^{-1}.
l = axial distance along the kiln, m.
MW = molecular mass, kg/kmol
Pi = partial pressure of species i, kPa
Ri = rate of reaction of species i, kmol/m^3. S
Sf = total surface area of the coke fines/unit length of the kiln, m.
T = temperature, K.
u = velocity, m/s.
V = volume of gas in the kiln, m^3.
X = mass fraction, kg/kg.
y = mole fraction, kmol/kmol.

Subscripts

b = coke bed.
c = coke or carbon
ch = tar
g = gas phase
l = liquid phase
o = initial
voc = volatile organic compound
v = volatile
w = water

References

1. ALBA, Private communication
2. Bui, R.T., Perron, J., Read, M.: Model-based optimization of the operation of the coke calcining kiln. Carbon 31(7), 1139–1147 (1993)

3. Dernedde, E., Charette, A., Bourgeois, T., Castonguay, L.: Kinetic Phenomena of the Volatiles in Ring Furnaces. In: Light Met. Pcoc. Tech, Sess. AIME 105th Annual Meeting, p. 589 (1986)
4. Elkanzi, E.M.: Simulation of the Coke Calcining Processes in Rotary Kilns. Chemical Product and Process Modeling 2(3), Article 20 (2007)
5. Howard, J.B., Williams, G.C., Fine, D.H.: Kinetics of Carbon Monoxide Oxidation in Post flame Gases. In: 14th International Symposium on Combustion, pp. 975–985 (1973)
6. Ibrahim, H.A., Ali, M.M.: Effect of the removal of sulphur and volatile matter on the true density of petroleum coke. Periodica Polytechnica Ser. Chem. 49(1), 19–24 (2005)
7. Li, K.W., Friday, J.R.: Simulation of Coke Calciners. Carbon 12, 225–231 (1974)
8. Lu, C.-W., Wu, Y.-J.: Experimental and theoretical investigations of rate coefficients of the reaction $S(^3P)+O_2$ in the temperature range 298-878 K). Journal of Chemical Physics 121(17), 8271–8278 (2004)
9. Lyons, J.W., Min, H.S., Parisot, P.F., Paul, J.F.: Experimentation with a Wet-Process Rotary Cement Kiln via the Analog Computer. Ind. Eng. Chem. Process. Des. Dev. 1(1), 29–33 (1962)
10. Martins, A., Marcio, O., Leandro, S., Franca, A.S.: Modeling and Simula-tion of Petroleum Coke Calcination in Rotary Kilns. Fuel 80, 1611–1622 (2001)
11. Perron, J., Bui, R.T., Nguyen, T.H.: Modelisation du four de calcination du coke de petrole: 2- simulation du procede. Can. J. Chem. Eng. 70, 1120–1131 (1992)
12. Menendez, J.A., Pis, J.J., Alvarez, R., Barriocanal, E., Fuente, E., Diez, M.A.: Characterization of Petroleum Coke as an Additive in Metallurgical Coke making Modification of Thermoplastic Properties of Coal. Energy & Fuels 10, 1262–1268 (1996)
13. Perron, J., Bui, R.T.: Rotary Cylinders: Solid transport Prediction by Dimensional Rheological Analysis. Can. J. Chem. Eng. 68, 61–68 (1990)
14. Srinivasan, R.J., Srirmulu, S., Kulasekaran, S.: Mathematical modeling of fluidized bed combustion- 2: combustion of gases. Fuel 77(9/10), 1033–1043 (1988)

Behavior of Elastomeric Seismic Isolators Varying Rubber Material and Pad Thickness: A Numerical Insight

Gabriele Milani[1] and Federico Milani[2]

[1] Politecnico di Milano, Piazza Leonardo da Vinci 32, 20133 Milan, Italy
[2] Chem.Co Consultant, Via J.F. Kennedy 2, 45030 Occhiobello (RO), Italy
gabriele.milani@polimi.it, federico-milani@libero.it

Abstract. A numerical approach for the determination of (a) the shear behavior under large displacements and (b) the compression elastic modulus of common parallelepiped elastomeric isolators is presented. Particular attention is devoted to the role played by the material used for the rubber pads and their thickness. For them, an experimental data fitting by means of both a nine constants Mooney-Rivlin and a five constants exponential law is utilized, within a Finite Element discretization of the isolator. Having at disposal a few experimental stretch-stress data points for each rubber compound in uniaxial tension, a cubic Bezier spline approach is firstly utilized, to generate numerically a large number of metadata containing the original experimental ones. Then, respectively the nine Mooney-Rivlin and five exponential law constitutive parameters are estimated through a least square approach. Once assessed the models, a full scale rectangular seismic isolator is analyzed when subjected to horizontal actions and normal compression, in order to provide estimates of the initial stiffness and the overall behavior of the isolator undergoing large deformations, using both models and for all the compounds considered. It is found that the global behavior may depend significantly on the material hypothesis assumed to model rubber and on pads thickness.

Keywords: Elastomeric Isolators, Rubber Typology, Compounds Performance, Stretch-strain Behavior under Large Deformations, Numerical Model Simulations, Finite Element Method, Pad Thickness.

1 Introduction

In the past, as a consequence of the high costs, elastomeric seismic isolation technology was very limited and mainly conceived for large, expensive buildings housing sensitive internal equipment, e.g. computer clusters, emergency operation centers, and hospitals. To extend this valuable earthquake-resistant strategy to common housing and commercial buildings, producers made several efforts to reduce the weight and cost of the isolators. Now they are available at relatively low prices, even for standard new buildings - Tsai & Lee [20], Hsiang-Chuan Tsai [9].

An elastomeric seismic isolator is a layered device constituted by thick rubber pads (10-30 mm) and thin reinforcing steel plates, laid between pads [13], see Fig. 1.

M.S. Obaidat et al. (eds.), *Simulation and Modeling Methodologies, Technologies and Applications*, Advances in Intelligent Systems and Computing 256,
DOI: 10.1007/978-3-319-03581-9_4, © Springer International Publishing Switzerland 2014

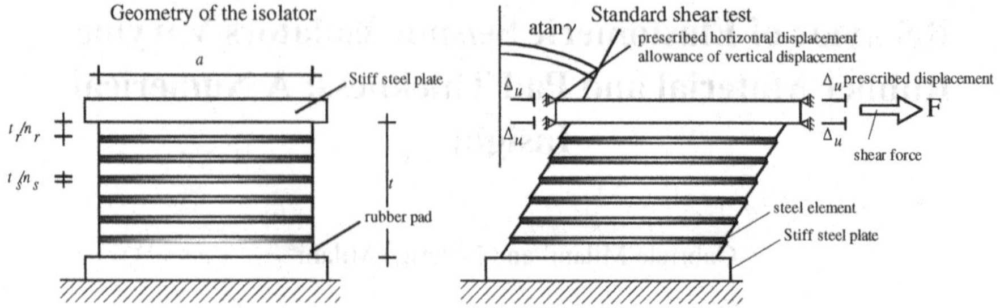

Fig. 1. Geometry of the elastomeric isolator

rubber pads may be realized with natural (NR) or artificial rubber (AR). NR is still the most diffused, but it will be superseded very soon in the built stock by AR, NR suffering from quick aging and being the industrial production capability limited. AR is usually neoprene, or less frequently, Braga et al. [4], EPDM (Ethylene Propylene Diene Monomer) and NBR-EPDM (Nitrile-Butadiene Rubber-EPDM) blend vulcanized with or without an increasing amount in weight of carbon black.

Thanks to rubber incompressibility [1,2] and the introduction of the thin steel sheets, a seismic isolator is extremely deformable for horizontal forces, but at the same time sufficiently stiff when loaded with vertical actions. This is essential in buildings subjected to seismic loads, where the main goal to achieve is to "isolate" (i.e. increase the period of the structure) the whole structure to the ground when seismic load acts and to sustain the vertical loads transferred to the foundation.

In this work, the influence of rubber pads mechanical properties and thickness on both vertical and horizontal stiffness of realistic seismic isolators is discussed. A number of different rubber blends available in the market are considered, to show that a different performance is achieved at a structural level when different materials for the pads are used. A numerical approach based on experimental data fitting is adopted to characterize the mechanical behaviour under tensile stretching and pure shear. Both a nine constants Mooney-Rivlin model [14] and an exponential law [1,2] are assumed for the evaluation of the energy density. Constants entering in both models are estimated by means of a standard least squares routine fitting experimental data available. Due to the insufficient number of experimental values at disposal, Bezier splines are utilized to numerically generate a large number of (meta)data.

Once that the unknown coefficients are evaluated for both models and for all the compounds analyzed, a standard parallelepiped seismic isolator is simulated using FEs in compression and in pure shear tests under large deformation. To perform the simulations, both a Mooney Rivlin and an Amin et al. [1] model, see also Milani & Milani [11,12] are implemented in an existing non commercial large deformation FE code.

Having at disposal an idealized mechanical behaviour for the pads to be used at a structural level, a second parameter that may influence the overall behaviour of the devices is investigated, namely the thickness of the single pads, Milani & Milani [12], which may vary the overall elastic compression modulus E_c of the isolator. E_c is evaluated by means of a full 3D Finite Element discretization, comparing results obtained with the model proposed with those provided by [19] model and Italian code [15] formulas, varying both first shape factor in the range of technical interest and compound used in the rubber pad.

From simulations results, generally, it is found that the most indicated compound is neoprene heavily loaded with C/B (Carbon Black), or an EPDM and NBR-EPDM blend with C/B in variable weight percentage. Both of them suffer from a sudden increase of the stiffness at relatively high stretches, meaning that a non-linear analysis of the building should be performed in this case.

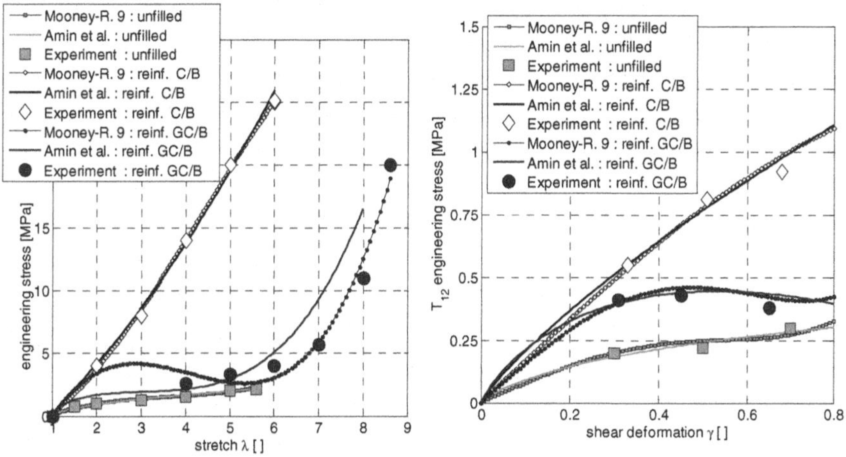

(a) (b)

Fig. 2. Unfilled, reinforced with carbon black (C/B) and with graphitized carbon black (GC/B) generic non crystallized rubber. –a: Uniaxial stretch-nominal stress diagram. –b: simple shear behaviour. Experimental data and numerical (Mooney Rivlin and 5 constants [1]) models response.

2 Pads Uniaxial and Shear Experimental Behavior at Finite Deformations with Different Materials

The most commonly performed mechanical test to charaterize rubber vulcanizates is the uniaxial extension of a strip to its breaking point. Fig. 2-a shows schematically the uniaxial stretch-stress behavior of an amorphous rubber incapable of crystallization under strain. The first curve (with green squares) is for a gum vulcanizate, the second (white diamonds) for a vulcanizate reinforced with a high-surface-area structure carbon black (C/B), and the latter for a vulcanizate reinforced with the same black after graphitization (GC/B). As it can be noticed, sample with C/B exhibits an initial slope of the stretch-strain curve much greater, typically due to the contribution of the filler. Such increase in initial stiffness varies varying the amount and typology of the filler and is a function of the state of subdivision of the filler and hence of its specific surface area. Once the effects of secondary agglomeration are overcome, several mechanisms remain, responsible of the stress rising faster than in the unfilled sample:

the most important is the strain amplification, since the rigid filler does not share in the deformation, Bueche [6]. It is clear that the inclusion of a rigid adhering filler in a soft matrix will cause the average local strain in the matrix to exceed the macroscopic strain. As a consequence, the rubber in the filled vulcanizate is highly strained and responds with a higher stress. Strain amplification also increases the mean rate at which the matrix is strained, leading to a further increase in stress.

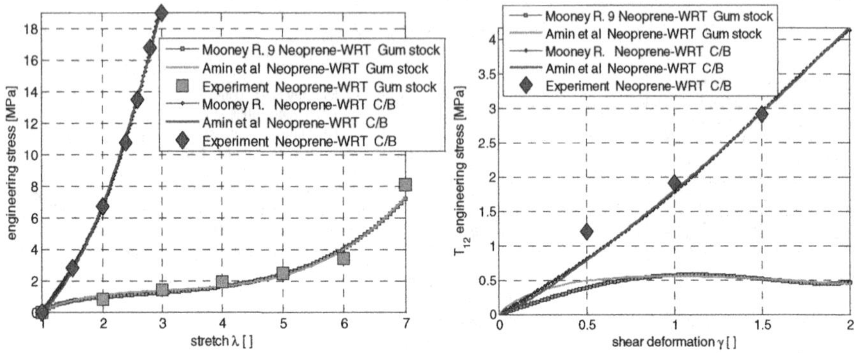

(a) (b)

Fig. 3. Neoprene WRT with C/B and Gum stock. –a: Uniaxial stretch-nominal stress diagram. –b: simple shear behaviour. Experimental data and numerical (Mooney Rivlin and 5 constants [1]) models response.

Stretch-stress data reported in Fig. 3-a are post-processed from experimental data collected in Studebaker & Beatty [17] and represent the typical uniaxial behavior of neoprene gum stock in comparison with neoprene loaded with carbon black at 50 phr. In general, in this case the behavior of the neoprene sample loaded with carbon black is sensibly stiffer, with a rather marked increase of the tensile strength and a decrease of elongation at break, which is typical of loaded elastomers.

Similar considerations may be repeated for shear tests, Fig. 3-b. Shear initial stiffness may depend strongly on the cristallization degree and on the amount of filler and unsaturation. However, to have at disposal shear experimental data is not always possible, see Fig. 2-b. When available, numerical models coefficients are obtained by least squares on all experimental data.

In Fig. 4-a, the uniaxial stretch-stress behavior of unfilled and filled EPDM is represented. Data are kept from Studebaker & Beatty [17]-. Here, it is worth noting that the filler amount used is exactly the same used for neoprene in Fig. 3. Tensile strength increases using carbon black, but, differently to neoprene, the elongation at break is almost the same of the unfilled compound, passing from 4.2 to 5.1: this is due to the low unsaturation of the EPDM rubber when compared with neoprene, natural rubber etc. Conversely, tensile strength is comparable to neoprene. Therefore, one of the advantages of EPDM compared with neoprene would be the more ductile behavior.

In Fig. 5-a, a comparison between stretch-stress curves for neoprene and a mixture of nitrile rubber (Elaprim-S354-EP) and EPDM (Dutral TER 9046) in the ratio of

70:30 in weight is represented for a standard tensile stress experimentation. Two experimental curves for Elaprim+Dutral are reported, to show the low experimental data scattering. The vulcanization conditions are nearly the same (160° for 30 and 20 minutes for neoprene and Elaprim+Dutral respectively). Both rubber compounds have been loaded with carbon black at 50 phr.

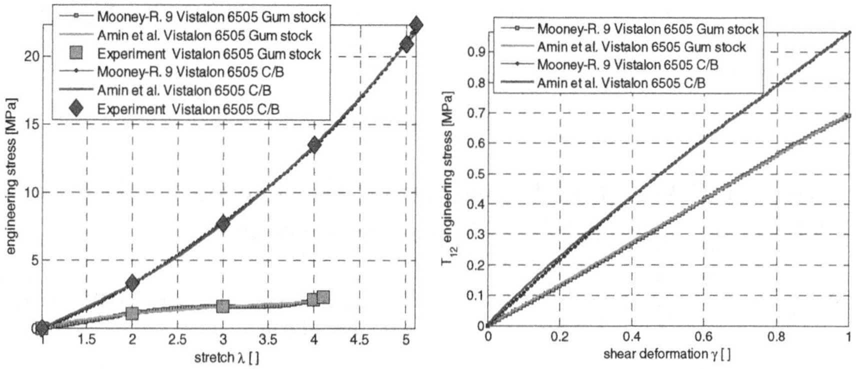

(a) (b)

Fig. 4. EPDM Vistalon 6505 Gum stock and Vistalon 6505 with C/B. –a: Uniaxial stretch-nominal stress diagram. –b: simple shear behaviour. Experimental data and numerical (Mooney Rivlin and 5 constants [1]) models response.

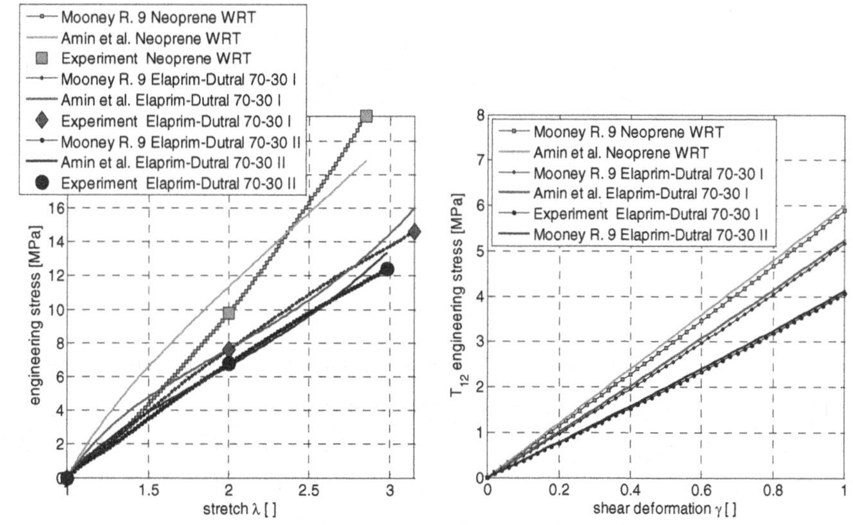

(a) (b)

Fig. 5. Neoprene WRT, Elaprim-Dutral 70-30 (1st test) and Elaprim-Dutral 70-30 (2nd test) compounds. –a: Uniaxial stretch-nominal stress diagram. –b: simple shear behaviour. Experimental data and numerical (Mooney Rivlin and 5 constants [1]) models response.

Neoprene behavior is nearly linear, with slightly higher tensile strength. The elongation at failure is comparable for both compounds. While generally it can be stated that neoprene mechanical performance is preferable, mainly for its quite linear response even at high levels of stretches, it is worth mentioning that nitrile rubber added with EPDM has some other advantages, as for instance the lower cost, lower time of vulcanization, lower specific weight for the items. In any case, limiting the analyses for stretches under 2 or shear distortion angles lower than 45°, Fig. 5-b, the mechanical behavior is, from an engineering point of view, the same, meaning that if a seismic isolator is designed to undergo deformations where rubber pads do not exceed this threshold stretch, nitrile rubber and EPDM may be considered as a valid alternative to neoprene.

3 Numerical Model for Rubber Pads: Nine Constants Mooney Rivlin and Amin et al. [1,2] Exponential Model

In an uniaxial test, we usually define the stretch as the ratio between the length in the deformed configuration divided by the length in the undeformed state. Let $\lambda_1 = \lambda$ be the stretch in the direction of elongation and $\sigma_1 = \sigma$ the corresponding stress. The other two principal stresses are zero, since no lateral forces are applied $\sigma_2 = \sigma_3 = 0$. For constancy of volume, the incompressibility condition $\lambda_1 \lambda_2 \lambda_3 = 1$ gives $\lambda_2 = \lambda_3 = 1/\sqrt{\lambda}$. The strain energy function, in the most general case, for the Mooney-Rivlin model is given by:

$$W = \sum_{r=0\,s=0}^{m,n} C_{rs} (I_1 - 3)^r (I_2 - 3)^s \quad C_{00} = 0 \tag{1}$$

where $I_1 = \lambda_1^2 + \lambda_2^2 + \lambda_3^2$ and $I_2 = \lambda_1^{-2} + \lambda_2^{-2} + \lambda_3^{-2}$.

When dealing with the six constants Amin et al. (2006) model, the strain energy function is given by:

$$W = C_2(I_2 - 3) + C_5(I_1 - 3) + \frac{C_3}{N+1}(I_1 - 3)^{N+1} + \frac{C_4}{M+1}(I_1 - 3)^{N+1} \tag{2}$$

where C_2, C_3, C_4, C_5, N, M are material parameters to be determined. Amin et al. [1] five constants model is obtained simply assuming C_2=0.

In uniaxial tension or compression equations it can be shown that the strain invariants are $I_1 = \lambda^2 + 2\lambda^{-1}$ and $I_2 = 2\lambda + \lambda^{-2}$. The engineering stress S' (force per unit unstrained area of cross-section) may be evaluated from energy density as follows:

$$S' = 2\left(1 - \frac{1}{\lambda^3}\right)\left(\lambda \frac{\partial W}{\partial I_1} + \frac{\partial W}{\partial I_2}\right) \tag{3}$$

To fully characterize a rubber compound in terms of Mooney-Rivlin stress-stretch curves, it is therefore necessary to perform suitable experimental mono-axial tests in pure traction and/or shear at large deformations on different rubber compounds. The relationship between the nominal stress and the corresponding stretch, after substituting the corresponding strain energy density function (1) or (2) into (3) can be generalized as follows:

$$S'(\lambda) = S'(C_k, \lambda) \tag{4}$$

where C_k are material constants in the strain energy density function, see Fig. 6-a. Equation (4) holds also in presence of experimental data available for pure shear tests, i.e. $T_{12}(\gamma) = T_{12}(C_k, \gamma)$, where γ identifies the shear strain, namely the distortion of

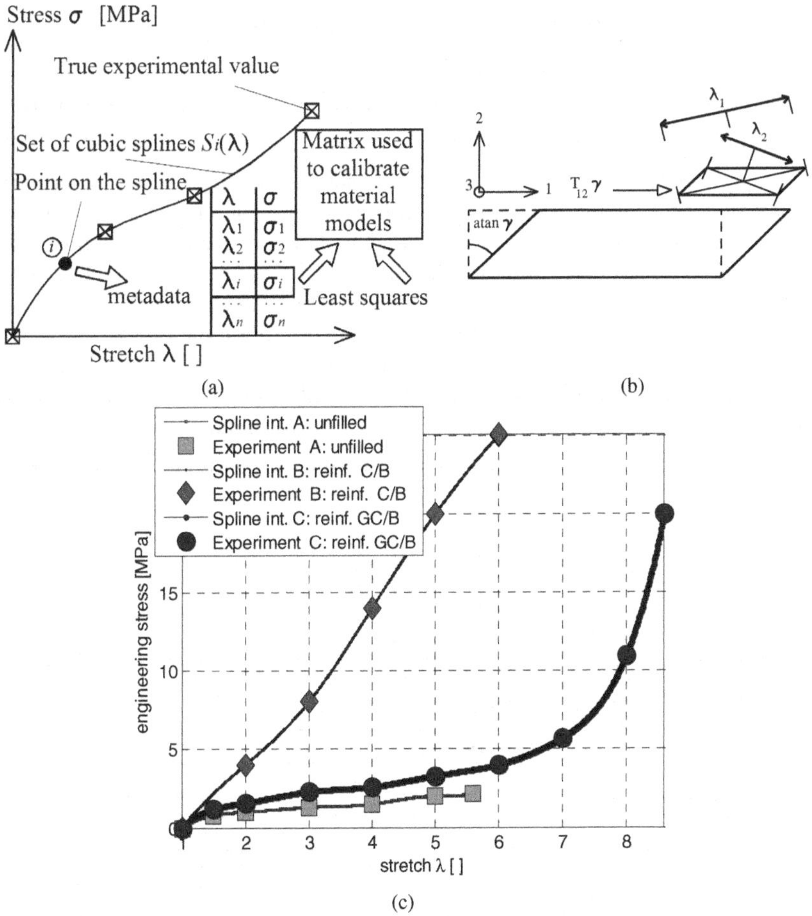

Fig. 6. Numerical procedure adopted to fit experimental data and to calibrate the material parameters model. -a: fitting of experimental data (squares) by means of a set of natural cubic splines and subsequent metadata extraction to use as input points to calibrate the material constants through least squares. –b: pure shear deformation: principal stretch directions, shear angle and T12 shear internal action. -c: splines fitting for the compounds represented in Fig. 2.

a rectangle subjected to pure shear, Fig. 6-b. C_k are determined by a (non) linear least squares fitting performed on experimental data and equation (4).

Assuming that there are $\overline{N}\left(\overline{N} \geq \overline{M}\right)$ pairs of experimental data (nominal stress and stretch, say $\overline{S}_i$ and $\overline{\lambda}_i$), we minimized the sum of squared differences between the calculated and the measured stress values, i.e.:

$$\min\left\{\sum_{s=1}^{N}\left(\overline{S}'_i - S'\left(\overline{\lambda}_i\right)\right)^2\right\} \tag{5}$$

Equation (5) differentiated with respect to C_k variables leads to a system of linear (in the Mooney Rivlin case) equations:

$$\sum_{s=1}^{N}\left(\overline{S}' - S'\left(\overline{\lambda}\right)\right)\left.\frac{\partial S'}{\partial C_k}\right|_{\lambda=\overline{\lambda}} = 0 \quad k = 1,\, 2,\, \overline{M} \tag{6}$$

which permits an evaluation of C_k constants. However, experimental data are usually insufficient ($\overline{N} \leq \overline{M}$) and therefore a numerical fitting is needed to collect (meta) data, which are assumed as reasonable approximation of experimental evidences. In order to avoid a polynomial fitting model, which is not unique and with an insufficient fitting performance, the actual experimental curve is approximated using a set of Bezier cubic splines and, subsequently, several intermediate points between the actual experimental data are numerically evaluated on the spline, to be used within a least squares procedure for material data calibration.

In pure shear deformation, see Fig. 6-b, and differently to uniaxial compression, the direction of applied displacement does not coincide with the direction of principal stretches; rather it involves a rotation of axes. Due to applied shear strain γ, the deformation gradient tensor $\mathbf{F}$ and the left Cauchy-Green deformation tensor $\mathbf{B}$ are described as:

$$\mathbf{F} = \begin{bmatrix} 1 & \gamma & 0 \\ 0 & 1 & 0 \\ 0 & 0 & 1 \end{bmatrix} \qquad \mathbf{B} = \begin{bmatrix} 1+\gamma^2 & \gamma & 0 \\ \gamma & 1 & 0 \\ 0 & 0 & 1 \end{bmatrix} \tag{7}$$

Consequently, the strain invariants are expressed as $I_1 = I_2 = 3 + \gamma^2$, $I_3 = 1$ and the expression for Cauchy stress becomes:

$$T_{12} = 2\gamma[\partial W / \partial I_1 + \partial W / \partial I_2] \tag{8}$$

The principal stretches associated with shear strain γ may be obtained as:

$$\lambda_1 = \sqrt{1 + \frac{\gamma^2}{2} + \gamma\sqrt{1 + \frac{\gamma^2}{4}}} \qquad \lambda_2 = \sqrt{1 + \frac{\gamma^2}{2} - \gamma\sqrt{1 + \frac{\gamma^2}{4}}} \tag{9}$$

λ_1 represents the principal tension stretch whereas λ_2 represents the principal compression stretch.

The shear stress in the six constants Amin et al. model (2006) is, for pure shear deformation, the following:

$$T_{12} = 2\gamma\left[C_5 + C_2 + C_4\gamma^{2M} + C_3\gamma^{2N}\right] \tag{10}$$

which reduces to $T_{12} = 2\gamma\left[C_5 + C_4\gamma^{2M} + C_3\gamma^{2N}\right]$ in the five constants model by Amin [1].

Where shear test data are available, parameters are obtained using a least squares routine on both tensile and shear tests experimental data, as envisaged in Amin et al. [2]. In absence of shear experimental data, parameters are evaluated only on the uniaxial behavior of the specimens. Resultant shear stress-deformation curves by Mooney-Rivlin and Amin et al. [1] models are in any case compared, to show how different is the response of the models when the fitting procedure is performed only on uniaxial tests.

In Fig. 6-c, for instance, the interpolation obtained with a cubic spline on experimental data of Fig. 2 is shown. As it is possible to notice, the fitting is almost perfect and allows to collect a number of 'metadata', which are defined hereafter as numerical stretch-stress points to use to calibrate rubber material properties and, in practice, will be used as they were a large set of experimental data available. This procedure is obviously necessary if a material model with many parameters has to be calibrated.

In figures from Fig. 2 to Fig. 5 a comparison among experimental mono-dimensional data (both tensile stress and shear) and curves provided by the two numerical models (nine constants Mooney-Rivlin and five constants [1] model) is represented. As expected, generally, the mono-axial behavior of polymers is well fitted by both models. On the other hand, shear behavior provided by both models is comparable and, in some cases, almost identical.

4 Influence of Pads Thickness on Elastic Compression Modulus

To evaluate the elastic compression modulus of a seismic isolator, a FE discretization is recommended. Due to the high vertical stiffness, linear elastic analyses are sufficient to determine precisely the compression modulus. Under small deformations rubbers are linearly elastic solids. Because of the high modulus of bulk compression, about 2000MN/m^2, compared to the shear modulus G, about 0,2-5MN/m^2 [18], they may be regarded as relatively incompressible. The elastic behavior under small strains can thus be described by a single elastic constant G, being Poisson's ratio very near to ½ and Young's modulus E equal to 3G with very good approximation.

In any case, elastic parameters to assign to single pads should be known in advance. In general, the so-called static modulus of a rubber compound is obtained in a standard stress-strain test in which the samples are extended at the rate of 20 in/min.

Hardness measures may help in the estimation of the elastic modulus, when a standard stress-strain characterization is not possible, takes too much time or it is too expensive. The test method ASTM D2240 [3] helps in the definition of a standard for hardness evaluation, standardizing the penetration of a specified indentor forced into the material under specified conditions.

In order to have the possibility to evaluated the relationship between hardness and Young's modulus, a semi empirical relation between the shore hardness and Young's modulus for elastomers has been derived by Gent (1978). This relation has the following form:

$$E = \frac{0.0981(56 + 7.62336 S_H)}{0.137505(254 - 2.54 S_H)} \tag{11}$$

where E is the Young's modulus in MPa and S is the shore hardness. This relation gives a value of E equal to infinite at S_H=100, but departs from experimental data for S_H lower than 40.

Another relation that fits experimental data slightly better is the following and is reported into British standards (BS 1950, BS 1957):

$$S_H = 100 erf\left(3.186 \times 10^{-4} \sqrt{E}\right) \tag{12}$$

where erf is the error function and E is in units of Pa. A first order estimate of the relation between shore D hardness and the elastic modulus for a conical indenter with a 15 degree cone is:

$$S_D = 100 - \left[20\left(-78.188 + \sqrt{6113.36 + 781.88E}\right)\right] / E \tag{13}$$

where S_D is the shore D hardness and E is in MPa.

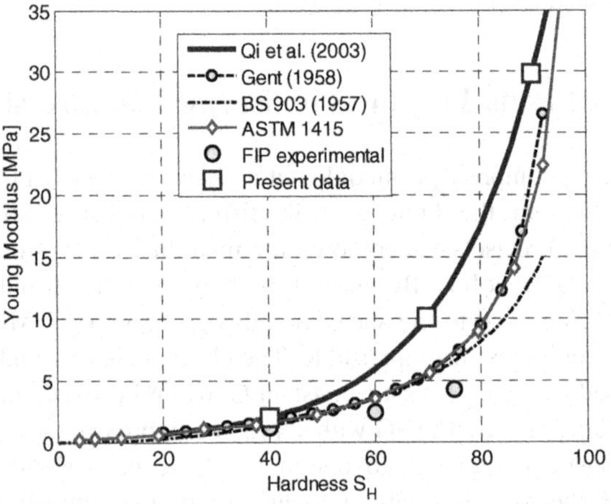

Fig. 7. Empirical dependence of the rubber elastic modulus in terms of international hardness. Squares denote elastic moduli used in the numerical simulations.

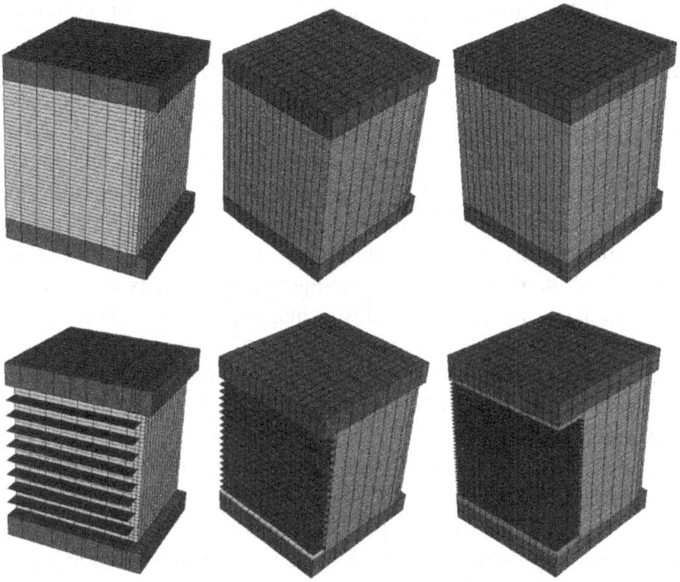

Fig. 8. FE discretization of the seismic isolators studied

A further recent linear relation between the shore hardness and the natural logarithm of Young's modulus is applicable over a large range of shore A and shore D hardness (Q_i, H) [16]. This relation has the form:

$$\ln(E) = 0.0235 S_H - 0.6403 \qquad (14)$$

where S_H=S_A for S_A between 20 and 80 and S_H=S_D +50 for S_D between 30 and 85, being S_A the shore A hardness, S_D the shore D hardness and E the Young's modulus in MPa.

Having at disposal hardness-Young modulus relationships of Fig. 7, they can be used in the FE model for the characterization of Ec compression modulus of the isolator.

The seismic isolator under study is a typical rectangular device which may be found in common building practice, Kelly [10], De Luca & Imbimbo [7], etc. The bearing, see Fig. 1 for the geometric dimensions, is composed by two steel plates, each 20 mm thick, between which rubber-steel elastomer are placed. One of the key parameters having a fundamental role in the determination of overall isolator compression elastic modulus E_c is the so called shape factor S (or primary shape factor), defined as the ratio between the loaded area and the lateral surface free to bulge. Since the shape factor refers to the single rubber layer, it represents a measure of the local slenderness of the elastomeric bearing. Experimental tests have shown that low shape factor bearings, characterized by values of S greater than 5 and less than 20, provide an isolation effect in both the horizontal and vertical directions whereas high shape factor bearings, characterized by values of S greater than 20, only provide a good isolation in the horizontal direction. It is even obvious that low values of the shape factor define thick rubber layers and, hence, provide bearings characterized by high deformability. As a rule, in seismic isolation applications the need to have a device with a high vertical stiffness and low shear stiffness requires that S assumes values greater than 5 and less than 30. The Finite Element model shown in Fig. 8 is used to model ¼ of the bearing subjected to compression. Three geometric cases corresponding to shape factors S around equal to 7, 15 and 30 are hereafter considered. In these cases, the thicknesses of the single pad are respectively equal to 15, 22.5 and 45 mm. Assuming in the first case a thickness of steel laminas equal to 1 mm, in the second 2 mm and in the third 3 mm, the number of steel plates to be used on such devices is respectively equal to 14, 10 and 5. Two types of elements are utilized, namely 8 noded plate and shell elements for thin laminas, and 20 nodes brick elements for the hyper-elastic material (rubber). For steel laminas, an isotropic elastic behavior has been assumed. Following literature data, we adopted a Young Modulus $E = 2x10^5$ MPa with Poisson ratio $v = 0.3$. For a shape factor S equal to 7 the number of rubber pads is 9 (2 pads with ½ thickness at the top and bottom), for S=15 is 18 and for S=30 is 36.

In Fig. 9-a, the trend of the initial compression elastic modulus E_c provided by the present FE approach is compared with those evaluated by means of the Tsai and Kelly [19] model and by the Italian code [15]. Deformed shapes of the bearings in pure compression are also represented. When dealing with the Italian code, E_c is evaluated as $E_c = \left(1/6G_{din}S^2 + 4/3E_b\right)^{-1}$, where G_{din} is the dynamic shear modulus of the isolator (hereafter assumed equal to rubber shear modulus in absence of experimental data available) and E_b is the rubber bulk modulus (hereafter assumed infinite, being rubber almost incompressible). For the sake of conciseness, only three different rubber compounds are tested, namely a Neoprene and an EPDM tested in pure tension by the authors [11] and a commercial Neoprene (DuPont) used ordinarily for seismic isolation. As it is possible to notice from the figure, a very good agreement is found between present FE results and Tsai & Kelly [19] model. Evident but in any case acceptable differences may be noticed for high shape factors between present model and Italian code predictions. However, here it is worth noting that data assumed for shear and bulk moduli are rather questionable and may strongly affect output

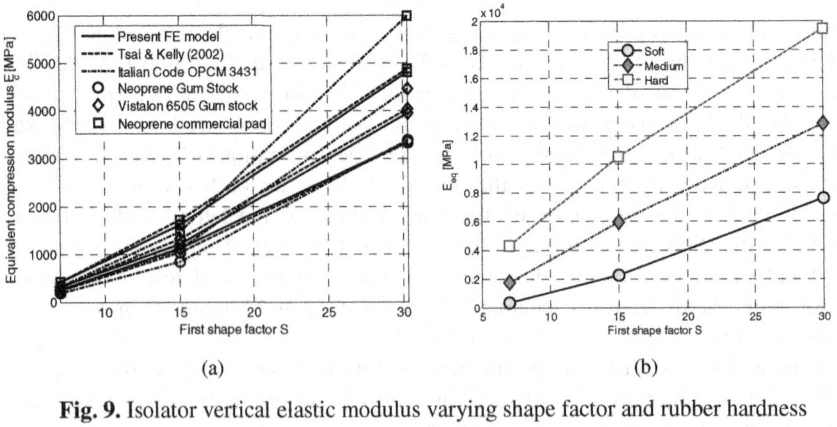

(a) (b)

Fig. 9. Isolator vertical elastic modulus varying shape factor and rubber hardness

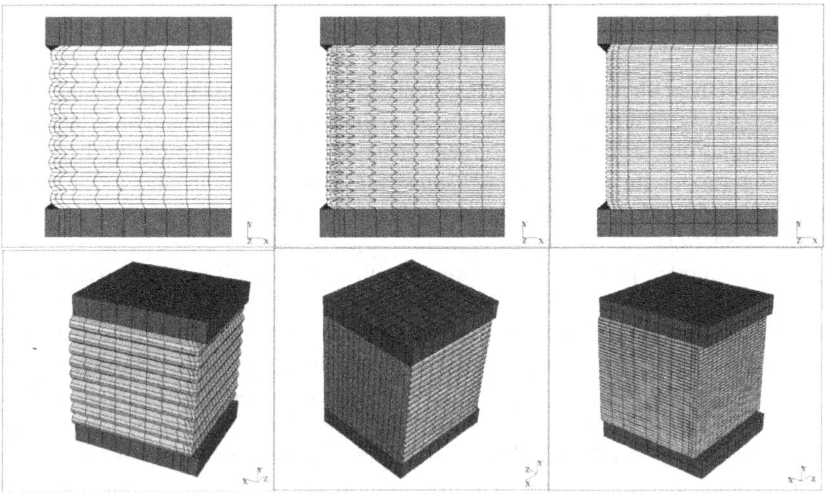

Fig. 10. Deformed shapes for vertical compression

numerical data for S>20. Finally, additional elastic analyses under small deformations are performed to characterize the vertical elastic modulus in compression at different values of the shape factor and for the three blends represented in Fig. 7 with square symbols, roughly corresponding to soft, medium and hard rubber.

As expected, elastic modulus increases with shape factor and is maximum for the hard blend. Usually, elastic moduli of a seismic isolator should range between 1500 and 7500 MPa, meaning that for a shape factor equal to 15, a medium or a soft blend should be used. Conversely, for higher shape factor, soft blend could have the beneficial role to progressively decrease vertical stiffness, whereas for lower shape factors hard blends are preferable.

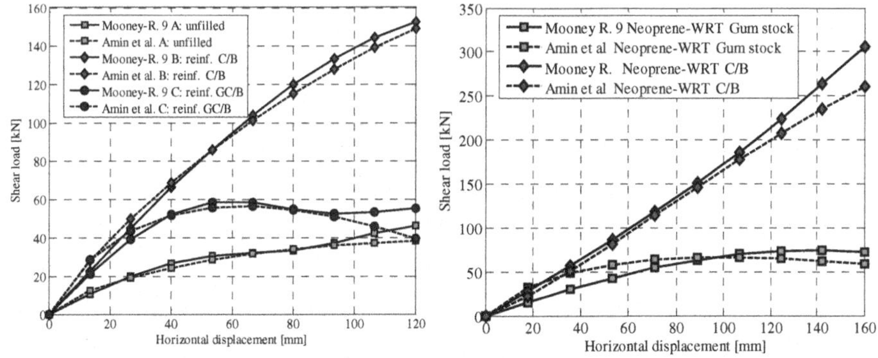

(a) (b)

Fig. 11. -a: Unfilled, reinforced with carbon black (C/B) and with graphitized carbon black (GC/B) generic non crystallized rubber. –b: EPDM Vistalon 6505 Gum stock and Vistalon 6505 with C/B. Seismic isolator global behavior under shear loads (total shear vs horizontal displacement curves obtained with a 9 constants Mooney Rivlin model and a 5 constants exponential model).

5 Shear Behavior of the Isolator

Lateral force (shear)-horizontal displacement behavior is predicted by means of the models previously described, at high levels of horizontal displacement. Clearly, for severe deformations, a large displacement analysis is required, under the hypothesis of modeling rubber through both a 9 constants Mooney-Rivlin and a 5 constants exponential model. Material constants are obtained using the procedure described in the previous Section.

Since the model proposed by Amin et al. [1] normally is not available in commercial codes, a Matlab based Open FEM software has been modified in order to implement an exponential law for the strain energy density function. The code obviously allows the usage of both brick and shell elements in large deformations.

In the model, a generic rubber (Fig. 2) and the neoprene with uniaxial behavior represented in Fig. 3 (filled with C/B) are considered, assuming the vertical displacement at the top edge of the stiff steel plate of the isolator not allowed. Since the global response of the isolator in terms of deformed shape is very similar in all the cases analyzed, a small sample of the huge amount of numerical results obtained are hereafter reported for the sake of conciseness. In particular, results in terms of applied shear at the top edge and corresponding horizontal displacement (which indirectly define the shear stiffness of the isolator) are depicted in Fig. 11 for the generic rubber (-a) of Fig. 2 and the neoprene (-b) of Fig. 3. As a rule, Mooney Rivlin and Amin et al. [1] models behave globally and generally in a similar way, providing comparable levels of horizontal load at assigned deformation.

This notwithstanding, it is worth noting that, in general, when hyperelastic constants are deduced exclusively from uniaxial tests fitting, some remarkable differences in shear tests between the models are possible. In addition, it is worth emphasizing that, in the case here analyzed, rubber pads are subjected to a complex state of stress depending on several factors, comprising the bending stiffness of the steel elements, the vertical pre-compression, etc., thus complicating further the prediction of their actual behavior.

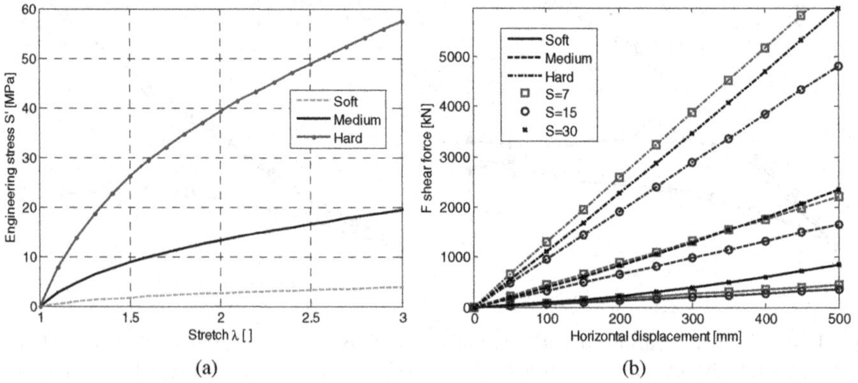

Fig. 12. Stretch-stress behaviour of a single rubber pad (left) and pure shear behaviour under large deformations (right) of a square isolator.

To study the shear behavior of the isolator as a function of rubber pads hardness, a two constants Mooney-Rivlin model is utilized, in order to put at disposal to practitioners a simple FE model that can be implemented in any FE commercial code.

The uniaxial stress-stretch behavior of the 2 constants Mooney-Rivlin model adopted to evaluate the shear behavior of a single pad in presence of soft, medium and hard rubber is depicted in Fig. 12-a. The resultant shear force F-horizontal displacement of the entire isolator is represented in Fig. 12-b. Such representation may be particularly useful for practical purposes, since such curves may be implemented at a structural level to study entire base-isolated buildings in the dynamic range. As it is possible to notice, the utilization of different hardness rubber pads in conjunction with slender or less slender isolators may considerably change the macroscopic response of the isolator and, hence, the effectiveness of the device inserted in a large case structure may be variable.

6 Conclusions

The important matter of the role played by the material used and the thickness of rubber pads within seismic isolators has been investigated by means of a comprehensive FE set of analyses in linear elasticity and large deformations. From the results obtained, it can be deduced that a proper calculation is needed when the seismic isolator is constituted by pads with large thickness. Hardness is finally a very practical key parameter to define rubber initial Young's modulus, whereas full stretch-stress uniaxial curves are needed within refined numerical models to describe rubber behavior under large deformations.

References

1. Amin, A.F.M.S., Alam, M.S., Okui, Y.: An improved hyperelasticity relation in modeling viscoelasticity response of natural and high damping rubbers in compression: experiments, parameter identification, and numerical verification. Mech. Mater. 34, 75–95 (2002)
2. Amin, A.F.M.S., Wiraguna, I., Bhuiyan, A.R., Okui, Y.: Hyperelasticity Model for Finite Element Analysis of Natural and High Damping Rubbers in Compression and Shear. Journal of Engineering Mechanics ASCE 132(1), 54–64 (2006)
3. ASTM D2240-91. Standard test method for fubber property- Durometer hardness. Annual book of ASTM standard, 388–391 (1992)
4. Braga, F., Dolce, M., Ferrigno, A., Laterza, M., Marotta, G., Masi, A., Nigro, D., Ponzo, F.: Development of new materials for seismic isolation and passive energy dissipation - part i: experimental tests on new compound elastomeric bearings. In: Proc. International Post-SMiRT Conference Seminar on Seismic Isolation, Passive Energy Dissipation and Active Control of Seismic Vibrations of Structures Taormina, Italy (1997)
5. British Standard 903 (1950, 1957). Methods of testing vulcanized rubber, Part 19 (1950) and Part A7 (1957)
6. Bueche, F.: Mullins effect and rubber-filler interaction. Journal of Applied Polymer Science 5(15), 271–281 (1961)
7. De Luca, A., Imbimbo, M.: F. E. stress analysis of rubber bearings under axial loads. Computers and Structures 68, 31–39 (1998)

8. Gent, A.N.: Rubber elasticity: basic concepts and behavior. In: Eirich, F.R. (ed.) Science and Technology Rubber. Academic Press, NY (1978)
9. Tsai, H.-C.: Compression analysis of rectangular elastic layers bonded between rigid plates. International Journal of Solids and Structures 42(11-12), 3395–3410 (2005)
10. Kelly, J.M.: Base Isolation of Structures. Design Guidelines. Holmes Consulting Group Ltd., Wellington (2001)
11. Milani, G., Milani, F.: Stretch-stress behavior of elastomeric seismic isolators with different rubber materials. A numerical insight. Journal of Engineering Mechanics ASCE 138(5), 416–429 (2012a)
12. Milani, G., Milani, F.: Elastomeric seismic isolators behavior at different pads thickness. In: Pina, N., Kacprzyk, J. (eds.) Proc. 2nd International Conference on Simulation and Modeling Methodologies, Technologies and Applications, Rome, IT, July 28-31 (2012b)
13. Moon, B.Y., Kang, G.J., Kang, B.S., Kim, G.S., Kelly, J.M.: Mechanical properties of seismic isolation system with fiber-reinforced bearing of strip type. International Applied Mechanics 39(10), 1231–1239 (2003)
14. Mooney, M.: A theory of large elastic deformation. J. Appl. Physics 2, 582–592 (1940)
15. OPCM 3431: Ulteriori modifiche ed integrazioni all'OPCM 3274/ 03 20/03/2003. Primi elementi in materia di criteri generali per la classificazione sismica del territorio nazionale e di normative tecniche per le costruzioni in zona sismica (2005) (in Italian)
16. Qi, H.J., Joyce, K., Boyce, M.C.: Durometer hardness and the stress-strain behavior of elastomeric materials. Rubber Chemistry and Technology 72(2), 419–435 (2003)
17. Studebaker, M.L., Beatty, J.R.: The rubber compound and its composition. In: Eirich, F.R. (ed.) Science and Technology of Rubber. Academic Press, NY (1978)
18. Tobolsky, A.V., Mark, H.F.: Polymer Science and Materials, ch. 13. Wiley, New York (1971)
19. Tsai, H.S., Kelly, J.M.: Stiffness Analysis of Fiber-Reinforced Rectangular Seismic Isolators. Journal of Engineering Mechanics ASCE 128(4), 462–470 (2002)
20. Tsai, H.S., Lee, C.C.: Tilting stiffness of elastic layers bonded between rigid plates. International Journal of Solids and Structures 36(17), 2485–2505 (1999)

Numerical Simulation of Coastal Flows in Open Multiply-Connected Irregular Domains

Yuri N. Skiba[1] and Denis M. Filatov[2]

[1] Centre for Atmospheric Sciences (CCA), National Autonomous University of Mexico (UNAM), Av. Universidad 3000, C.P. 04510, Mexico City, Mexico

[2] Centre for Computing Research (CIC), National Polytechnic Institute (IPN), Av. Juan de Dios Batiz s/n, C.P. 07738, Mexico City, Mexico

skiba@unam.mx, denisfilatov@gmail.com

Abstract. We develop a numerical method for the simulation of coastal flows in multiply-connected domains with irregular boundaries that may contain both closed and open segments. The governing equations are the shallow-water model. Our method involves splitting of the original nonlinear operator by physical processes and by coordinates. Specially constructed finite-difference approximations provide second-order unconditionally stable schemes that conserve the mass and the total energy of the discrete inviscid unforced shallow-water system, while the potential enstrophy results to be bounded, oscillating in time within a narrow range. This allows numerical simulation of coastal flows adequate both from the mathematical and physical standpoints. Several numerical experiments, including those with complex boundaries, demonstrate the skilfulness of the method.

Keywords: Coastal Shallow-water Flows, Conservative Finite Difference Schemes, Multiply-connected Domains, Irregular Boundaries, Operator Splitting.

1 Introduction

When studying a 3D fluid dynamics problem in which typical horizontal scales are much larger than the vertical ones—say, the vertical component of the velocity field is rather small compared to the horizontal ones, or horizontal movements of the fluid are normally much larger than the vertical ones—it is often useful to reduce the original problem, usually described by the Navier-Stokes equations, to a 2D approximation. This leads to a shallow-water model [19, 10, 6].

Shallow-water equations (SWEs) naturally arise in the researches of global atmospheric circulation, tidal waves, river flows, tsunamis, among others [5]. In the spherical coordinates (λ, φ) the shallow-water equations for an ideal unforced fluid can be written as [18]

$$\frac{\partial U}{\partial t} + \frac{1}{R\cos\varphi}\frac{1}{2}\left[\left(\frac{\partial uU}{\partial\lambda} + u\frac{\partial U}{\partial\lambda}\right) + \left(\frac{\partial vU\cos\varphi}{\partial\varphi} + v\cos\varphi\frac{\partial U}{\partial\varphi}\right)\right] - \left(f + \frac{u}{R}\tan\varphi\right)V = -\frac{gz}{R\cos\varphi}\frac{\partial h}{\partial\lambda}, \qquad (1)$$

M.S. Obaidat et al. (eds.), *Simulation and Modeling Methodologies, Technologies and Applications*, Advances in Intelligent Systems and Computing 256,
DOI: 10.1007/978-3-319-03581-9_5, © Springer International Publishing Switzerland 2014

$$\frac{\partial V}{\partial t}+\frac{1}{R\cos\varphi}\frac{1}{2}\left[\left(\frac{\partial uV}{\partial\lambda}+u\frac{\partial V}{\partial\lambda}\right)+\left(\frac{\partial vV\cos\varphi}{\partial\varphi}+v\cos\varphi\frac{\partial V}{\partial\varphi}\right)\right]$$
$$+\left(f+\frac{u}{R}\tan\varphi\right)U=-\frac{gz}{R}\frac{\partial h}{\partial\varphi}, \qquad (2)$$

$$\frac{\partial h}{\partial t}+\frac{1}{R\cos\varphi}\left[\frac{\partial zU}{\partial\lambda}+\frac{\partial zV\cos\varphi}{\partial\varphi}\right]=0. \qquad (3)$$

Here $U\equiv uz$, $V\equiv vz$, where $u=u(\lambda,\varphi,t)$ and $v=v(\lambda,\varphi,t)$ are the fluid's velocity components, $H=H(\lambda,\varphi,t)$ is the fluid's depth, $z\equiv\sqrt{H}$, $f=f(\varphi)$ is the Coriolis acceleration due to the rotation of the sphere, R is the radius of the sphere, $h=h(\lambda,\varphi,t)$ is the free surface height, g is the gravitational acceleration. Besides, $h=H+h_T$, where $h_T=h_T(\lambda,\varphi)$ is the bottom topography. We shall study (1)-(3) in a bounded domain D on a sphere with an arbitrary piecewise smooth boundary Γ, assuming that λ is the longitude (positive eastward) and φ is the latitude (positive northward).

As we are dealing with a boundary value problem, system (1)-(3) has to be equipped with boundary conditions.

The question of imposing correct boundary conditions for SWEs is not trivial. Many independent research papers have been dedicated to this issue for the last several decades [19, 9, 20, 1]. Depending on the type of the boundary—inflow, outflow or closed—as well as on the particular physical application, one or another set of boundary conditions should be used. Following [1], we represent the boundary as $\Gamma=\Gamma_{\mathrm{o}}\cup\Gamma_{\mathrm{c}}$, where Γ_{o} is the open part of the boundary, while Γ_{c} is its closed part. Such a representation of the boundary simulates a bay-like domain, where the coastline corresponds to the closed part Γ_{c}, while the influence of the ocean is modelled via the open segment Γ_{o}. Yet, the open segment is divided into the inflow $\Gamma_{\mathrm{inf}}:=\{(\lambda,\varphi)\in\Gamma:\mathbf{n}\cdot\mathbf{u}<0\}$ and outflow $\Gamma_{\mathrm{out}}:=\{(\lambda,\varphi)\in\Gamma:\mathbf{n}\cdot\mathbf{u}>0\}$. Here $\mathbf{n}$ is the outward unit normal to Γ, $\mathbf{u}=(u,v)^{\mathrm{T}}$. On the closed part we put

$$\mathbf{n}\cdot\mathbf{u}=0, \qquad (4)$$

on the inflow we assume

$$\tau\cdot\mathbf{u}=0, \qquad h=h_{(\Gamma)} \qquad (5)$$

and on the outflow it holds

$$h=h_{(\Gamma)}, \qquad (6)$$

where τ is the tangent vector to Γ, whereas $h_{(\Gamma)}$ is a given function defined on the boundary [1].

From the mathematical standpoint unforced inviscid SWEs are based on several conservation laws. In particular, the mass

$$M(t)=\int_D HdD, \qquad (7)$$

the total energy

$$E(t) = \frac{1}{2} \int_D \left[\left(u^2 + v^2\right) H + g \left(h^2 - h_T^2\right) \right] dD \tag{8}$$

and the potential enstrophy

$$J(t) = \frac{1}{2} \int_D H \left(\frac{\zeta + f}{H} \right)^2 dD, \tag{9}$$

where

$$\zeta = \frac{1}{R \cos \varphi} \left(\frac{\partial v}{\partial \lambda} - \frac{\partial u \cos \varphi}{\partial \varphi} \right), \tag{10}$$

are kept constant in time for a closed shallow-water system [20, 6]. In the numerical simulation of shallow-water flows one should use the finite difference schemes which preserve the discrete analogues of the integral invariants of motion (7)-(9) as accurately as possible. It is crucial that for many finite difference schemes the discrete analogues of the mass, total energy and potential enstrophy are usually not invariant in time, so the numerical method can be unstable and the resulting simulation becomes inaccurate [20]. This emphasises the importance of using conservative difference schemes while modelling fluid dynamics phenomena.

In the last forty years there have been suggested several finite difference schemes that conserve some or other integral characteristics of the shallow-water equations [12, 2, 4, 11, 3, 7, 13]. In all these works, however, only semi–discrete (i.e., discrete only in space, but still continuous in time) conservative schemes are constructed. After using an explicit time discretisation those schemes stop being conservative. Besides, while aiming to achieve the desired full conservatism (see, e.g., [14]), when all the discrete analogues of the integral invariants of motion are conserved, some methods require rather complicated spatial grids (e.g., triangular, hexagonal, etc.), which makes it difficult to employ those methods in a computational domain with a boundary of an arbitrary shape; alternatively, it may result in a resource-intensive numerical algorithm.

In this work we suggest a new efficient method for the numerical simulation of shallow-water flows in domains of complex geometries. The method is based on our earlier research devoted to the modelling of atmospheric waves with SWEs [16–18]. The method involves operator splitting of the original equations by physical processes and by coordinates. Careful subsequent discretisation of the split 1D systems coupled with the Crank-Nicolson approximation of the spatial terms yields a fully discrete (i.e., discrete both in time and in space) finite difference shallow-water model that, in case of an inviscid and unforced fluid, exactly conserves the mass and the total energy, while the potential enstrophy is bounded, oscillating in time within a narrow band. Due to the prior splitting the model is extremely efficient, since it is implemented as systems of linear algebraic equations with tri– and five–diagonal matrices. Furthermore, the model can straightforwardly be realised on high-performance parallel computers without any significant modifications in the original single-threaded algorithm.

The paper is organised as follows. In Section 2 we give the mathematical foundations of the suggested shallow-water model. In Section 3 we test the model with several numerical experiments aimed to simulate shallow-water flows in a bay-like domain with a complex boundary. We also test a modified model, taking into account fluid viscosity and external forcing for providing more realistic simulation. In Section 4 we give a conclusion.

2 Governing Equations of the Fully Discrete Conservative Shallow-Water Model

Rewrite the shallow-water equations (1)-(3) in the operator form

$$\frac{\partial \psi}{\partial t} + \mathbf{A}(\psi) = 0, \tag{11}$$

where $\mathbf{A}(\psi)$ is the shallow-water nonlinear operator, while $\psi = (U, V, h\sqrt{g})^{\mathrm{T}}$ is the unknown vector. Now represent the operator $\mathbf{A}(\psi)$ as a sum of three simpler operators, nonlinear $\mathbf{A_1}$, $\mathbf{A_2}$ and linear $\mathbf{A_3}$

$$\mathbf{A}(\psi) = \mathbf{A_1}(\psi) + \mathbf{A_2}(\psi) + \mathbf{A_3}\psi. \tag{12}$$

Let (t_n, t_{n+1}) be a sufficiently small time interval with a step τ $(t_{n+1} = t_n + \tau)$. Applying in (t_n, t_{n+1}) operator splitting to (11), we approximate it by the three simpler problems

$$\frac{\partial \psi_1}{\partial t} + \mathbf{A_1}(\psi_1) = 0, \tag{13}$$

$$\frac{\partial \psi_2}{\partial t} + \mathbf{A_2}(\psi_2) = 0, \tag{14}$$

$$\frac{\partial \psi_3}{\partial t} + \mathbf{A_3}\psi_3 = 0. \tag{15}$$

According to the method of splitting, these problems are to be solved one after another, so that the solution to (11) from the previous time interval (t_{n-1}, t_n) is the initial condition for (13): $\psi_1(t_n) = \psi(t_n)$, then $\psi_2(t_n) = \psi_1(t_{n+1})$ and finally $\psi_3(t_n) = \psi_2(t_{n+1})$. Therefore, the solution to (11) at the moment t_{n+1} is approximated by the solution $\psi_3(t_{n+1})$ [8].

Operators $\mathbf{A_1}$, $\mathbf{A_2}$, $\mathbf{A_3}$ can be defined in different ways. In our work equation (13) has the form

$$\frac{\partial U}{\partial t} + \frac{1}{R\cos\varphi}\frac{1}{2}\left[\frac{\partial uU}{\partial \lambda} + u\frac{\partial U}{\partial \lambda}\right] = -\frac{gz}{R\cos\varphi}\frac{\partial h}{\partial \lambda}, \tag{16}$$

$$\frac{\partial V}{\partial t} + \frac{1}{R\cos\varphi}\frac{1}{2}\left[\frac{\partial uV}{\partial \lambda} + u\frac{\partial V}{\partial \lambda}\right] = 0, \tag{17}$$

$$\frac{\partial h}{\partial t} + \frac{1}{R\cos\varphi}\frac{\partial zU}{\partial \lambda} = 0, \tag{18}$$

for (14) we take

$$\frac{\partial U}{\partial t} + \frac{1}{R\cos\varphi}\frac{1}{2}\left[\frac{\partial vU\cos\varphi}{\partial \varphi} + v\cos\varphi\frac{\partial U}{\partial \varphi}\right] = 0, \tag{19}$$

$$\frac{\partial V}{\partial t} + \frac{1}{R\cos\varphi}\frac{1}{2}\left[\frac{\partial vV\cos\varphi}{\partial\varphi} + v\cos\varphi\frac{\partial V}{\partial\varphi}\right] = -\frac{gz}{R}\frac{\partial h}{\partial\varphi}, \tag{20}$$

$$\frac{\partial h}{\partial t} + \frac{\partial zV\cos\varphi}{\partial\varphi} = 0, \tag{21}$$

and for (15)—

$$\frac{\partial U}{\partial t} - \left(f + \frac{u}{R}\tan\varphi\right)V = 0, \tag{22}$$

$$\frac{\partial V}{\partial t} + \left(f + \frac{u}{R}\tan\varphi\right)U = 0. \tag{23}$$

This choice of $\mathbf{A_i}$'s corresponds to the splitting by physical processes (transport and rotation) and by coordinates (λ and φ). The latter means that while solving (16)-(18) in λ, the coordinate φ is left fixed; and vice versa for (19)-(21).

Introducing the grid $\{(\lambda_k, \varphi_l) \in D : \lambda_{k+1} = \lambda_k + \Delta\lambda, \varphi_{l+1} = \varphi_l + \Delta\varphi\}$, we approximate systems (16)-(18) and (19)-(21) by the central second-order finite difference schemes, so that eventually in λ we obtain (the subindex l, in the φ–direction, is fixed, as well as omitted for simplicity)

$$\frac{U_k^{n+1} - U_k^n}{\tau} + \frac{1}{2c_l}\left[\frac{\overline{u}_{k+1}U_{k+1} - \overline{u}_{k-1}U_{k-1}}{2\Delta\lambda} + \overline{u}_k\frac{U_{k+1} - U_{k-1}}{2\Delta\lambda}\right] = -\frac{g\overline{z}_k}{c_l}\frac{h_{k+1} - h_{k-1}}{2\Delta\lambda}, \tag{24}$$

$$\frac{V_k^{n+1} - V_k^n}{\tau} + \frac{1}{2c_l}\left[\frac{\overline{u}_{k+1}V_{k+1} - \overline{u}_{k-1}V_{k-1}}{2\Delta\lambda} + \overline{u}_k\frac{V_{k+1} - V_{k-1}}{2\Delta\lambda}\right] = 0, \tag{25}$$

$$\frac{h_k^{n+1} - h_k^n}{\tau} + \frac{1}{c_l}\frac{\overline{z}_{k+1}U_{k+1} - \overline{z}_{k-1}U_{k-1}}{2\Delta\lambda} = 0, \tag{26}$$

while in φ we get (the subindex k, in λ, is fixed and omitted too)

$$\frac{U_l^{n+1} - U_l^n}{\tau} + \frac{1}{2c_l}\left[\frac{\overline{v}_{l+1}U_{l+1}c_+ - \overline{v}_{l-1}U_{l-1}c_-}{2\Delta\varphi} + \overline{v}_l\cos\varphi_l\frac{U_{l+1} - U_{l-1}}{2\Delta\varphi}\right] = 0, \tag{27}$$

$$\frac{V_l^{n+1} - V_l^n}{\tau} + \frac{1}{2c_l}\left[\frac{\overline{v}_{l+1}V_{l+1}c_+ - \overline{v}_{l-1}V_{l-1}c_-}{2\Delta\varphi} + \overline{v}_l\cos\varphi_l\frac{V_{l+1} - V_{l-1}}{2\Delta\varphi}\right] = -\frac{g\overline{z}_l}{R}\frac{h_{l+1} - h_{l-1}}{2\Delta\varphi}, \tag{28}$$

$$\frac{h_l^{n+1} - h_l^n}{\tau} + \frac{1}{c_l}\frac{\overline{z}_{l+1}V_{l+1}c_+ - \overline{z}_{l-1}V_{l-1}c_-}{2\Delta\varphi} = 0. \tag{29}$$

Here, in a standard manner, $w_{kl}^n = w(\lambda_k, \varphi_l, t_n)$, where $w = \{U, V, h\}$; besides, we denoted $c_l \equiv R\cos\varphi_l$ and $c_\pm \equiv \cos\varphi_{l\pm1}$. In turn, the rotation problem (22)-(23) has the form

$$\frac{U_{kl}^{n+1} - U_{kl}^n}{\tau} - \left(f_l + \frac{\overline{u}_{kl}}{R}\tan\varphi_l\right)V_{kl} = 0, \tag{30}$$

$$\frac{V_{kl}^{n+1} - V_{kl}^n}{\tau} + \left(f_l + \frac{\overline{u}_{kl}}{R}\tan\varphi_l\right)U_{kl} = 0. \tag{31}$$

The functions U_{kl}, V_{kl} in the presented schemes are defined via the Crank-Nicolson approximation as $U_{kl} = \left(U_{kl}^n + U_{kl}^{n+1}\right)/2$, $V_{kl} = \left(V_{kl}^n + V_{kl}^{n+1}\right)/2$. As for the overlined functions $\overline{u}_{kl}$, $\overline{v}_{kl}$ and $\overline{z}_{kl}$, they can be chosen in an arbitrary manner [16]. For instance, the choice $\overline{w}_{kl} = w_{kl}^n$, where $w = \{u, v, z\}$, will yield linear second-order finite difference schemes, whereas the choice $\overline{w}_{kl} = w_{kl}$ coupled with the corresponding Crank-Nicolson approximations for w_{kl} will produce nonlinear schemes.

The developed schemes have several essential advantages.

First, the coordinate splitting allows simple parallelisation of the numerical algorithm without any significant modifications of the single-threaded code. Indeed, say, when solving (24)-(26), all the calculations along the longitude at different φ_l's can be done in parallel; analogously, for (27)-(29) the calculations at different λ_k's are naturally parallelisable. Finally, equations (30)-(31) can be reduced to explicit formulas with respect to U_{kl}^{n+1}, V_{kl}^{n+1} [17].

Second, the simple 1D longitudinal and latitudinal spatial stencils used do not impose any restrictions on the shape of the boundary Γ. Therefore, the developed schemes can be employed for the simulation of shallow-water flows in computational domains of complex geometries.

Third, the developed schemes are mass– and total-energy–conserving for the inviscid unforced shallow-water model in a closed basin ($\Gamma = \Gamma_c$). To show this, consider, e.g., (24)-(26). The boundary condition will be $U|_\Gamma = 0$, which can be approximated as

$$\frac{1}{2}\left(U_0 + U_1\right) = 0, \tag{32}$$

$$\frac{1}{2}\left(U_K + U_{K+1}\right) = 0, \tag{33}$$

where the nodes $k = 1$ and $k = K$ are inside the domain D, while the nodes $k = 0$ and $k = K + 1$ are out of D (i.e., fictitious). Multiplying (26) by $\tau R\Delta\lambda$, summing over all the $k = \overline{1, K}$'s and taking into account the boundary conditions (32)-(33), we find that the spatial term vanishes, so that

$$M_l^{n+1} = R\Delta\lambda \sum_k H_{kl}^{n+1} = \cdots = R\Delta\lambda \sum_k H_{kl}^n = M_l^n, \tag{34}$$

which proves that the mass conserves in λ (at a fixed l). Further, multiplying (24) by $\tau R\Delta\lambda U_{kl}$, (25) by $\tau R\Delta\lambda V_{kl}$ and (26) by $\tau R\Delta\lambda g h_{kl}$, summing over the internal k's and taking into account (32)-(33), we obtain

$$\begin{aligned} E_l^{n+1} &= R\Delta\lambda \sum_k \frac{1}{2}\left(\left[U_{kl}^{n+1}\right]^2 + \left[V_{kl}^{n+1}\right]^2 + g\left(\left[h_{kl}^{n+1}\right]^2 - \left[h_{T,kl}\right]^2\right)\right) = \cdots \\ &= R\Delta\lambda \sum_k \frac{1}{2}\left(\left[U_{kl}^{n}\right]^2 + \left[V_{kl}^{n}\right]^2 + g\left(\left[h_{kl}^{n}\right]^2 - \left[h_{T,kl}\right]^2\right)\right) = E_l^n, \end{aligned} \tag{35}$$

that is the energy conserves in λ at $\varphi = \varphi_l$ too. Similar results can be obtained with respect to the second coordinate, φ (problem (27)-(29)), while the Coriolis problem (30)-(31) does not affect the conservation laws. Note that to establish both the mass and the energy conservation we used the divergence of the spatial terms in (24)-(26)

and (27)-(29) [14]. The conservation of the total energy guarantees that the constructed finite difference schemes are absolutely stable [8].

Fourth, from (24)-(26), (27)-(29) it follows that under the choice $\overline{w}_{kl} = w_{kl}^n$, where $w = \{u, v, z\}$, the resulting finite difference schemes are systems of linear algebraic equations with either tri– or five–diagonal matrices. Obviously, fast direct (i.e., non-iterative) linear solvers can be used for their solution, so that the exact conservation of the mass and the total energy will not be violated.

3 Numerical Experiments

For testing the developed model we consider two problems. In the first problem the SWEs are a closed system, so that we are able to verify the mass and the total energy conservation laws; besides, the ranges of the variation of the potential enstrophy are analysed. In the second experiment the problem is complicated by introducing a complex boundary with open and closed segments, a nonzero viscosity, as well as a nonzero source function imitating the wind. Such a setup simulates wind-driven shallow-water flows in a bay.

3.1 Rectangular Domain Test

In this experiment we consider the simplest case: for the computational domain we take the spherical rectangle $D = \{(\lambda, \varphi) : \lambda \in (12.10, 12.65), \varphi \in (45.16, 45.60)\}$ with a closed boundary $\Gamma = \Gamma_c$. This will allow to verify whether the mass and the total energy of an inviscid unforced fluid are exactly conserved in the numerical simulation. For the initial velocity field we take $u = v = 0$, while the free surface height is a hat-like function (Fig. 1). The gridsteps are $\Delta\lambda \approx \Delta\varphi \approx 0.015°$, $\tau = 1.44$ min.

In Fig. 2 we plot a graph of the discrete analogues of the potential enstrophy (9). The mass and the total energy are not shown, as they are trivial constants, that is those invariants are conserved exacly, while the behaviour of the potential enstrophy is stable—it is oscillating within a narrow band, with a drastically small maximum relative error about 0.32%, without unbounded growth or decay. This confirms the theoretical calculations (34)-(35), as well as demonstrates that the developed schemes allow obtaining physically adequate simulation results.

3.2 Irregular Domain Test

Having a numerical shallow-water model that conserves the mass and the total energy in the absence of sources and sinks of energy, now consider a more complex problem.

For the computational domain we choose the region shown in Fig. 3. Unlike the previous problem, the boundary is now of an arbitrary shape; besides, there are several on-shore parts surrounded by water which represent small isles. The boundary Γ is divided into the closed and open segments: $\Gamma_o = \{\lambda \in (12.32, 12.65), \varphi = 45.16\} \cup \{\lambda = 12.65, \varphi \in (45.16, 45.50)\}$, $\Gamma_c = \Gamma \backslash \Gamma_o$. This setup aims to simulate flows that may occur in the Bay of Venice.

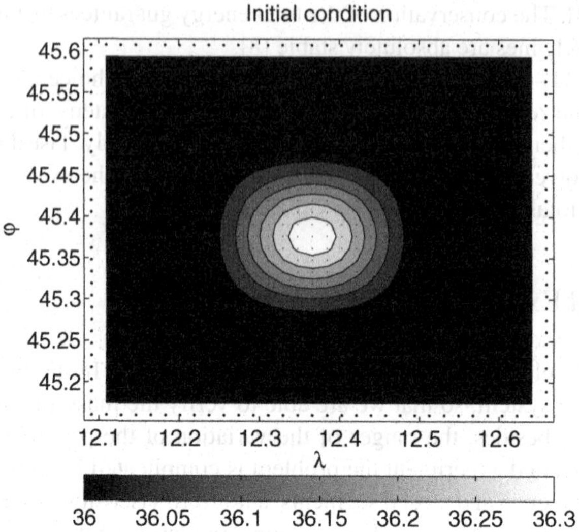

Fig. 1. Rectangular Domain Test: Initial condition (the free surface height is shown in meters; the markers '.' denote the fictitious nodes outside the domain)

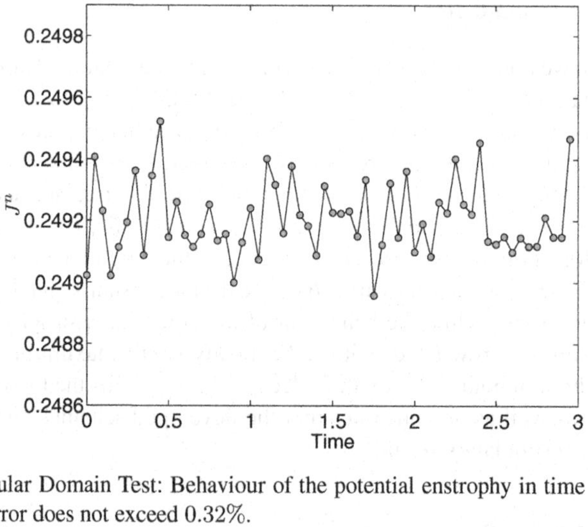

Fig. 2. Rectangular Domain Test: Behaviour of the potential enstrophy in time (in days). Maximum relative error does not exceed 0.32%.

In order to make the flows more realistic, terms responsible for fluid viscosity are also added into the first two equations of the shallow-water system. Specifically, on the right-hand side of (11) we add the vector $\mathbf{D}\psi$, where

$$\mathbf{D} = \begin{pmatrix} d_{11} & 0 & 0 \\ 0 & d_{22} & 0 \\ 0 & 0 & 0 \end{pmatrix}, \tag{36}$$

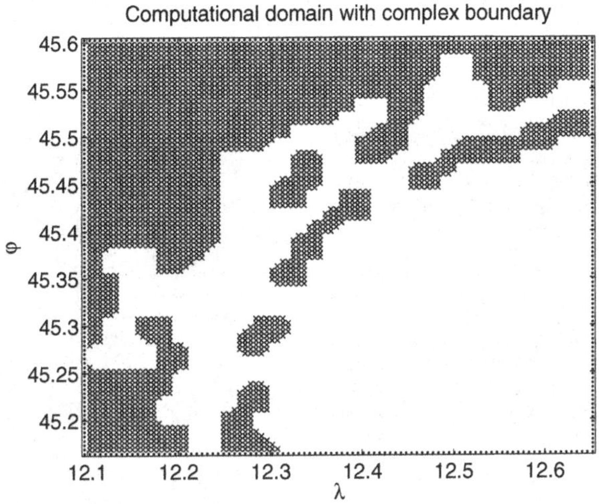

Fig. 3. Irregular Domain Test: The computational domain (white area) with onshore parts and interior isles (grey areas)

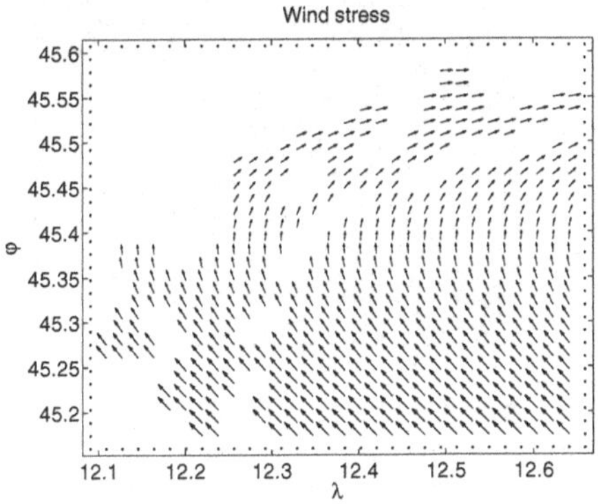

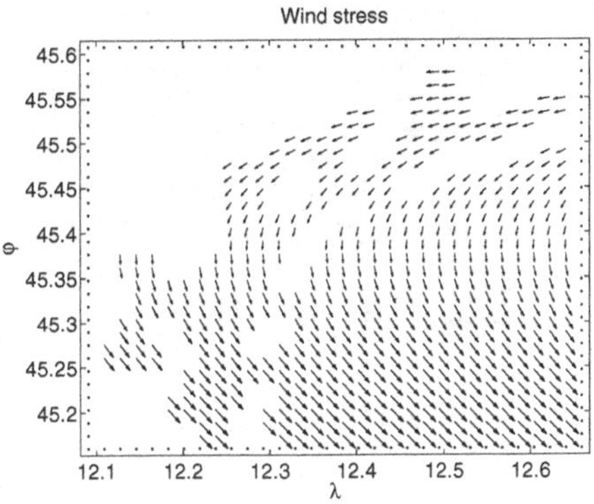

Fig. 4. Irregular Domain Test: Field of the wind stress at $t = 0.25$ (left) and $t = 0.75$ (right)

where

$$d_{11} = d_{22} = \frac{D}{R^2 \cos^2 \varphi} \frac{\partial^2}{\partial \lambda^2} + \frac{D}{R^2 \cos \varphi} \frac{\partial}{\partial \varphi} \left(\cos \varphi \frac{\partial}{\partial \varphi} \right). \tag{37}$$

Here D is the viscosity coefficient. However, addition of the viscosity terms into (24)-(25) and (27)-(28) requires a modification of boundary conditions (4)-(6). Following [1], we use the boundary conditions

$$\mathbf{n} \cdot \mathbf{u} = 0, \qquad Dh \frac{\partial \mathbf{u}}{\partial \mathbf{n}} \tau = 0, \tag{38}$$

$$\left(|\mathbf{n} \cdot \mathbf{u}| - \mathbf{n} \cdot \mathbf{u}\right)\left(h - h_{(\Gamma)}\right) = 0, \tag{39}$$

$$Dh \frac{\partial \mathbf{u}}{\partial \mathbf{n}} = 0. \tag{40}$$

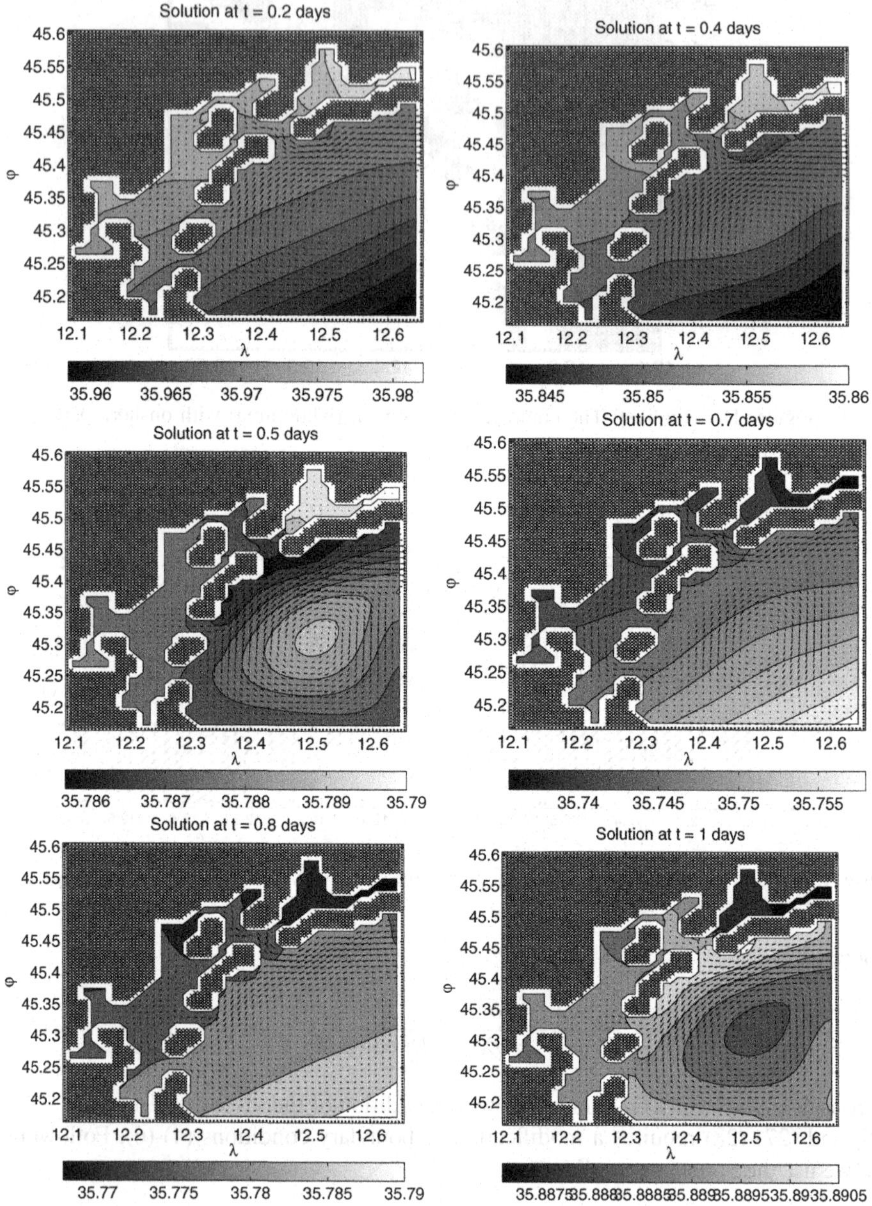

Fig. 5. Irregular Domain Test: Numerical solution at several time moments (the solution is reduced to a coarser grid $\Delta\lambda \approx \Delta\varphi \approx 0.01°$ for better visualisation; the fluid's depth is shown by colour, while the velocity field is shown by arrows)

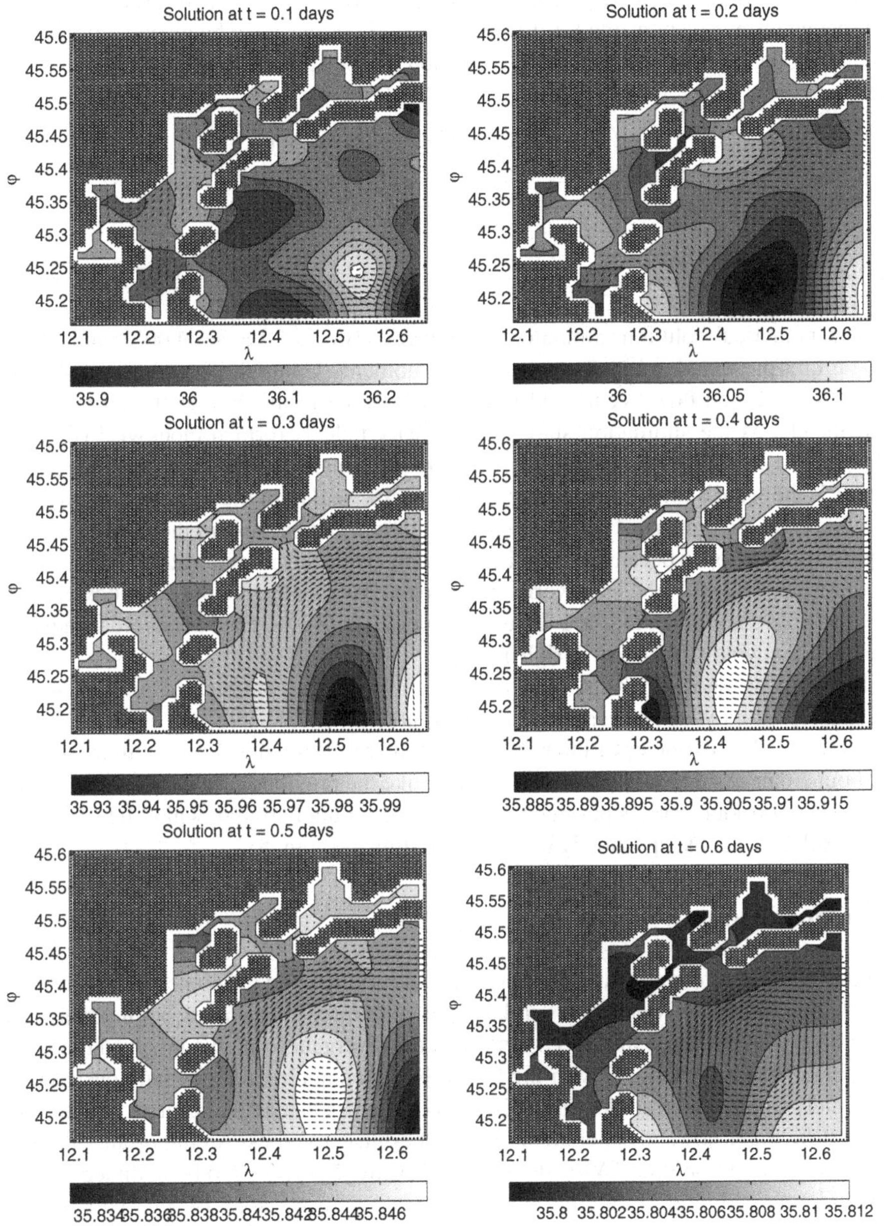

Fig. 6. Irregular Domain Test with Nonzero Bottom Topography: Numerical solution at several time moments

Condition (38) is for $\mathbf{u}$ on the closed segment of the boundary (see also [15]), while (39) and (40) are for h and $\mathbf{u}$ on the open segment, respectively. Condition (39) is supposed to consist of two parts: the first term fires on the outflow (when $|\mathbf{n}\cdot\mathbf{u}| = \mathbf{n}\cdot\mathbf{u}$), whereas the second term is responsible for the inflow ($h_{(\Gamma)}$ is supposed to be given a priori).

Finally, on the right-hand side of (11) we add a wind stress of the form $\mathbf{W}\psi \sin 2\pi t$, where

$$\mathbf{W} \sim \begin{pmatrix} -\cos\frac{\pi(\varphi-\varphi_{\min})}{L_\varphi} & 0 & 0 \\ 0 & \cos\frac{\pi(\varphi-\varphi_{\min})}{2L_\varphi} & 0 \\ 0 & 0 & 0 \end{pmatrix}, \tag{41}$$

while $L_\varphi = \varphi_{\max} - \varphi_{\min}$. The wind stress field at $t = 0.25$ and $t = 0.75$ is shown in Fig. 4.

The numerical solution computed on the grid $\Delta\lambda \approx \Delta\varphi \approx 0.005°$ and $\tau = 1.44$ min is presented in Fig. 5 at several time moments. Comparison with Fig. 4 shows that a wind-driven flow occurs and is then developing in the computational domain. Specifically, as the simulation starts, the velocity field is formed clockwise (Fig. 5, $t = 0.2, 0.4$), in accordance with the wind stress at small times (Fig. 4, left). Later, at $t = 0.5$, the wind's direction changes to anticlockwise due to the term $\sin 2\pi t$, which is reflected in the numerical solution with a little time gap because of the fluid's inertia, especially in the open ocean far from the coastline: while the coastal waters change their flows at $t \approx 0.5 - 0.7$, the large vortex in the open bay begins rotating anticlockwise at $t \approx 0.8$ (Fig. 5). Finally, at $t = 1$ the entire velocity field is aligned in accordance with the late-time wind stress (Fig. 4, right).

In Fig. 6 we show the numerical solution to a slightly complicated problem—the bottom topography is now a smooth hat-like function with the epicentre at $\lambda_c = 12.51$, $\varphi_c = 45.24$. The diameter of the bottom's irregularity is about $0.1°$, maximum height is about 1.6 m. As it is seen from the figure, unlike the previous case the presence of an underwater obstacle causes permutations in the depth field around, while the structure of the velocity field keeps mostly unchanged in time (cf. Fig. 5).

4 Conclusions

A new fully discrete mass– and total-energy–conserving finite difference model for the simulation of shallow-water flows in bay-like domains with complex boundaries was developed. Having taken the SWEs written in the divergent form, we involved the idea of operator splitting coupled with the Crank-Nicolson approximation and constructed absolutely stable second-order finite difference schemes that allow accurate simulation of shallow-water flows in spherical domains of arbitrary shapes. An important integral invariant of motion of the SWEs, the potential enstrophy, proved to be bounded for an inviscid unforced fluid, oscillating in time within a narrow range. Hence, the numerical solution is mathematically accurate and provides physically adequate results. Due to the method of splitting the developed model can straightforwardly be implemented for distributed simulation of shallow-water flows on high-performance parallel computers. Numerical experiments with a simple inviscid unforced closed shallow-water system and with a viscous open wind-driven shallow-water model with zero and nonzero bottom topography, simulating real situations, nicely confirmed the skills of the new method.

Acknowledgements. This work was supported by Mexican System of Researchers (SNI) grants 14539 and 26073. It is part of the project PAPIIT DGAPA UNAM IN104811-3.

References

1. Agoshkov, V.I., Saleri, F.: Recent Developments in the Numerical Simulation of Shallow Water Equations. Part III: Boundary Conditions and Finite Element Approximations in the River Flow Calculations. Math. Modelling 8, 3–24 (1996)
2. Arakawa, A., Lamb, V.R.: A Potential Enstrophy and Energy Conserving Scheme for the Shallow-Water Equation. Mon. Wea. Rev. 109, 18–36 (1981)
3. Bouchut, F., Le Sommer, J., Zeitlin, V.: Frontal Geostrophic Adjustment and Nonlinear Wave Phenomena in One-dimensional Rotating Shallow Water. Part II: High-Resolution Numerical Simulations. J. Fluid Mech. 514, 35–63 (2004)
4. Heikes, R., Randall, D.A.: Numerical Integration of the Shallow-Water Equations on a Twisted Icosahedral Grid. Part I: Basic Design and Results of Tests. Mon. Wea. Rev. 123, 1862–1880 (1995)
5. Jirka, G.H., Uijttewaal, W.S.J.: Shallow Flows. Taylor & Francis, London (2004)
6. Kundu, P.K., Cohen, I.M., Dowling, D.R.: Fluid Mecanics, 5th edn. Academic Press, London (2012)
7. LeVeque, R.J., George, D.L.: High-Resolution Finite Volume Methods for the Shallow-Water Equations with Bathymetry and Dry States. In: Yeh, H., Liu, P.L., Synolakis, C.E. (eds.) Advanced Numerical Models for Simulating Tsunami Waves and Runup, pp. 43–73. World Scientific Publishing, Singapore (2007)
8. Marchuk, G.I.: Methods of Computational Mathematics. Springer, Berlin (1982)
9. Oliger, J., Sündstrom, A.: Theoretical and Practical Aspects of Some Initial Boundary Value Problems in Fluid Dynamics. SIAM J. Appl. Anal. 35, 419–446 (1978)
10. Pedlosky, J.: Geophysical Fluid Dynamics, 2nd edn. Springer (1987)
11. Ringler, T.D., Randall, D.A.: A Potential Enstrophy and Energy Conserving Numerical Scheme for Solution of the Shallow-Water Equations on a Geodesic Grid. Mon. Wea. Rev. 130, 1397–1410 (2002)
12. Sadourny, R.: The Dynamics of Finite-Difference Models of the Shallow-Water Equations. J. Atmos. Sci. 32, 680–689 (1975)
13. Salmon, R.: A Shallow Water Model Conserving Energy and Potential Enstrophy in the Presence of Boundaries. J. Mar. Res. 67, 1–36 (2009)
14. Shokin, Y.I.: Completely Conservative Difference Schemes. In: de Vahl Devis, G., Fletcher, C. (eds.) Computational Fluid Dynamics, pp. 135–155. Elsevier, Amsterdam (1988)
15. Simonnet, E., Ghil, M., Ide, K., Temam, R., Wang, S.: Low-Frequency Variability in Shallow-Water Models of the Wind-Driven Ocean Circulation. Part I: Steady-State Solution. J. Phys. Ocean. 33, 712–728 (2003)
16. Skiba, Y.N.: Total Energy and Mass Conserving Finite Difference Schemes for the Shallow-Water Equations. Russ. Meteorol. Hydrology 2, 35–43 (1995)
17. Skiba, Y.N., Filatov, D.M.: Conservative Arbitrary Order Finite Difference Schemes for Shallow-Water Flows. J. Comput. Appl. Math. 218, 579–591 (2008)

18. Skiba, Y.N., Filatov, D.M.: Simulation of Soliton-like Waves Generated by Topography with Conservative Fully Discrete Shallow-Water Arbitrary-Order Schemes. Internat. J. Numer. Methods Heat Fluid Flow 19, 982–1007 (2009)
19. Vol'tsynger, N.E., Pyaskovskiy, R.V.: The Theory of Shallow Water. Gidrometeoizdat, St. Petersburg (1977)
20. Vreugdenhil, C.B.: Numerical Methods for Shallow-Water Flow. Kluwer Academic, Dordrecht (1994)

System Dynamics and Agent-Based Simulation for Prospective Health Technology Assessments*

Anatoli Djanatliev[1], Peter Kolominsky-Rabas[2], Bernd M. Hofmann[3], Axel Aisenbrey[3], and Reinhard German[1]

[1] Department of Computer Science 7, University of Erlangen-Nuremberg, Germany
[2] Interdisciplinary Centre for Public Health, University of Erlangen-Nuremberg, Germany
[3] Siemens AG, Healthcare Sector, Germany
{anatoli.djanatliev,german}@cs.fau.de

Abstract. Healthcare innovations open new treatment possibilities for patients and offer potentials to increase their life quality. But it is also possible that a new product will have negative influences on patient's life quality, if it has not been checked before. To prevent latter cases three already established methods can be used in order to assess healthcare technologies and to inform regulatory agencies. But these tools share a common problem. They can only be applied when a product is already developed and high costs have been already produced. This work describes Prospective Health Technology Assessment. This approach uses hybrid simulation techniques in order to learn about the impacts of a new innovation before a product has been developed. System Dynamics is used to perform high-level simulation and Agent-Based Simulation allows to model individual behavior of persons.

Keywords: Healthcare, Simulation, Hybrid Simulation, System Dynamics, Agent-based Simulation, Prospective Health, Technology Assessment, Ischemic Stroke.

1 Introduction

The global market for medical technology products is growing rapidly due to an increasing demand of innovative healthcare technologies. Presumably, some reasons of this notice are the increasing life expectancy of the population, new technical opportunities and an increasing complexity of the healthcare delivery.

A significant consequence of this trend is that the healthcare industry has to handle even faster trade-off decisions before developing new innovations in medicine. To be more profitable a new product must have a short development phase, low costs and it is also important that the expected revenue can be reached. In addition, health technologies are safety-critical, as the life quality can depend on the product's quality. Following this fact, proven evidence is necessary and regulatory barriers have to be overcome. Usually, such activities are time-consuming and their results are difficult to predict. Often there is also no proven evidence available, so the development of new innovations can degenerate into a risky, non-profitable project.

* On behalf of the ProHTA Research Group.

M.S. Obaidat et al. (eds.), *Simulation and Modeling Methodologies, Technologies and Applications*, Advances in Intelligent Systems and Computing 256,
DOI: 10.1007/978-3-319-03581-9_6,

The main difference between the healthcare domain and other safety-critical areas is that many involved stakeholders with different interests have to be satisfied before a product can be placed on the market. Some examples of them are patients, healthcare providers such as hospitals and outpatient departments, manufacturers in healthcare, technology developers, insurance companies, governments, regulatory institutions and academia.

According to the described situation, assessments of health technologies are very complex and advanced assessment methods are crucial to prevent disinvestments and to make the product development more effective. All players must be considered for this process to reach an overall credibility of the product's quality and to detect all kinds of problems as early as possible.

Three assessment methodologies for healthcare technologies have been introduced until now. Health Technology Assessment, Early Health Technology Assessment and Horizon Scanning. Health Technology Assessment (HTA) is an approach used after the market launch of a product when evidence is available from studies. In accordance with Goodman [10], the main field of HTA is to inform, among others, regulatory agencies and lawmakers about permission decisions for a commercialization of a regarded innovation. Early Health Technology Assessment (Early HTA) tends to be an appropriate tool in cases where the product is already in the development phase but only limited evidence is available. Some examples for such evidence levels are animal testing, early clinical experience, or data from previous technology generations [17]. The impact of Horizon Scanning (HS) is the comparative assessment of similar technologies to observe trends of possible disinvestments and impacts of new healthcare technologies. Horizon Scanning Systems focus on health technologies that are ready to enter the market, i.e. in early post-marketing phases [9].

In many cases the technological progress is the key factor for new products and there is only a little consideration of possible future consequences. But new innovations must reach positive assessment results before their impact can be seen by all stakeholders as reasonable. Hence, an enormous number of disinvestments can be expected, as established assessment methods (HTA, Early HTA and HS) are just used for products that already have passed the expensive design and development process.

For that reason we call for advanced foresight healthcare assessment methods which are applicable early before high efforts and investments were made. This is where our approach ProHTA can take an important role, because simulation techniques can be applied to assess new innovations prospectively.

2 Related Work

There are already a couple of publications which are presenting the use of simulation for the healthcare domain. Most of them are considering healthcare processes and are modeled by the Discrete Event Simulation approach. Brailsford [3] discussed System Dynamics (SD) as tool for healthcare modelers. The author presented several examples and depicted some reasons for the growth in SD popularity. Following this, some advantages of SD are lower data requirements, the sight of the big-picture and in particular very fast simulation runs.

In order to benefit from high-level abstractions as well as to allow more detailed modeling possibilities, hybrid simulation techniques are gaining acceptance. Heath et al. [11] discussed several challenges and successes of cross-paradigm modeling. In particular, Discrete Event Simulation, System Dynamics and Agent-Based Simulation approaches were considered in detail. Though hybrid simulation is not precisely defined yet, there are however software packages that allow multi-paradigm modeling.

3 Prospective Health Technology Assessment

Prospective Health Technology Assessment (ProHTA) is a new approach that extends the tool environment of healthcare assessment methodologies and fills the gap, mentioned previously. For this reason the project is located within the Medical Valley EMN (European Metropolitan Region Nuremberg) and is a part of the Centre of Excellence for Medical Technology.

The innovative intention of ProHTA is the assessment of health technologies from many perspectives, early before the development of a medical innovation begins. It helps to detect potentials for process optimization and allows learning about the influence of a new technology on the established health system structures. Innovation's cost-effectiveness has to be prospectively calculated as well as the impact on the patient's health.

Figure 1 summarizes the main prospective evaluation processes. In this respect, two questions are central to ProHTA:

1. What are the changes that result from the launch of a new technology? - ProHTA will be able to simulate and assess the effects of changes to processes, which are introduced e.g. by a new technology. ProHTA will project the effects of innovative health technologies on the quality and costs of health care.
2. What does a technology need to be like in order to have a specific effect? - If certain input requirements of a desired change are specified, the ProHTA tools will enable outcome related conclusions concerning the required changes to the process (hypothesis development). ProHTA will envisage the effects of the potential efficiency enhancement on the health care system.

Especially the second question allows ProHTA to examine the existing health system and to find bottlenecks and weaknesses of currently applied practices. This can lead to new ideas for health technologies that are particularly based on desired effects and not only on technical opportunities. In the course of the method creation of ProHTA a new scientific service platform has to be developed targeting on the scope of the just described project.

The regarded challenges and questions are handled together by an interdisciplinary team consisting of experts from the areas of Public Health, Health Technology Assessment, Clinical Medicine, Health Economics and Outcomes Research, Medical Informatics, Knowledge Management as well as Modeling and Simulation. Furthermore, two representatives of the healthcare industry also participate in the project.

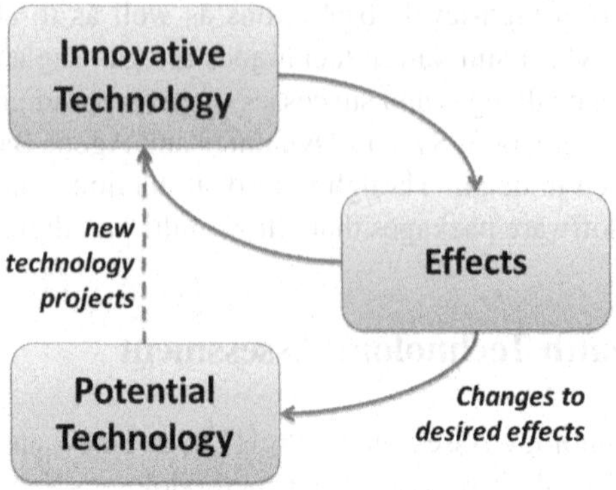

Fig. 1. ProHTA overview

In this paper we discuss the use of hybrid simulation approaches, consisting of System Dynamics (SD) and Agent-Based Simulation (ABS), for the new foresight assessment method ProHTA. As the complexity of the considered domain and the interdisciplinary co-working is enormous, a structured methodology towards a hybrid simulation model has been developed. The current focus of the approach is the assessment of innovative technologies in management of acute stroke. We use Mobile Stroke Units (MSU) within a Metropolitan area as an innovative health technology to show exemplary how hybrid simulation techniques can be applied to the impact of ProHTA. Within the discussion section we look ahead to the future work, including the validation of our approach and its application to oncology diseases.

4 Conceptual Modeling

As already mentioned, many challenges have to be mastered during the project realization phase. To achieve an effective and intensive co-working of all experts with different backgrounds, a structured methodology is essential.

For that reason a dedicated *Conceptual Modeling Process (CMP)* for ProHTA was defined, classifying the main steps towards a hybrid simulation model, beginning at the domain experts' knowledge collection. It is largely based on current work of a research group from the University of Warwick and Loughborough University [19,15].

As hybrid simulation builds the focus of this paper and the CMP will be the topic of another publication, we introduce in the following only some parts of it, as it is important for the simulation.

The CMP distinguishes in particular between *domain experts* and *technical experts* and defines fields of activity and intermediate interfaces for all of them. Furthermore, the reality is summarized within the *Domain World* and the abstraction of it is contained in the, so called, *Model World*. The significant artifact of the Domain World is the non-formal specification of the problem, represented by the *Conceptual Domain Model (CDM)*. This result is used by an iterative formalization process as input for the Model World, according to create the *Formal Conceptual Model (FCM)* which serves as a basis for the simulation model and knowledge base.

5 Stroke Therapy by Mobile Stroke Units

Stroke is one of the leading causes of death and disability and absorbs a considerable proportion of healthcare budgets [13]. In Germany an increase of 1.5 million stroke cases is estimated in the next decade [14]. Due to an ageing population, increasing incidence values could be expected in the future, as stroke appears mainly at higher age groups. The costs for treatment and rehabilitation in Germany are estimated to be totally about 58 billion of euros in the period 2006-2015 [14].

One of the most frequent form of stroke is the *ischemic stroke*, caused by an oclusion of cerebral arteries. Currently, an approved method for the treatment is the *intravenous thrombolysis* by recombinant tissue plasminogen activator (rtPA) [7]. However, even in specialized hospitals (Stroke Units), only 7 to 10 percent of patients are treated by this effective method [12]. The main reason for such a bad rate is the applicability of the thrombolysis only within 4.5 hours after the obstruction's occurrence and the time, elapsed during the transport to the hospital [18]. A significant increase of the rate from 3 percent to 7 percent had been observed after the extension of the recommended time window from 3 hours to 4.5 hours in 2008 [18].

In case of an affection, nearly 1.9 million of neurons can die per minute [16]. Hence, the "time is brain" concept has to be applied to the extremely time-critical treatment of stroke. Following this fact, particularly innovations in stroke treatment are necessary which are able to reduce the call-to-therapy-decision time.

Two German research groups from Saarland and Berlin [20,6] are working with their partners on methods targeting at the transfer of the thrombolytic inpatient treatment to the pretreatment phase. In that case the therapy can be applied before the time-intensive transfer to the hospital and an increase of the number of patients, treated by thrombolysis can be expected. The main problem of this idea is the important exclusion of an intracerebral haemorrhage, before applying rtPA. This can be done by laboratory analyses and Computer Tomography (CT), usually installed in hospitals. To prevent a loss of crucial time, both research groups developed similar prototypes, the *Mobile Stroke Unit (MSU)* and the *Stroke-Einsatz-Mobil (STEMO)*. Such a vehicle extends the standard emergency equipment by a CT and further tools for a rapid diagnostic decision onsite at stroke occurrence location.

In an early phase, first trials have shown that a shortened call-to-therapy-decision-time of approximately 35 minutes is not a vision [16]. To inspect other relevant effects (e.g. long-term cost-effectiveness, application in other regions) more trials are necessary, so that further time and cost investments have to be made. This is where ProHTA can provide an early analysis by hybrid simulation.

6 Hybrid Simulation

For the purpose of the Prospective Health Technology Assessment simulation has been identified to be an appropriate tool. Customized large scale simulation models are crucial to handle complex questions in situations where an innovative technology possibly hasn't been developed yet. Depending on the point of view, ProHTA must be able to answer specialized questions (e.g. effectiveness of a new innovation) as well as to calculate global and long-term consequences. Hence, large scale models are necessary to fulfill these requirements. Furthermore, separate data management and data quality components are crucial to handle a couple of different input data formats in an efficient way [2].

We need methods to create models on an abstract, macroscopic level (e.g. economic flows, population dynamics) where data about details is not available, or even is not necessary. In our approach System Dynamics (SD) [8] was qualified as an appropriate tool for this kind of simulation. To model individual aspects of patient's workflows and its behavior, the Agent-Based Simulation (ABS) approach has been selected. New technological innovations can also be modeled by the ABS approach to allow behavioral changes, according to inspect their impacts on output values.

Each assessment scenario is individual. For this reason we propose to combine the SD and the ABS approach into a common hybrid simulation using an adequate ratio according to the desired simulation level and high simulation performance. This trade-off decision can be handled by following a top-down approach. In that case the problem environment can be modeled as deeply as needed by abstract methods and a special simulation part will be framed-out and realized by the ABS approach.

The idea of hybrid simulation is getting more and more popular in current research [11,4]. There are many software packages that allow to create simulations by SD, Discrete Event Simulation (DES) or ABS models, but most of them support only one of the presented approaches. AnyLogic [21] offers the power to combine different simulation paradigms within a common simulation environment. This is why this tool is qualified to create models within the scope of ProHTA.

6.1 Modular Environment

In accordance to the Conceptual Modeling Process, the CDM represents a non-formal description of the collected experts' knowledge. The FCM is central to develop simulation models within the context of ProHTA.

We have to deal with different, interacting simulation parts, such as money flows, healthcare structures, disease, and population development. To make models reusable for further assessment scenarios and to master the complexity, modularization is important. Further advantages of a modular structure are *domain-focused modeling* (domain experts focus on modeling of such parts that are mostly covered by their knowledge), *validation* and *maintenance*. It is not easy to find optimal units in healthcare, as there are usually not well-formed boundaries available. For this reason it is usually an iterative process until optimal modules can be defined.

Chahal and Eldabi [5] proposed a *process environmental format* for hybrid simulations. In our hybrid simulation approach for ProHTA, SD models can be used to build the environment of a simulation. An important simulation part can be modeled by ABS approaches and used as core within the process environmental format. Figure 2 depicts important modules within the scope of ProHTA.

Population Dynamics summarizes models and parameters that handle with the development of demographic structures, e.g. birth rate, mortality, immigration and emigration.

Disease Dynamics includes generic model parts that deal with illness parameters, e.g. incidence, prevalence, case fatality rate and remission.

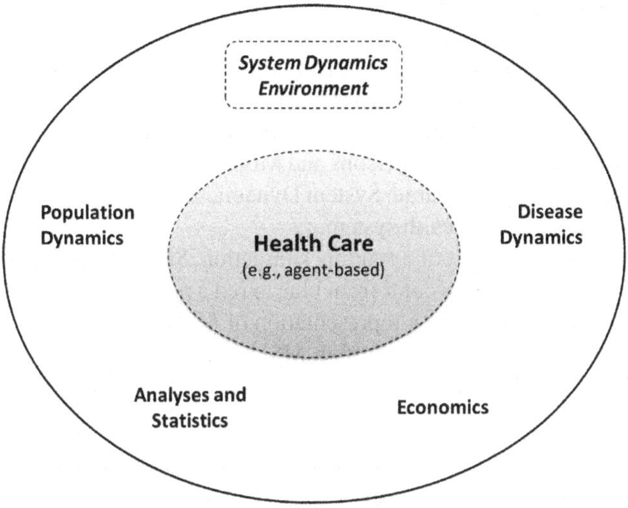

Fig. 2. Overview of the ProHTA modules

Economics combines the dynamics of money flows of the health system. Statutory and private health insurance can be modeled there as well as the long-term care system.

Health Care is a module where especially workflows of prevention, pre-treatment, treatment and post-treatment of a regarded disease can be modeled.

6.2 Scenario Description

Before we set the focus on our model, the scenario with Mobile Stroke Units within a Metropolitan region will be shortly presented in the following:

- Stroke scenario with usage of a predefined number of 10 MSUs.
- A German Metropolitan region, represented by Berlin, with approximately 3.4 mill. people is modeled using data from the statistical calculations of [1].
- The regional distribution of people is done by information about the district density.
- MSUs are randomly distributed within the region boundaries.
- People can get stroke and call the Rescue Service (RS).
- RS decides whether an MSU can be sent (e.g. there are free MSUs available) to the affected patient.
- During the affection phase, people pass through diagnostic and therapy workflows and lose life quality according to a therapeutic effectiveness.
- In case of using an MSU, thrombolytic therapy can be started onsite at patient's location, after a mobile CT and laboratory diagnose were performed.
- Calculated costs are drawn from financial budgets.
- Demographical information is used to reproduce the population development.
- Prevalence is used to separate the population in affected and not-affected parts initially.
- Incidence is used as a dynamic affection rate.

For assessment purposes the same scenario without MSUs will be run afterwards. It allows to make the effects of a new intervention (MSUs) visible.

6.3 Simulation Model

Overview. The simulation is started within an initialization phase and proceeds with a forecasting phase after a couple of data has been produced.

There are two types of agents modeled within the hybrid simulation of our MSU scenario, the individual behavior of persons and Mobile Stroke Units. Demographic development is reproduced by a separate System Dynamics model (Population Dynamics) as well as money flows of the health system.

Figure 3 depicts a screenshot of a running simulation. Small points represent person agents (number 1). Changing the color from blue to red a patient becomes affected. Yellow bullets with black borders are a representation of MSU agents (number 2). During simulation runs they are colored in red, if an MSU is busy to see an overall utilization of the new technology. MSUs pick up patients and move them to the next hospital.

Agents. In our previous models the whole population had to be represented by agents. As this is a large performance drawback, we scaled down the Berlin population that is not a proper method. Currently, Disease Dynamics is responsible to calculate the number of new affections and an agent is only generated afterwards. So the main advance is that the number of person agents can be reduced extremely without scaling and the simulation is running faster.

Age, gender and life quality (LQ) are important attributes of a person and are sampled initially. LQ is represented by a value between 0 and 100. Dependent on the therapy success, the patient loses more or less LQ points. In case of the stroke use case LQ represents the self-reliance of a person.

After an agent has been created, a cognition process starts. Within this phase time elapses until the person decides to call the emergency service. While performing the call, the agent enters the *preTreatment* state and a dedicated workflow is being instantiated dynamically. After finishing pre-treatment, the agent traverses hospital workflows. Finally, the agent remains noticed for a while in the post treatment phase. The just described process can be depicted in Figure 4 from the left hand behavior state chart of the person agent.

The state chart of the agent type MSU is depicted on the right hand of the Figure 4 and includes two states (*free*, *busy*). The variable *timeApprToScene* is used to simulate the delay of an MSU until arrival at patient's location. In a further step it can be used to reproduce traffic by an additional simulation part.

Each step that produces costs (e.g. special treatments, materials, staff employment) collects them in a global variable within the economics System Dynamics model. The calculated number is pulled off from the corresponding budgets once-a-year.

Pre-Treatment. After the expiration of the cognition time the person enters the *pre-Treatment* state and starts the workflow of the pre-treatment phase. According to the workflow an emergency call is modeled first. After talking to the dispatcher a first diagnosis is made and a rescue service is alarmed to move to the patient. Due to the assessment of MSUs, the dispatcher searches for free Mobile Stroke Units. The more of them are initially available, the higher is the probability to get one free. Otherwise,

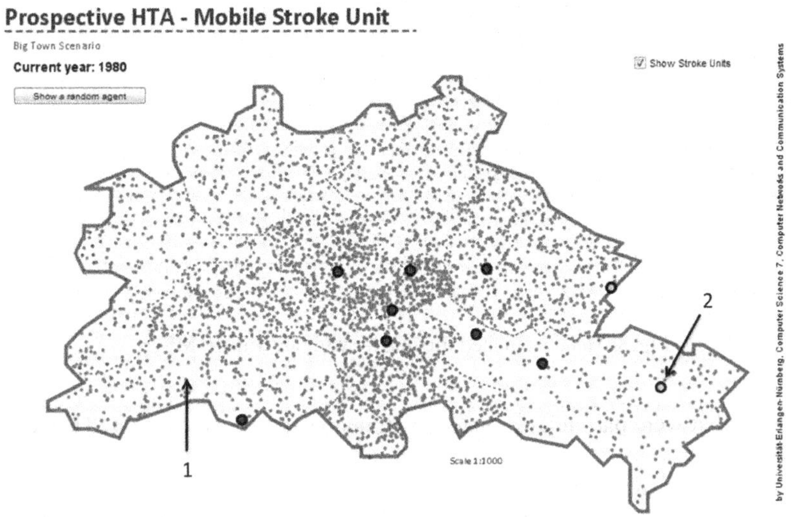

Fig. 3. Mobile Stroke Units within a Metropolitan Scenario

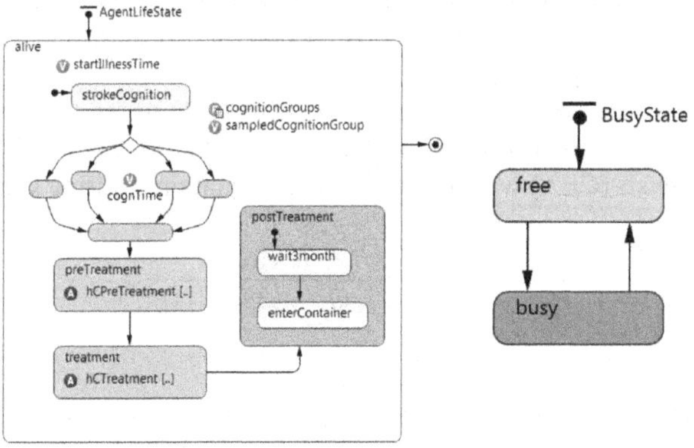

Fig. 4. Behavior state charts of agent person (left) and MSU (right)

a normal emergency doctor will be sent. In case of MSU usage, a CT exam can be performed at the patient's location in the pre-treatment phase, as described in section 5.

Treatment. Due to the arrival in the hospital the treatment phase starts by an instantiation of the hospital workflows. Diagnosis and therapy steps which were already performed during the pre-treatment phase (in case of MSU usage) are skipped and a "happy path", modeled by our domain experts, is traversed by the patient.

6.4 Simulation Results

In our exemplary use-case of Mobile Stroke Units we noticed some trends due to the implementation of 10 MSUs in a metropolitan area. According to the used input

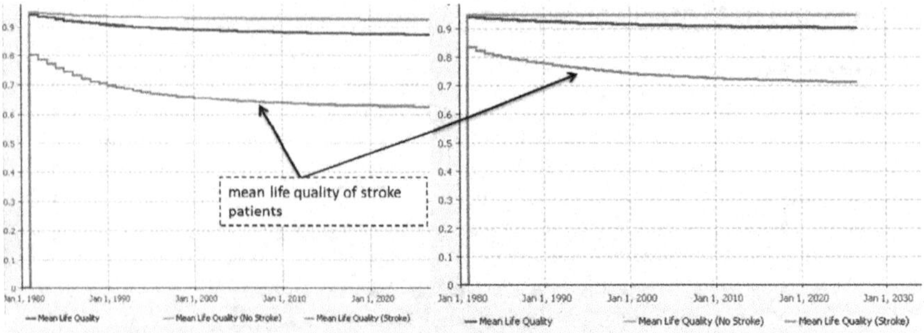

Fig. 5. Mean estimated life quality without/with 10 MSUs

parameters that were partially estimated by domain experts, we monitored some of the expected results.

Figure 5 shows two plots of the estimated mean life quality. On the left hand we can see graphs that were calculated by a simulation run without MSU usage; on the right hand a scenario with 10 Mobile Stroke Units was applied. According to the results we can notice that persons with stroke affection achieve a higher mean life quality after the launch of the regarded innovation. The main reason for this is the possibility to apply the thrombolytic therapy immediately at the stroke occurrence location and an intervention can be done within the important time-window of 4.5 hours.

In our model we also aggregated costs caused by MSU usage to show exemplary economic assessment calculations. To check if a new technology is reasonable, a Cost-Effectiveness-Analysis (CEA) was computed by the equation 1 during simulation runs.

$$CEA = \frac{K_{MSU} - K_{NoMSU}}{\Delta LQ_{NoMSU} - \Delta LQ_{MSU}} \tag{1}$$

We regarded life quality as effectiveness parameter to determine the effect of MSUs. Within the equation the following variables are defined as follows: $K_{MSU} - K_{NoMSU}$ (cost difference between MSU usage and conventional therapy workflows), ΔLQ_{NoMSU} and ΔLQ_{MSU} are mean life quality losses of stroke patients without/with MSU implementation.

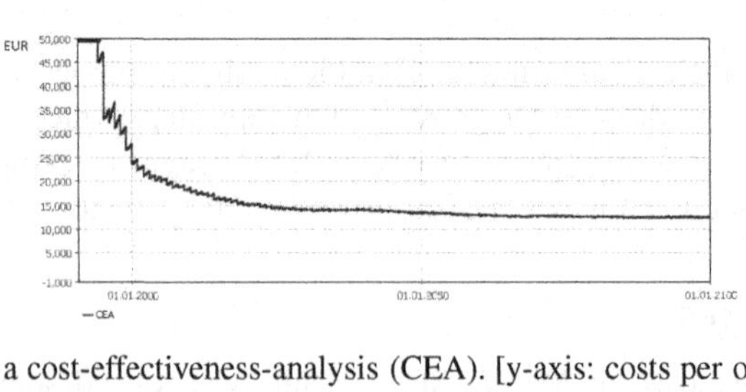

Fig. 6. Example of a cost-effectiveness-analysis (CEA). [y-axis: costs per one life quality point, x-axis: time].

Graph 6 shows an example result of a Cost-Effectiveness-Analysis. The plot suggests that an increase of the mean quality of life of the population of one point comes with a cost of approximately 12,000 EUR. It also shows the effects of the initialization phase which have been alleviated in the forecasting phase starting around the year 2010.

7 Conclusions

We identified hybrid simulation, consisting of System Dynamics and Agent-Based Simulation, to be an appropriate modeling approach for large scale simulations. This method enables to solve problems on a macroscopic level, e.g. disease dynamics in a global context, as well as on a detailed, behavioral microscopic level. A further benefit of this approach within the scope of ProHTA is the possibility to build models top-down, beginning from high abstractions and going more in detail to frame-out individual workflows. To make models capable for reuse and to master complexity, we designed a modular and generic environment that is also eligible for flexible model changes.

To evaluate our methods, an exemplary use-case scenario of an innovative stroke treatment approach, represented by Mobile Stroke Units within a Metropolitan Scenario, had been implemented, using the simulation software AnyLogic [21]. This tool is predestined for multi-method simulation paradigms. The project procedure strictly followed the CMP and an overall expert-credibility has been achieved.

There are still many challenges to master in the future. MSU usage shall be defined regionally, as it it not realistic to use a free MSU immediately. A further complex task will be the application of our hybrid simulation approach to other diseases, especially within the domain of personalized medicine. As the ProHTA research group includes oncology experts, cancer diseases will be the focus of further work.

Acknowledgements. Prospective Health Technology Assessment (ProHTA) is funded by the German Federal Ministry of Education and Research (BMBF) as part of the National Cluster of Excellence Medical Technology - Medical Valley EMN (Project grant No. 01EX1013B).

References

1. Amt für Statistik Berlin-Brandenburg: Die kleine Berlin-Statistik (2011)
2. Baumgärtel, P., Lenz, R.: Towards data and data quality management for large scale healthcare simulations. In: Conchon, E., Correia, C., Fred, A., Gamboa, H. (eds.) Proceedings of the International Conference on Health Informatics, pp. 275–280. SciTePress - Science and Technology Publications (2012) ISBN: 978-989-8425-88-1
3. Brailsford, S.C.: System Dynamics: What's in it for Healthcare Simulation Modelers. In: Proceedings of the 2008 Winter Simulation Conference, pp. 1478–1483 (2008)
4. Brailsford, S.C., Desai, S.M., Viana, J.: Towards the holy grail: Combining system dynamics and discrete-event simulation in healthcare. In: Proceedings of the 2010 Winter Simulation Conference, pp. 2293–2303 (2010)
5. Chahal, K., Eldabi, T.: Applicability of hybrid simulation to different modes of governance in UK healthcare. In: Proceedings of the 2008 Winter Simulation Conference, pp. 1469–1477 (2008)

6. Ebinger, M., Rozanski, M., Waldschmidt, C., Weber, J., Wendt, M., Winter, B., Kellner, P., Baumann, A., Malzahn, U., Heuschmann, P., Fiebach, J., Endres, M., Audebert, H.: PHANTOM-S: the prehospital acute neurological therapy and optimization of medical care in stroke patients - study. International Journal of Stroke (February 2, 2012)
7. Fassbender, K., Walter, S., Liu, Y., Muehlhauser, F., Ragoschke, A., Kuehl, S., Mielke, O.: Mobile Stroke Unit for Hyperacute Stroke Treatment. Stroke 34(6), 44e (2003)
8. Forrester, J.W.: Industrial Dynamics. Pegasus Communications, Waltham and MA (1964)
9. Geiger-Gritsch, S.: Horizon Scanning in Oncology: Concept Development for the Preparation of a Horizon Scanning System in Austria. Ludwig Boltzmann Institute for Health Technology Assessment. (2008)
10. Goodman, C.S.: HTA 101: Introduction to Health Technology Assessment. National Information Center on Health Services Research and Health Care Technology (NICHSR). The Lewin Group (2004)
11. Heath, S.K., Brailsford, S.C., Buss, A., Macal, C.M.: Cross-paradigm simulation modeling: challenges and successes. In: Proceedings of the 2011 Winter Simulation Conference, pp. 2788–2802 (2011)
12. Heuschmann, P., Busse, O., Wagner, M., Endres, M., Villringer, A., Röther, J., Kolominsky-Rabas, P., Berger, K.: Frequency and Care of Stroke in Germany. Aktuelle Neurologie 37(07), 333–340 (2010)
13. Kjellström, T., Norrving, B., Shatchkute, A.: Helsingborg Declaration 2006 on European Stroke Strategies. Cerebrovascular Diseases 23(2-3), 229–241 (2007)
14. Kolominsky-Rabas, P.L., Heuschmann, P., Marschall, D., Emmert, M., Baltzer, N., Neundoerfer, B., Schöffski, O., Krobot, K.J.: Lifetime Cost of Ischemic Stroke in Germany: Results and National Projections From a Population-Based Stroke Registry: The Erlangen Stroke Project. Stroke 37(5), 1179–1183 (2006)
15. Kotiadis, K., Robinson, S.: Conceptual modelling: Knowledge acquisition and model abstraction. In: Proceedings of the 2008 Winter Simulation Conference, pp. 951–958 (2008)
16. Kuehn, A., Grunwald, I.: Das Konzept der Mobilen Stroke Unit (MSU). Der Radiologe 51(4), 261–262 (2011)
17. Pietzsch, J.B., Paté-Cornell, M.E.: Early technology assessment of new medical devices. International Journal of Technology Assessment in Health Care 24(01), 36–44 (2008)
18. Purrucker, J., Veltkamp, R.: Thrombolysis Beyond the 3-h Time Window. Aktuelle Neurologie 38(09), 494–504 (2011)
19. Robinson, S.: Choosing the right model: Conceptual modeling for simulation. In: Proceedings of the 2011 Winter Simulation Conference, pp. 1428–1440 (2011)
20. Walter, S., Kostpopoulos, P., Haass, A., Helwig, S., Keller, I., Licina, T., Schlechtriemen, T., Roth, C., Papanagiotou, P., Zimmer, A., Vierra, J., Körner, H., Schmidt, K., Romann, M.S., Alexandrou, M., Yilmaz, U., Grunwald, I., Kubulus, D., Lesmeister, M., Ziegeler, S., Pattar, A., Golinski, M., Liu, Y., Volk, T., Bertsch, T., Reith, W., Fassbender, K., Noor, A.M.: Bringing the Hospital to the Patient: First Treatment of Stroke Patients at the Emergency Site. PLoS One 5(10), e13758 (2010)
21. XJ Technologies Company Ltd.: AnyLogic -, `http://www.xjtek.com`

Simple and Efficient Algorithms to Get a Finer Resolution in a Stochastic Discrete Time Agent-Based Simulation*

Chia-Tung Kuo, Da-Wei Wang, and Tsan-sheng Hsu**

Institute of Information Science,
Academia Sinica, Taipei 115, Taiwan
{ctkuo,wdw,tshsu}@iis.sinica.edu.tw

Abstract. A conceptually simple approach on adjusting the time step to a finer one is proposed and an efficient two-level sampling algorithm for it is presented. Our approach enables the modelers to divide each original time step into any integral number of equally spaced sub-steps, and the original model and the finer one can be formally shown to be equivalent with some basic assumptions.

The main idea behind the two-level sampling algorithm is to make "big decision" first and make subsequent "small decisions" when necessary, i.e., first to decide if an event occurred in original time scale then refine the occurrence time to a finer scale if it does occur.

The approach is used on a stochastic model for epidemic spread and show that the refined model produces expected results. The computational resources needed for the refined model increases only marginally using the two-level sampling algorithm together with some implementation techniques which are also highlighted in the paper.

Approach proposed in this paper can be adapted to be used in merging continuous time steps into a super step to save simulation time.

1 Introduction

The stochastic discrete time simulation model is a useful and efficient way in simulating agent-based activities and has gained significant popularity in modelling many dynamical biological or physical systems in recent years, such as the dynamics of epidemic spread [10,8]. Often time, such simulations can give insights into problems where traditional models are too complicated and analytic results are very difficult or currently impossible to obtain [10,4]. This high level abstraction, however, introduces artifacts which do not pertain to real world behavior, namely, the discretization of time. In particular, event simultaneity whereby multiple distinct events occur at exactly the same time may be due to an insufficiently detailed discrete-time model [11]. The additional parameter, the size of the time step, can potentially have significant impact on the results of simulation without the awareness of the modeller [1]. Buss and Rowaei investigated

* An earlier extended abstract of this paper appears in [7].

** Corresponding Author. Supported in part by NSC of Taiwan Grants NSC100-2221-E-001-011-MY3.

M.S. Obaidat et al. (eds.), *Simulation and Modeling Methodologies, Technologies and Applications*, 97
Advances in Intelligent Systems and Computing 256,
DOI: 10.1007/978-3-319-03581-9_7, © Springer International Publishing Switzerland 2014

time advancement mechanism and the role of time step size in a somewhat different context [1,2], and found no systematic studies had been done with its effects. Moreover, the choice of time step size plays a role in the efficiency of the simulation as the occurrence of events need to be checked more frequently for smaller time steps. Accordingly, the size of the time step should be carefully selected to match the real world phenomenon to be modelled as realistically as possible, and without too much sacrifice of simulation time. Other than for its realistic nature, this choice of time step size is often influenced by the empirical data we have. For example, in modelling the epidemic spread, the unit of measurements for latent and infectious periods also affect the choice of time step size [5]. Specifically, the unit of measurements for latent and infectious periods should be smaller than or equal to the length of a time step to make sure that no event advances two steps in one simulated time step.

We have developed a simple, yet efficient, technique based on reasonable assumptions to split each time step into any integral number of equally spaced sub-steps with small increase in simulation time in a stochastic discrete time agent based simulation system for epidemic spread we developed earlier [10,9]. More precisely, we first modify the probabilities of events according to the basic probability theory such that they correspond to events occurring in a smaller time frame. This allows the introduction of certain types of events in a smaller time scale. Then we introduce a structurally different implementation model that significantly outperforms the straightforward step-by-step linear model. The experiments show that our improved implementation produces stochastically identical results as the straightforward implementation with significantly less time. Furthermore, our results also show that, given the same set of possible events, simulation with a finer time scale causes events to occur slightly earlier. This is well justified since we assign certain probability of occurrence to each event in a smaller time interval and update the system state more frequently.

This paper is organized as follows. Section 2 gives a simplified view of the basic model of a stochastic discrete time agent-based simulation. Section 3 describes techniques of our algorithms to obtain the same simulation in a finer time resolution. Section 4 describes potential implementation issues and their solutions. Section 5 describes efficient implementation techniques for our algorithms. Section 6 demonstrates our experimental results and provides a discussion of these results. Section 7 remarks on how to make the time step size larger. Section 8 concludes our paper, describes the general applicability of our approach and discusses its limitations, and points out directions for future work.

2 Simulation Model

In this section, we describe the basic model of a stochastic discrete time agent-based simulation system, with specific reference to the simulation system developed by [10] following approach in [4]. Algorithm 1 provides a high level description of the agent-based simulation model of an epidemic spread.

The time step refers to the indexing variable in the outermost loop. This step size characterizes the unit of time in which the system progresses. In other words, the system state must remain constant between successive steps; thus, all events must occur with

Algorithm 1. Stochastic Agent-based Simulation Model.

1: **for all** time step T from *beginning* to *end_of_simulation* **do**
2:     **for all** infectious agent I **do**
3:         **for all** susceptible agent S in contact with I during T **do**
4:             **if** I infects S successfully **then**
5:                 update status of S
6:             **end if**
7:         **end for**
8:     **end for**
9: **end for**

periods of integral multiples of this time step size. In line 2, we identify the agents that may change the system state in the current time step. In the inner loop, for each of these agents, we find all interacting agents in the same step, and decide whether an event between agents actually occurs. There are often times scheduled events other than among interacting agents that may take place; these could be dealt with similarly and we focus our attention on events described in Algorithm 1.

In our simulation of an epidemic spread, each time step is one half-day. This is largely due to the facts that agents (people) are in contact with different other agents during daytime and nighttime, and also state transitions take place in multiple of half-days. An agent is classified as one of the susceptible, infectious, and recovered (or removed); in each step, an infectious agent may infect susceptible agents in contact according to a transmission probability, which is dependent upon the interacting agents' ages and contact locations. Once an infectious agent successfully infects a susceptible agent, the susceptible will be assigned a latent period and an infectious period, and become infectious at the end of the half-day step. These latent and infectious periods are drawn from two pre-specified discrete random variables, derived from observed data on the epidemic to be modelled.

Besides the obvious possibility where finer time scale result is desired, there are still at least two potential issues with this current model regarding the size of the time step. First, if we wish to model a disease for which the latent and infectious periods are better modelled with finer time unit, the current system could not easily accommodate this change without major revision efforts. Second, if we wish to (indeed we do) record not only who is infected, but also the infector, then the artificial simultaneity (See Section 1) may be introduced: two or more infectious agents may infect the same susceptible agent in one step, and thus some mechanism is required to decide the infector precisely.

3 Adjustment of Time Step Size

3.1 Finer Step Model

In the original model, each pair of infectious (I) and susceptible (S) in contact are associated with a transmission probability P^{IS}. In each half-day step, we iterate through all pairs of infectious and susceptible in contact, and for each pair, decide whether S is infected by I with probability P^{IS}. Now, for each P^{IS}, we derive a k-hour transmission probability, p_k^{IS}, that satisfies

$$(1 - p_k^{IS})^{\frac{12}{k}} = 1 - P^{IS} \tag{1}$$

where k is a factor of 12 (we use the term, *granularity* of the system, to denote the smallest unit of time interval in which an event to be modelled could take place). The probability p_k^{IS} is derived such that the overall probability of S getting infected by I does not change if S is decided for infection with probability p_k^{IS} every k hour(s), provided that no change in state occurs in each half-day step. Notice that this derivation also makes the assumptions that the probability of transmitting a disease is uniform in the half-day step and independent among each smaller sub-steps. That is, we use the same p_k^{IS} for all sub-steps, instead of a number of different (conditional) probabilities. These may be debatable assumptions, depending upon what events are being modelled.

The probability p_k^{IS} is derived between each IS-pair; we would like the probabilities of transmitting a disease (between any pair) in each step to be the same as the original probability, P^{IS} in cases of multiple pairs of infectious and susceptible agents in contact. Now we show that this is indeed the case. More formally, the probability of each susceptible agent S getting infected remains the same as long as the duration of contact between each IS-pair is unchanged and a multiple of 12-hour. Notice that it suffices to demonstrate the case where there are more than one infectious agents in contact with only one susceptible agent, as susceptible agents do not influence each other. This is an immediate result from the assumed independence of infection events by different infectious agents and the commutative property of multiplication. Suppose there are n infectious agents, $I_1, ..., I_n$ and one susceptible agent S in contact in some arbitrary half-day step. Let S_t and S^t denote the events S gets infected at t and S gets infected by t, respectively. Also, we use $\neg$ to denote logical negation. Then

$$\begin{aligned}
\Pr\{S \text{ infected}\} &= 1 - \Pr\{S \text{ not infected}\} \\
&= 1 - \Pi_{t_j=jk,\ j\in\mathbb{N}}^{12} \Pr\{\neg S_{t_j} | \neg S^{t_{j-1}}\} \\
&= 1 - \Pi_{t_j=jk\ j\in\mathbb{N}}^{12} \Pi_{i=1}^{n} \Pr\{\neg S_{t_j} \text{ by } I_i | \neg S^{t_{j-1}}\} \\
&= 1 - \Pi_{i=1}^{n} \Pi_{t_j=jk,\ j\in\mathbb{N}}^{12} (1 - p_k^{I_i S}) \\
&= 1 - \Pi_{i=1}^{n} (1 - P^{I_i S})
\end{aligned}$$

Notice that in the derivation above, we split one half-day into $12/k$ sub-steps of k hour(s) each. The same approach could be used for splitting a time step of any size into any integral number of equally spaced sub-steps.

This refinement does not introduce any conceptually new artefact into the model. All it does is to perform the simulation with shorter time step size, and in each time step, the probabilities for events to occur are altered. Specifically, in Algorithm 1, we substitute a step with a smaller sub-step in the outermost loop, and use p_k^{IS} instead of P^{IS} when deciding infection (line 4). Regarding the two concerns we have at the end Section 2, the first is solved as we can now model any events that take place with periods greater than or equal to the *granularity* of the system (k hours in this example). This approach does not deal with the second concern directly. However, by reducing the time step size (and thus the transmission probability in a step), the chances of simultaneous events could be reduced significantly.

Now a new issue concerning efficiency is introduced. Typically, in a large scale agent-based simulation system, the number of possible interactions among agents or other events (line 2 in Algorithm 1) is very large in each step. Therefore, after applying this

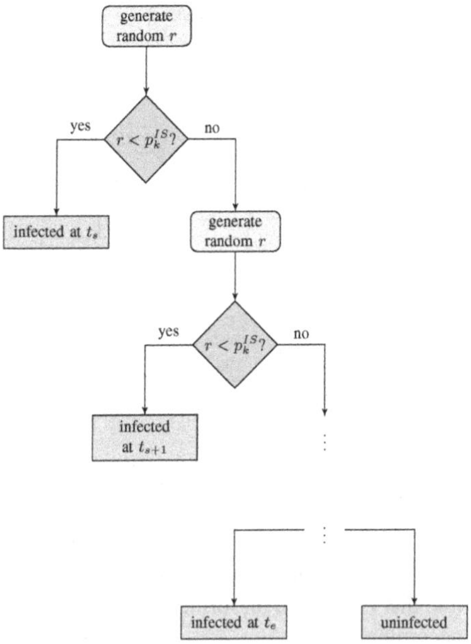

Fig. 1. A naive sequential infection decision process sub-step by sub-step. One full step T consists of sub-steps $t_1, t_2, ..., t_\alpha$, and I starts infecting at $t_s \in T$ and recovers at $t_e \in T$ where t_e is later than t_s. Note that r is drawn uniformly random from $(0, 1)$.

technique to reduce the step size, we will examine, in each step, a long list of possible events, of which most will not take place due to the reduced probabilities. In response to this efficiency issue, in Section 3.2 we introduce techniques in implementation, which allow the system to run almost as fast as with the coarser time step, but achieve the benefits produced by the finer time step.

3.2 Our Techniques

Our goal in this section is to implement the refined model more efficiently. For the ease of description, we refer to the original time step (e.g., half-day in the model above) as *step*, and the finer time step (e.g., k hour(s)) as *sub-step*, and also maintain the use of P and p_k to denote the transmission probabilities in each step and sub-step, respectively. The challenge is that we wish to achieve the effect of advancing the system every sub-step unit of time, but we do not want to examine all possible transmission events at such high frequency. The main idea here is to make the "big decision" first, and make subsequent "small decisions" only if the first result turns out "favorable". First, we "select" a possible transmission event with the aggregate probability, P^{IS}. This determines whether this agent gets infected in a step. Following that, we decide which sub-step this event actually occurs according to probabilities, $\frac{p_k^{IS}}{P^{IS}}$, $\frac{p_k^{IS}(1-p_k^{IS})}{P^{IS}}$,

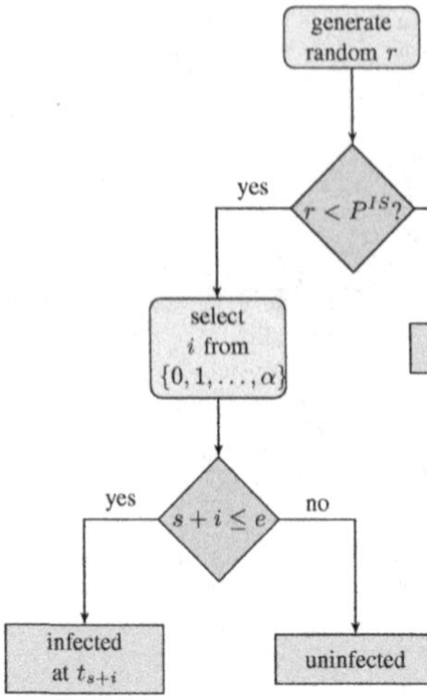

Fig. 2. Schematic picture of infection decision processes in our refined implementation. One full step T consists of sub-steps $t_1, t_2, ..., t_\alpha$, and I starts infecting at $t_s \in T$ and recovers at $t_e \in T$ where t_e is later than t_s. Note that r is drawn uniformly random from $(0, 1)$ and i is drawn from the computed distribution accordingly.

$\frac{p_k^{IS}(1-p_k^{IS})^2}{P^{IS}}, \ldots$. Flow charts of infection decisions between an IS-pair are shown in Figures 1 and 2.

From Figures 1 and 2, it is straightforward to verify that in our modified implementation, the probability of an infectious agent transmitting the disease to a susceptible agent in contact in a step is the same as in the original model. In cases of multiple IS-pairs, an argument similar to the one in Section 3.1 could show that the probabilities of transmission do not change, provided the length of duration for any IS-pair in contact is fixed. There is, however, one more premise for this implementation to work: the contacts among agents in a step must be known before we start the step. This is necessary since we iterate through infectious agents only every step (line 12 in Algorithm 2) , instead of every sub-step as in Section 3.1. This assumption is often easily satisfied as in many agent-based simulations, the possible interactions among agents are determined by the pre-initialized properties of the agents (e.g., the contact locations in which agents reside).

The argument above shows that this improved implementation should, in principle, achieve the same result as the straightforward sub-step by sub-step implementation (Algorithm 1).

4 Potential Issues and Their Solutions

We point out three issues (may be more for more complicated models) for which extra care should be taken in practical implementation, in order to get the expected result.

4.1 Issue I: State Transition within One Step

Notice in the actual model with period being a sub-step, an event may become ready or may be removed between two successive sub-steps within a step. For example, in the model of epidemic spread, an infectious agent I may recover, or a susceptible agent S may turn infectious in any sub-step, so that contact with I no longer results in new infections, or new transmission routes become possible within a step. In these cases, assuming the knowledge of when events are ready or removed is known before each step, we can perform a simple lookup and filter out those events that we have determined its occurrence in a sub-step before it is ready or after it has been removed.

4.2 Issue II: Artificial Simultaneity

It is possible that two events could each occur with some probabilities, but they could not occur both in the same sub-step. That is, the occurrence of one prevents the occurrence of the other one. This issue pertains to the problem of "artificial simultaneity" in Section 1 where some arbitrary ordering is needed. For example, a susceptible agent S cannot get infected from two infectious agents I_1 and I_2 in two different sub-steps of the same step. To conform to the original model (Algorithm 1) with time period being a sub-step, we must ignore the occurrence of all conflicting events in a step except the one which occurs in the earliest sub-step (There could still be more than one occurring in the earliest sub-step; in this case, an arbitrary selection is made. But the chances of such cases are significantly reduced as indicated in Section 3.1).

4.3 Issue III: New Events Created within One Step

The occurrence of an event in a sub-step may introduce new events that are possible to occur immediately starting from the next sub-step. To deal with these cases, a list of these possible new events must be maintained in a step, and each event in this list is to be decided for its occurrence and cleared in the current step. This iterative examination will eventually terminate as fewer events will be added for the remaining sub-steps and probabilities of occurrence are smaller as well. If a later examined event $\mathcal{A}$ takes place and prevents the occurrence of an earlier decided event $\mathcal{B}$, which occurs later than $\mathcal{A}$ in time (as illustrated in Section 4.2), we must update the system accordingly to take event $\mathcal{A}$ and ignore event $\mathcal{B}$.

For example, consider the case where in some step T, an infectious agent I_1 successfully infects a susceptible agent S_1 at sub-step t_1 and another susceptible agent S_2 at a later sub-step t_2 ($t_1 < t_2$, but they belong to the same step T). Assume S_1 turns infectious immediately and successfully infects S_2 in sub-step t_3 where $t_1 < t_3 < t_2$. Then we must update the infection time of S_2 to be t_3 in stead of t_2. In practical implementation, the list of these newly triggered events within the same step should be sorted in order of the sub-steps of occurrence to avoid a long sequence of updates in the cases of the example above.

5 Implementation Techniques

Algorithm 2 gives a high-level description of the practical implementation of the system that could achieve the same results as in Algorithm 1 with finer step size.

Algorithm 2. Refined Stochastic Agent-based Simulation Model.

1: **for all** time step T from *beginning* to *end_of_simulation* **do**
2: initialize an empty sorted list $\mathcal{L} = \emptyset$
3: **for all** infectious agent I **do**
4: `TryToInfect` $(I, T, \mathcal{L})$
5: **end for**
6: **while** $\mathcal{L}$ is not empty **do**
7: $I_{new} \leftarrow$ remove the head of $\mathcal{L}$
8: `TryToInfect` $(I_{new}, T, \mathcal{L})$
9: **end while**
10: **end for**

11: **procedure** TRYTOINFECT(I,T,$\mathcal{L}$)
12: **for all** susceptible agent S in contact with I during T **do**
13: **if** I is still infectious and infects S successfully in sub-step t, and S has not been infected before t **then**
14: update status of S
15: **if** S turns infectious within the current step T **then**
16: Add S to $\mathcal{L}$
17: **end if**
18: **end if**
19: **end for**
20: **end procedure**

5.1 List Management

Note that in step 2 of Algorithm 2, $\mathcal{L}$ keeps a sorted list of events that have happened so far where each event has a key that is the index of the sub-step it occurred. The reason to keep the list sorted is for the checking of a proper timing sequence of events in step 13 to be efficient. Hence we need a good data structure so that the insertion of an event S into $\mathcal{L}$ in step 16 can be done equally efficient.

Since the range of key values in $\mathcal{L}$ is a small constant say within 20, we could use a simple hash table with linked lists of events of the same key values. Note that in the actual running environment, the list is likely to be short. So we use standard C++ STL library of a balanced binary search tree that supports insertion in $O(\log |\mathcal{L}|)$ time.

5.2 Index Location

The other implementation issue is how to speed up the process of finding the right sub-step as illustrated in Figure 2. Note that our approach could in principle divide each step into any integral number, α, of sub-steps according to equation (1). However, in deciding the sub-step in which the an event takes place, the average number of comparisons needed is $\sim \alpha/2$. This could possibly slow down the process observably if α is large. If we know the probability of occurrence for an event in the beginning, we could use a classical method in sampling from a discrete random variable with finite number of outcomes, the *alias method* [12]. It requires $O(\alpha)$ preprocessing time to build up *alias tables* in the beginning, and each subsequent sampling takes at most 2 comparisons. This technique may not fit all simulation models as some events are associated with probabilities computed in the run time, or there are too many possible distinct events and storing a table for each one is not cost effective. In the case of our simulation of epidemic spread, the transmission probability between an IS-pair depends on the location of contact and the age groups of two agents; hence, we only need one table for each distinct triple of *(contact location, infectious agent's age group, susceptible agent's age group)*, and typically the number of such triples is not too large. In our system, we follow the straightforward implementation described in [6] since α is usually a small constant smaller than 12.

6 Experimental Results and Discussion

In this section, we wish to demonstrate that our refined model indeed achieves stochastically identical results as the original model with a shorter time period, and significantly reduces the simulation time. Also, we give a reasonable account for the observed differences in results from simulations of different time step sizes.

We build the two simulation systems (denoted by ALG_1 and ALG_2 for Algorithm 1 and Algorithm 2, respectively) by modifying a simulation system for epidemic spread developed by [10], keeping all parameters as original except the ones we explicitly wish to manipulate. Below, we perform 100 baseline simulations for each system with a particular granularity, and report the average results for both the simulation outputs and the simulation time consumed.

Figure 3 shows the epidemic curves (daily new cases) produced by the two systems, ALG_1 (blue) and ALG_2 (red), when granularity (GR) is set to 1 (+), 6 (o), and 12 (x) hour(s), respectively. Each pair of curves corresponding to the same granularity overlap well, showing that this modified implementation indeed reproduces the result of the original model. Moreover, the leftward shift of epidemic curves for finer granularities is also expected. This is a result of our assumption that the probability of transmission in each smaller step is uniformly distributed, and this causes the expectation of infection time to become earlier. Similar phenomenon is also observed if we shorten the latent period since it does not affect infectivity while causing a earlier spread of disease. Figure 4 shows the epidemic curves produced by the original system with varied latent periods.

Table 1 shows the average simulation times (SimTime in seconds) and attack rates (AR) for each of the two systems run on a workstation with 8 Intel Xeon X5365 CPUs and 32GB RAM. The consistency in attack rates confirms our argument that the overall probabilities of infection are not altered; the huge reduction in simulation time demonstrates the effectiveness of our proposed model for practical implementation.

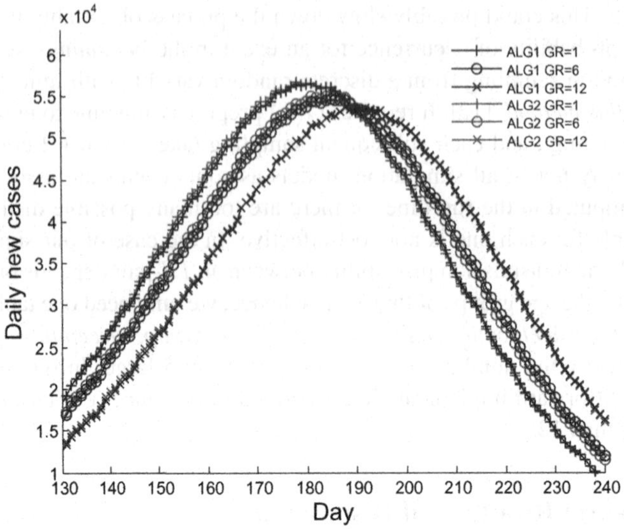

Fig. 3. Epidemic curves by ALG_1 and ALG_2 with different *granularities*

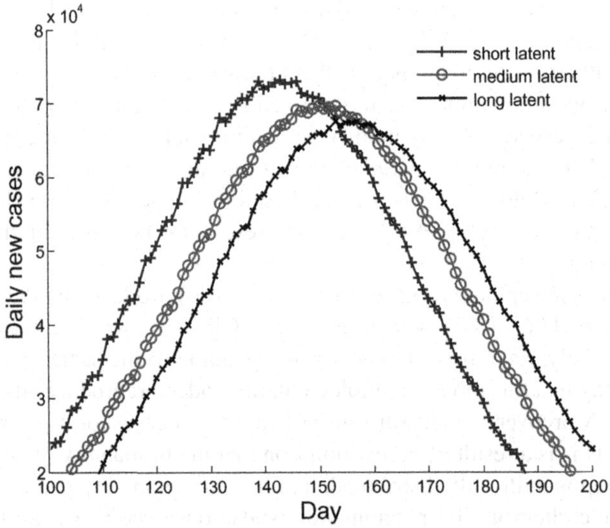

Fig. 4. Epidemic curves for different lengths of latent periods

7 Making the Time Step Size Larger

In this paper, we describe a general technique to adjust the time step size of an already-built discrete time simulation with a system set time step size. We demonstrate our techniques by making an epidemic spread simulation system to have a finer time step size with the same computational efficiency.

Table 1. Average attack rates and simulation times for the two systems ALG_1 and ALG_2 with different granularities

GR (hr)	AR (ALG_1)	AR (ALG_2)	SimTime (ALG_1)	SimTime (ALG_2)
1	0.312	0.312	971	134
2	0.312	0.312	504	133
3	0.312	0.312	351	133
4	0.312	0.312	276	133
6	0.312	0.312	203	133
12	0.311	0.311	127	130

Our technique, in theory, can be used to change the time step size larger if the original event occurring probability P^{IS} is very small. The system will run in less time steps, namely, less time consuming. That is,we may need to set a new time step size that is a factor of the original T that satisfies

$$(1 - P^{IS})^k = 1 - \hat{P}_k^{IS} \tag{2}$$

where k is a positive integer.

We call the merged time step a *super step*. We then use a similar formula to decide in which step of a super step an event happens. That is, it happens in the ith step of a super step with a probability of $(1 - P^{IS})^i \cdot P^{IS}$.

In our case of the epidemic spread simulation, the day time event occurring probability P_{day}^{IS} and the night time event occurring probability P_{night}^{IS} are different. In the case of merging time steps of unequal occurring probabilities such as in our example, we use a similar technique to find the occurring step index in a super step. For example, if we are merging the day and night periods, the the probability of an event happened in the day time is P_{day}^{IS}, and the probability of it happened in the night time is $(1 - P_{day}^{IS}) \cdot P_{night}^{IS}$ with

$$(1 - P_{day}^{IS}) \cdot (1 - P_{night}^{IS}) = 1 - \hat{P}_2^{IS}. \tag{3}$$

8 Concluding Remarks

We proposed a simple approach in adjusting the time step size in a stochastic discrete time agent-based simulation model based on reasonable assumptions. This provides flexibility in modelling events with finer time scales; such flexibility is often desired when finer empirical observations were to be incorporated into simulations. Furthermore, we described a structurally different model for practical implementation, which achieves identical result significantly faster.

To demonstrate this, we modified a simulation system for epidemic spread, developed by [10], constructed the proposed simulation model in both ways (Algorithm 1 and Algorithm 2), and through experiments, showed that they indeed computed the same results with the later one having a huge reduction in simulation time.

We note that many simulation systems have parameters that have been calibrated against historic data, and it is important to re-scale these parameters appropriately to make sure they correspond well to the finer scaled system. One of the most important parameters is the probability of an event to be occurred since it has something to do with the time step size. Our studies have also provided a simple solution to automatically calculate this occurring probability.

8.1 Applicability and Limitations

We briefly describe the context in which this approach is applicable, as well as its current limitations. This approach was initially designed to handle situations in epidemic spread where the time scale of the disease's natural history is finer than the time step size in the original model. This was achieved by certain assumptions and basic probability theory, as shown in Section 2. Algorithm 2 was then developed to improve the efficiency of the straightforward model (Algorithm 1) with a small step size and small interacting probabilities among agents. As a result, it should be easily applicable to other agent-based simulations when a similar factor affecting agents' interactions needs to be measured in finer time scale; that is agents could become active (e.g., infectious) or inactive (e.g., recovered or isolated) in such time scale.

It is crucial to notice that the sub-step in which agents become inactive must be known before the start of each step (the case of turning active in the middle of a step could be handled as described in Section 4.3). It is not a problem in our example of epidemic spread simulation as the length of latent and infectious periods of an agent (thus the information of when it recovers) is determined when the agent gets infected. This, however, may pose a problem for other kinds of simulations. It may require tracing back of when an agent becomes inactive, and undoing all its interactions thereafter.

8.2 Possible Future Work

The implementation technique introduced in this paper may also be applied to purely increase the efficiency without any effect on results; we could view the original time step as the *sub-step* and run the system with an enlarged *step* following Algorithm 2 to produce the same result more efficiently. This may, however, introduce difficulties in determining the possible events to occur in an enlarged step. In the example of the epidemic spread simulation, two agents may be in contact only in daytime and not nighttime (see step 3 in Algorithm 1; in an enlarged step, attention must be paid to such circumstances. It is also likely that other calibrated parameters need to be rescaled appropriately when such refinement is employed. We will work on overcoming the limitations mentioned above and in Section 8.1, and try to apply such techniques to a larger variety of general simulations.

Another interesting direction is to compare the effect of time step size in discrete time agent-based simulations with the more traditional approaches to simulation modelling, such as event scheduling, and also the modelling with differential equations, commonly seen in mathematical epidemiology [3].

Acknowledgements. We thank anonymous reviewers for their comments and suggestions. Also we sincerely thank Steven Riley (s.riley@imperial.ac.uk) for his valuable comments throughout this study.

References

1. Buss, A., Al Rowaei, A.: A comparison of the accuracy of discrete event and discrete time. In: Proceedings of the 2010 Winter Simulation Conference (2010)
2. Buss, A., Al Rowaei, A.: The effects of time advance mechanism on simple agent behaviors in combat simulations. In: Proceedings of the 2011 Winter Simulation Conference (2011)
3. Diekmann, O., Heesterbeek, J.A.P.: Mathematical Epidemiology of Infectious Diseases: Model Building, Analysis, and Interpretation. John Wiley and Sons Ltd. (2000)
4. Germann, T.C., Kadau Jr., K., Longini, I.M., Macken, C.A.: Mitigation strategies for pandemic influenza in the united states. In: Proceedings of the National Academy of Sciences, vol. 103, pp. 5935–5940 (April 2006)
5. Kelker, D.: A random walk epidemic simulation. Journal of the American Statistical Association 68(344), 821–823 (1973)
6. Kronmal, R.A., Peterson Jr., A.V.: On the alias method for generating random variables from a discrete distribution. The American Statistician 33(4), 214–218 (1979)
7. Kuo, C.-T., Wang, D.-W.: T.-s. Hsu. A simple efficient technique to adjust time step size in a stochastic discrete time agent-based simulation. In: Proceedings of the 2nd International Conference on Simulation and Modeling Methodologies, Technologies and Applications (SIMULTECH), pp. 42–48. SciTePress (2012) ISBN 978-989-8565-20-4
8. Riley, S.: Large-scale spatial-transmission models of infectious disease. Science 316(5627), 1298–1301 (2007)
9. Tsai, M.-T., Chern, T.-C., Chung, J.-H., Hsueh, C.-W., Kuo, H.-S., Liau, C.-J., Riley, S., Shen, B.-J., Wang, D.-W., Shen, C.-H., Hsu, T.-S.: Efficient simulation of the spatial transmission dynamics of influenza. PLoS One 5(11), e13292 (2010)
10. Tsai, M.-T., Wang, D.-W., Liau, C.-J., Hsu, T.-s.: Heterogeneous subset sampling. In: Thai, M.T., Sahni, S. (eds.) COCOON 2010. LNCS, vol. 6196, pp. 500–509. Springer, Heidelberg (2010)
11. Vangheluwe, H.: Discrete event modelling and simulation. Lecture Notes, CS522 McGill University (December 2001)
12. Walker, A.J.: An efficient method for generating discrete random variables with general distributions. ACM Transactions on Mathematical Software 3(3), 253–256 (1977)

Numerical Study of Turbulent Boundary-Layer Flow Induced by a Sphere Above a Flat Plate

Hui Zhao, Anyang Wei, Kun Luo, and Jianren Fan

State Key Laboratory of Clean Energy Utilization, Zhejiang University, Hangzhou 310027, PR China
{hzhao,wally08,zjulk,fanjr}@zju.edu.cn

Abstract. The flow past a three-dimensional obstacle on a flat plate is one of the key problems in the boundary-layer flows, which shows a significant value in industry applications. A direct numerical study of flow past a sphere above a flat plate is investigated. The immersed boundary (IB) method with multiple-direct forcing scheme is used to couple the solid sphere with fluid. The detail information of flow field and vortex structure is obtained. The velocity and pressure distributions are illuminated, and the recirculation region with the length of which is twice as much as the sphere diameter is observed in the downstream of the sphere. The effects of the sphere on the boundary layer are also explored, including the velocity defect, the turbulence intensity and the Reynolds stresses.

Keywords. Direct Numerical Simulation, Immersed Boundary Method, Boundary-layer Flow, Sphere, Flat Plate.

1 Introduction

Boundary-layer flows widely exist both in nature and in many practical engineering areas, such as the chemistry engineering, the aeronautic and the aircraft engineering, the energy engineering, the wind engineering and so on. To obtain enough data is of extreme necessity for helping us gain a better knowledge about the boundary-layer flow. In boundary-layer flows, the flow past a three-dimensional obstacle on the flat plate is one of the key problems, which shows a significant value in industry applications, including surface roughness, artistic structures, gas tanks, cooling towers and some types of vehicles. Based on the first study of the flow past a blunt obstacle placed on a flat plate by Schlichting [1], Klemin et al. [2] investigated the drag of a sphere placed on a ground plate. In recent decades, flows past spheres placed in boundary-layer on a flat plate have attracted lots of researchers' attention.

Okamoto [3] experimentally studied the turbulent shear layer behind a sphere placed in a boundary-layer on a flat plate. In the experiment, the sphere surface pressure distribution, velocity and pressure behind the sphere were measured; various results on the drag and lift coefficients were shown; recirculation length, generation positions of the arch vortices and horse-shoe vortex were observed. It was found that the drag coefficient was larger than that of a sphere alone without the influence of the

M.S. Obaidat et al. (eds.), *Simulation and Modeling Methodologies, Technologies and Applications*, Advances in Intelligent Systems and Computing 256,
DOI: 10.1007/978-3-319-03581-9_8, © Springer International Publishing Switzerland 2014

boundary-layer on the plate, and the wall wake behind the sphere became low and spread transversely with increasing downstream distance. The conclusion summarized the variation trend of the thickness of the shear layer, the streamwise velocity defect, and the spanwise turbulence intensity.

Takayuki [4] investigated the flow around individual spheres placed at various heights above a flat plate. With the smoke-wire method and the surface oil-flow pattern method, the behaviors of the flow and vortices were visualized. The surface pressure distributions on the sphere and the plate were measured, meanwhile empirical correlations for the drag and lift coefficients were derived. The ranges of S/d (Here, S was the gap between the plate and the sphere, d was the diameter of the sphere.) in which horseshoe and arch vortices were formed and the vortices shed from the sphere were pointed out.

Recently, with the blooming of the computational capability, it is possible to conduct direct numerical simulation (DNS) studies of two-phase turbulent boundary-layer flows. Zeng et al. [5] considered fully resolved DNS to investigate a turbulent channel flow over an isolated particle with finite size, in which the spectral element methodology (SEM) was used. A fully developed turbulent channel flow was applied as inflow and a simple convective outflow boundary condition was applied at the outlet. Two different particle locations were considered within the channel, one was in the buffer region and the other was along the channel centerplate. The study summarized instantaneous forces, time-averaged mean streamwise forces, drag forces, lift forces and vortex shedding with different particle diameters and locations.

The immersed boundary (IB) method was originally proposed by Peskin [6] to investigate the fluid dynamics of blood flow in human heart. Following the lead of Peskin [6], variants of IB method have been developed to study the interaction between the solid boundary and the fluid [7-11]. This method has been applied to several different areas such as flexible boundaries in fluid flow, roughness elements passed by fluid, particle sedimentation, wearing process and so on [6,12-21].

In the IB method the fluid is represented in an Eulerian coordinate frame while the irregular structures in a Lagrangian coordinate frame. Luo et al. [16] proposed a DNS method jointed with a combined multiple-direct forcing and immersed boundary (MDF/IB) method. The MDF/IB method guaranteed the no-slip boundary condition at the immersed boundary. Using the joint application proposed by Luo et al, Zhou et al. [15] simulated a flow past an isolated three-dimensional hemispherical roughness element. In his study, the evolutional process of the discrete hairpin vortex and the formation of two secondary vortex structures were captured. It also demonstrated that the numerical methods mentioned above can be extended to the simulation of the transitional boundary-layer flows induced by randomly distributed three-dimensional roughness elements.

Although a lot of methods have been employed to study the flow past a three-dimensional obstacle placed on the flat plate, investigations on the interaction between spheres and boundary layer on flat plate, which are significant for engineering applications, are still rare, especially for the moderate Reynolds number. Numerical simulations performed with MDF/IB method also were not common in this field.

The present study outlines the flow field with a sphere above a flat plate using direct numerical simulation with the MDF/IB method. The influence of the sphere on the boundary layer is explored. The paper is organized as follows. The next section introduces the numerical method including the basic governing equations, the IB method with multi-direct forcing, and the computational domain of this numerical experiment. The results including vortex structure, velocity distribution, turbulence intensity and Reynolds stresses are evaluated in the following section. Finally the conclusions are made.

2 Numerical Method

2.1 Governing Equations

In the computational domain Ω, the dimensionless governing equations for incompressible viscous flows are:

$$\nabla \cdot \boldsymbol{u} = 0, \tag{1}$$

$$\frac{\partial \boldsymbol{u}}{\partial t} + \boldsymbol{u} \cdot \nabla \boldsymbol{u} = -\nabla p + \frac{1}{Re} \nabla^2 \boldsymbol{u} + \boldsymbol{f}. \tag{2}$$

Here, $\boldsymbol{u}$ is the dimensionless velocity of fluid, p is the dimensionless pressure, Re is the Reynolds number defined as $\mathrm{Re} = \rho u_c l / \mu$, where ρ is the characteristic density of fluid, u_c is the characteristic velocity, l is the characteristic length of flow field and μ is the viscosity of fluid. $\boldsymbol{f}$ is external force exerted on the flow field which is the mutual interaction force between fluid and immersed boundary, and is expressed as following:

$$\boldsymbol{f}(x) = \int_{\Omega} \boldsymbol{F}_k(X_k) \cdot \delta(x - X_k) \, \mathrm{d}X_k, \tag{3}$$

where $\delta(x - X_k)$ is the Dirac delta function, X_k is the position of the Lagrangian points set at the immersed boundary, x is position of the computational Eulerian mesh and $\boldsymbol{F}_k(X_k)$ is the force exerted on the Lagrangian point X_k.

2.2 Multi-direct Forcing Immersed Boundary Method

In order to ensure that the no-slip boundary condition of the velocity at the immersed boundary could be satisfied, direct forcing [10] $\boldsymbol{F}_k(X_k)$ is introduced to make the velocity on the Lagrangian points approaching the velocity of the no-slip boundary. Wang et. al. [13] described the derivation and the algorithm in detail, which is briefly introduced here.

The time-discretized form of Eq. (2) could be written as following:

$$\begin{aligned} \frac{\boldsymbol{u}^{n+1} - \boldsymbol{u}^n}{\Delta t} &= -\left(\boldsymbol{u} \cdot \nabla \boldsymbol{u} + \nabla p - \frac{1}{\mathrm{Re}} \nabla^2 \boldsymbol{u} \right) + \boldsymbol{f} \\ &= \boldsymbol{rhs} + \boldsymbol{f}, \end{aligned} \tag{4}$$

where ***rhs*** represents a group of the convective, pressure and viscous terms. Then we can get the expression of the interaction force:

$$\boldsymbol{f} = \frac{\boldsymbol{u}^{n+1} - \boldsymbol{u}^{n}}{\Delta t} - \boldsymbol{rhs}, \tag{5}$$

where n and $n+1$ represent two different time steps. An immediate velocity $\hat{\boldsymbol{u}}$ is introduced, which is a temporary parameter satisfying the governing equations of the pure flow field:

$$\frac{\hat{\boldsymbol{u}} - \boldsymbol{u}^{n}}{\Delta t} - \boldsymbol{rhs} = 0, \tag{6}$$

then one can get:

$$\boldsymbol{f} = \frac{\boldsymbol{u}^{n+1} - \hat{\boldsymbol{u}}}{\Delta t}. \tag{7}$$

Eqs. (1) and (2) are enforced throughout the domain Ω, so we can get the expressions at the Lagrangian force points X_k and the Eulerian grid nodes x respectively :

$$\boldsymbol{F}_k(X_k) = \frac{\boldsymbol{U}_k^{n+1} - \hat{\boldsymbol{U}}_k}{\Delta t}, \tag{8}$$

$$\boldsymbol{f}(x) = \frac{\boldsymbol{u}^{n+1}(x) - \hat{\boldsymbol{u}}(x)}{\Delta t}, \tag{9}$$

and this dimensionless external force $\boldsymbol{f}$ is expressed as:

$$\boldsymbol{f}(x) = \sum_{k=1}^{N_L} \boldsymbol{F}_k(X_k)\delta(x - X_k)\Delta V_k, \tag{10}$$

where $\delta(x - X_k)$ is the Dirac-delta function [22,23]. N_L represents the total Lagrangian points on the immersed boundary. ΔV_k is the volume with each force point such that the union of all these volumes forms a thin shell (with the thickness equal to one mesh size) around each particle [24].

Therefore the force exerted on the k th Lagrangian point at the immersed boundary is:

$$\boldsymbol{F}_k(X_k) = \frac{\boldsymbol{U}_k^{n+1} - \hat{\boldsymbol{U}}_k}{\Delta t} = \frac{\boldsymbol{U}_{d-}\hat{\boldsymbol{U}}_k}{\Delta t}. \tag{11}$$

If this direct forcing is exerted by l+1 times, the intermediate velocity $\hat{\boldsymbol{U}}_k$ could be much closer to the desired velocity $\boldsymbol{U}_d$ .Then $\boldsymbol{F}_k(X_k)$ could be expressed as:

$$\boldsymbol{F}_k^{l+1}(X_k) = \frac{\boldsymbol{U}_d - \hat{\boldsymbol{U}}_k^{l}}{\Delta t}. \tag{12}$$

At the same time, to spread the force from the Lagrangian points to the Eulerian grids, the two-way coupling between Lagrangian points and Eulerian grids could be achieved through the Dirac delta function [22,23]. However, during this process, different schemes of discrete delta function can lead to different results, as demonstrated by Griffith and Peskin [23]. While the velocities on the Lagrangian points may not satisfy the no-slip boundary condition very well during the process of interpolation to obtain the simulated velocity on the Lagrangian points and extrapolation to spread the forcing effect to its surrounding Eulerian grids. Therefore, the multi-direct forcing technique is proposed to remedy the problem [16].

According to Luo et al. [16], when the direct forcing is exerted by *NF* times in a time step, the velocity at the immersed boundary can get close enough to the desired velocity under the no-slip condition. The interaction force between fluid and Lagrangian points could be described as:

$$\boldsymbol{F}_k\left(X_k\right)=\sum_{i=1}^{NF}\boldsymbol{F}_k^i\left(X_k\right). \tag{13}$$

The method mentioned above is called Multi-direct Forcing, which has been validated by Luo et al. [16], Wang et al. [13] and Zhou et al. [15] to ensure adequate reliability of the method. Here, *NF* is a crucial parameter of the multi-direct method which decides the accuracy of the simulation to some extent. The value of *NF* can be chosen arbitrarily, but for retaining the computational efficiency of the method it should preferably be kept low. In consideration of the calculation accuracy, the value of *NF* is chosen as 50.

2.3 Computational Domain

The geometrical parameters of the domain are X × Y × Z = 74.55 mm × 14.1 mm × 10.5 mm, which can be seen in Fig.1, and the sphere center is placed at the origin of coordinates. A uniform grid spacing of 50 μm is used, and the total grid number is 92,897,280. Parameters for the sphere and fluid are listed in Table 1. The Reynolds number based on the sphere diameter is 4171.

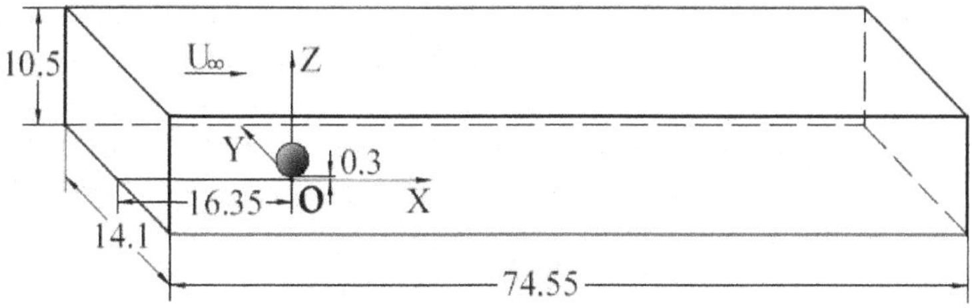

Fig. 1. Schematic view of the computational domain

Table 1. Summary of Prediction Conditions

Parameter	Value
Sphere diameter, D (mm)	3
Air density, ρ(kg/m3)	1.205
Air viscosity, μ (kg/m/s)	1.82×10^{-5}
Free stream velocity, U (m/s)	21

The distance of the gap between the bottom of the sphere and the plate is 0.1D, which is equivalent to 0.3 mm. According to the expression $N_L \geq \pi/3\left(12r_c^2/h^2+1\right)$ [24], the amount of Lagrangian points is 9520. An expression for the velocity distribution over a smooth wall is obtained from the wall up to the freestream [25]. In this paper, it is applied to calculate the streamwise velocity at the inlet, while the convective velocity boundary condition [26] is applied to the outlet.

3 Results and Discussion

The structure of the vortex is observed in Fig.2, from which horseshoe vortex could not be find obviously, but hairpin vortex shed from the sphere and form forest vortices. Because of the existence of the sphere, the original laminar boundary layer is impacted, and vortices move upward are generated from the plate. The vortex structure become disorderly where X / D > 7, and "break-up" phenomenon could be observed near the outlet where X / D > 16.

The instantaneous velocities at the center section Y / D = 0 at four different time steps are presented in Fig.3, *t* represents the dimensionless time. The streamwise velocity distribution of the boundary layer is visualized clearly. It can be seen from the illustrations that there is a typical laminar flow velocity distribution in the upstream of the sphere. Then the flow is separated and the under part flows through the gap between the sphere and the plate. Because the across area reduces suddenly, a high velocity region is formed, extending to the recirculation region behind the sphere. The upper part climbs upward along the sphere. Boundary layer separation takes place on the

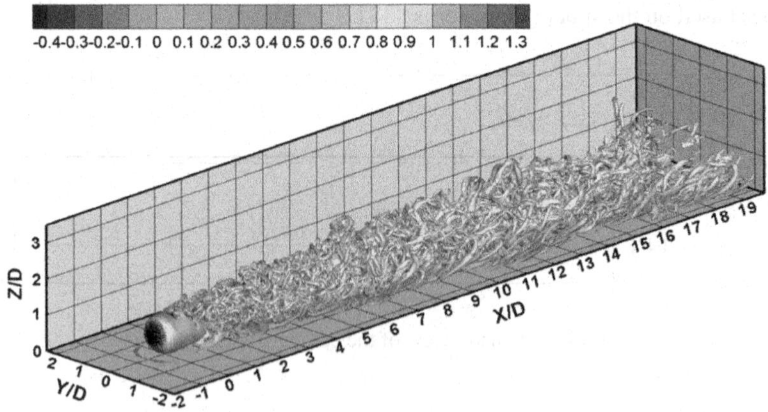

Fig. 2. The vortex structure

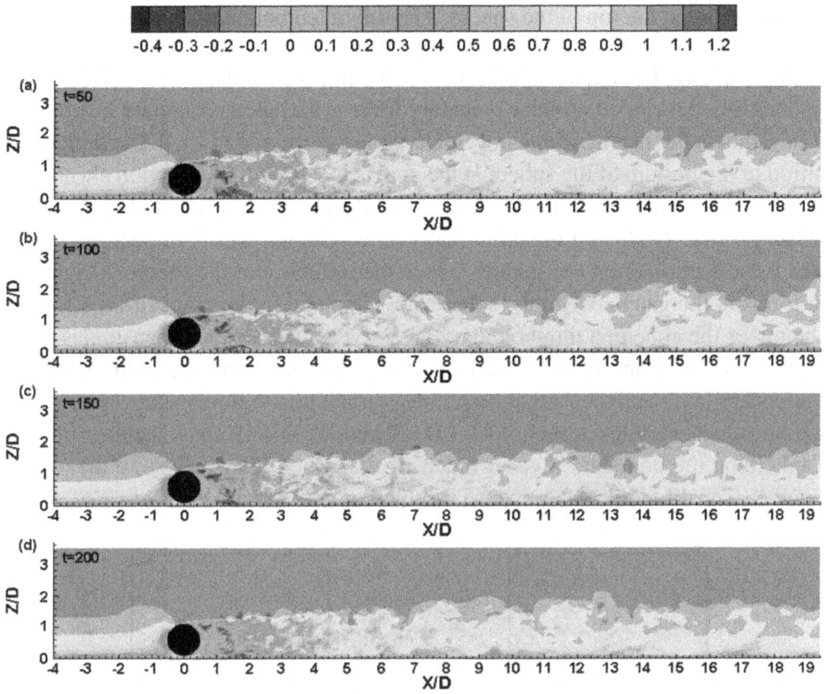

Fig. 3. The streamwise velocity distribution of the boundary layer

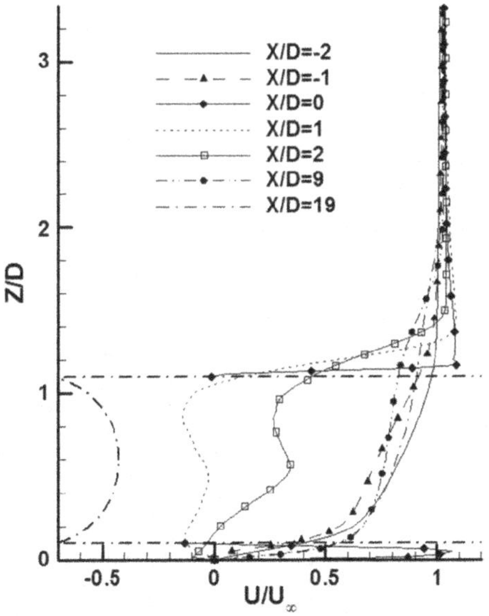

Fig. 4. Profiles of mean velocity in the center section Y / D = 0

separation point at the top of the sphere. The separated boundary layer sharply thickens along the flow, and under the separated boundary layer, a recirculation region is formed behind the sphere. Behind the sphere, the profile of the boundary layer velocity changes from the fully developed laminar boundary layer to turbulent boundary layer.

Fig.4 shows the profile of mean velocity at the section Y / D = 0.The dimensionless position of the bottom of the sphere at the Z-axis is 0.075. At the section ahead of X / D = - 2, a typical laminar boundary layer velocity profile is presented as the entrance velocity profile. There is a breaking part in the velocity profile of section X / D = 0, which is the section cross the sphere. Due to the influence of the sphere, a high velocity area forms in the gap between the sphere and the plate. In the range of 1.1 < Z / D < 1.3 at the top of the sphere, a thin boundary layer exists, where the dimensionless velocity sharply increases from 0 to 1.2. At the section X / D = 1, because of the separation of the boundary layer and the formation of the recirculation region, mean velocity presents negative values. Profiles at sections X / D = 9 and X / D = 19 show that the influence of the sphere on the boundary layer is much weaker when X gets close to the outlet, and the velocity profile indicates a typical turbulent layer velocity profile.

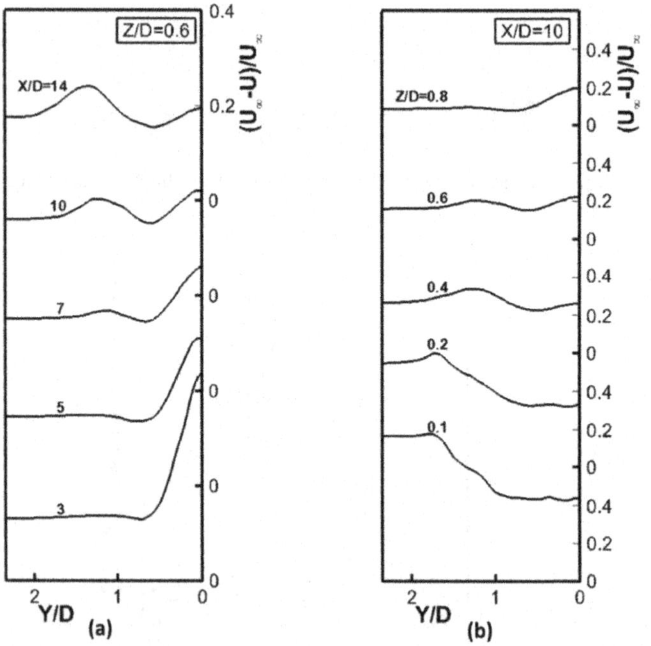

Fig. 5. Profiles of mean velocity defect, (a) in the horizontal center section Z / D = 0.6; (b) in the vertical section X / D = 0

In order to study the velocity defect of the wake behind the sphere, the mean velocity defect at sections Z / D = 0.6 and X / D = 10 are shown in Fig.5 (a) and Fig.5 (b), respectively. Fig.5 (a) indicates that, in the range of Y / D < 0.7, the peak defect is mainly affected by the recirculation region behind the sphere. With the increasing downstream

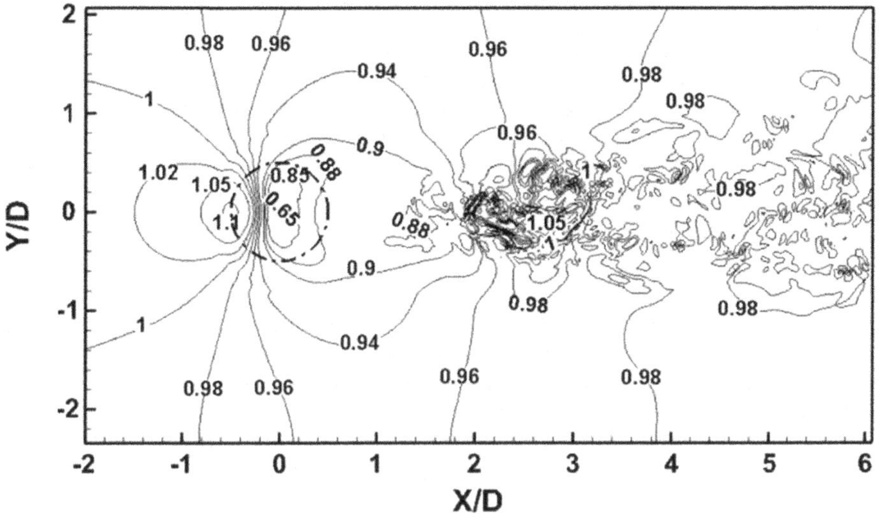

Fig. 6. The pressure distribution on the plate

distance, the center peak defect decreases. The spanwise peak defect shifts in the Y-direction, which takes place at Y / D = 1.1 for X / D = 7 and at Y / D = 1.4 for X / D = 14. According to Fig.5 (b), the peak velocity defect decreases and almost vanishes at Z / D = 0.8 with the increasing vertical distance. While with the increasing upward distance, its position closed to the center when Z / D < 0.6, and moved away from the center when Z / D > 0.6. In the spanwise direction, the velocity defect decreases faster when Y / D > 0.5 than the center behind the sphere.

The pressure distribution on the plate can be observed in Fig.6. There are two regions of high pressure, i.e. in the upstream of the sphere and behind the recirculation region. Behind the sphere, the pressure of the recirculation region is reduced rapidly because of a strong downwash behind the sphere. The length of the recirculation region between the two high pressure regions is considered to be twice as much as the diameter of the sphere.

The turbulence intensities are shown in Figs.7-9. Because of the existence of the sphere, the region in the vertical direction is divided into three zones naturally, according to the turbulence intensity. In the range of 0 < Z / D < 0.1, the turbulence intensities in the X-direction and the Y-direction decrease rapidly with the vertical distance increasing, but in the Z-direction it increases. In the range of 0.1 < Z / D < 1.1, turbulence intensity is about 0.1 in all directions and decreases with streamwise distance increasing. The value of turbulence intensity gradually approaches to the value of freestream in the range of Z / D > 1.1. The change is gentler as the streamwise distance increases. Thus it can be seen that the turbulence intensity increased in the shear layer.

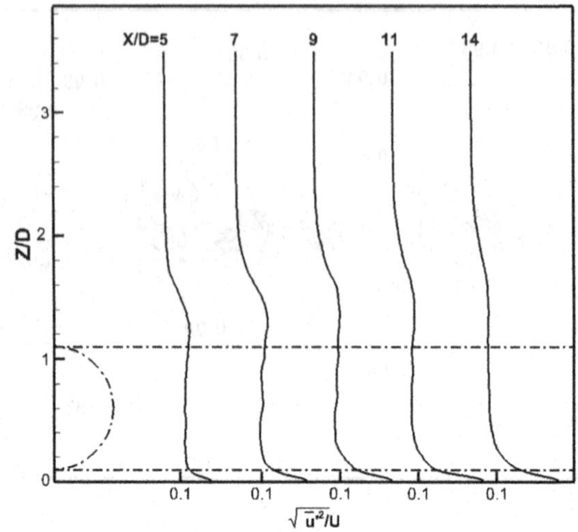

Fig. 7. Profiles of turbulence intensity in X-direction in center section Y / D = 0

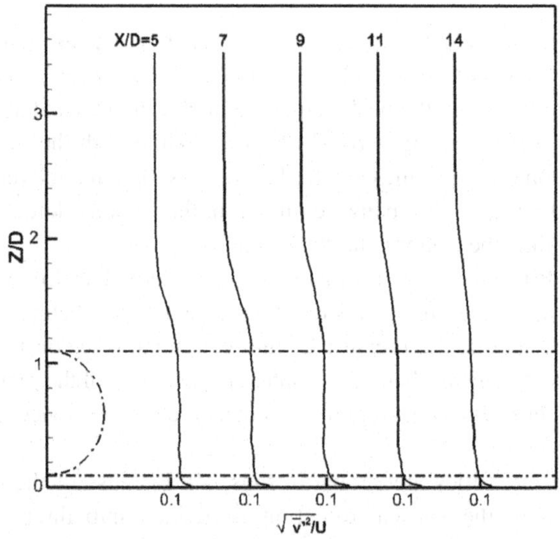

Fig. 8. Profiles of turbulence intensity in Y-direction in center section Y / D = 0

The three components of the turbulence intensities in the horizontal center section (Z/D=0.6) and the vertical section where X/D=10 are shown in Fig.10 and Fig.11. The position of the peak turbulence intensity moves in a manner similar to peak velocity defect. While Z/D<0.8, the turbulence intensity at the streamwise is larger than the lateral and vertical comes turbulence intensities. At Z/D=1.2, the turbulence be almost isotropic.

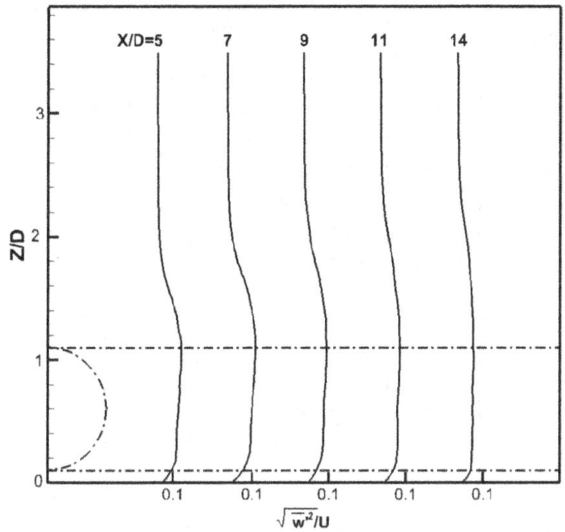

Fig. 9. Profiles of turbulence intensity in Z-direction in center section Y / D = 0

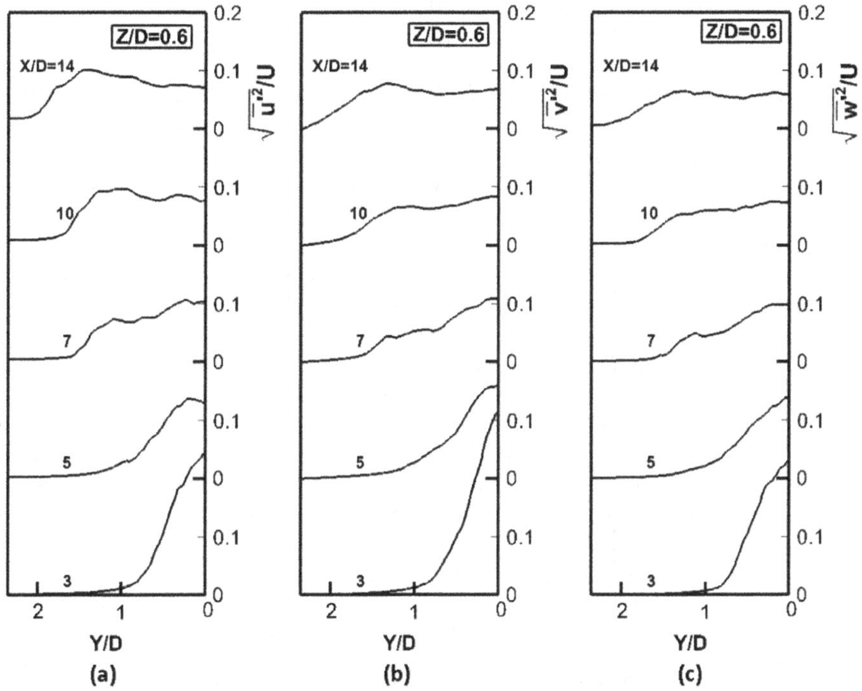

Fig. 10. Profiles of turbulence intensity at Z/D=0.6 in X-direction, Y-direction and Z-direction

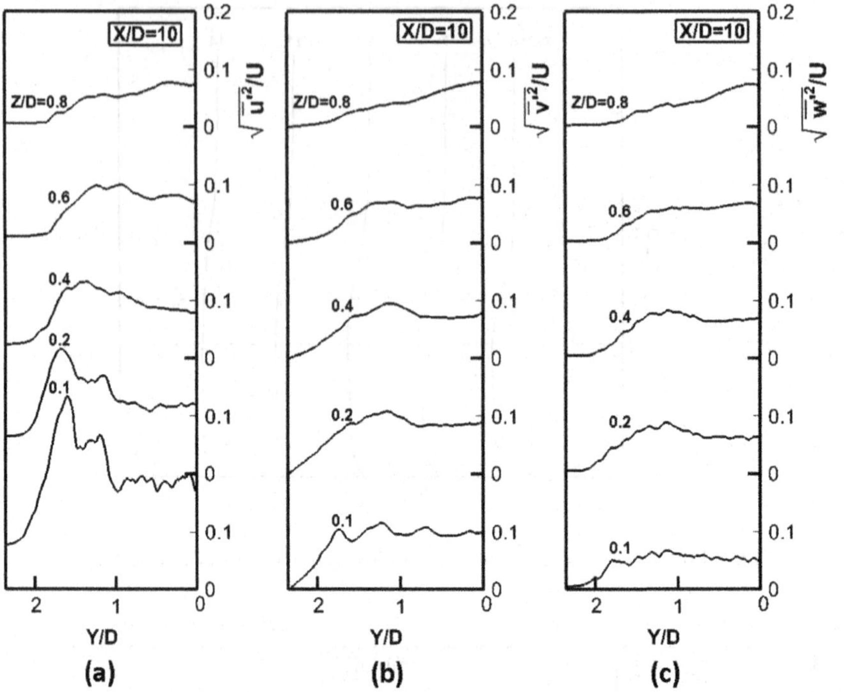

Fig. 11. Profiles of turbulence intensity at X / D = 10 in X-direction, Y-direction and Z-direction

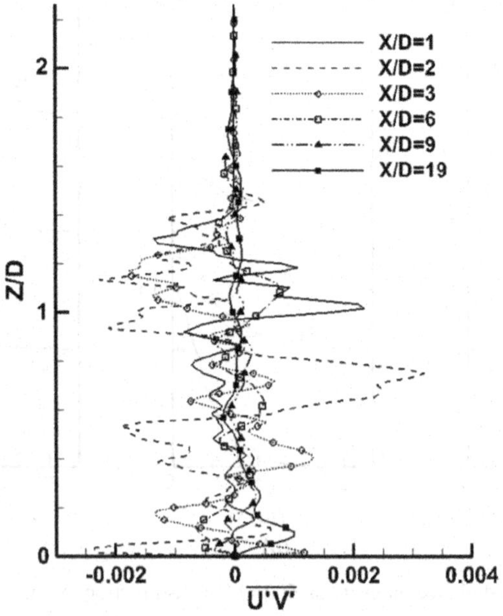

Fig. 12. Profiles of Reynolds stress $\overline{u'v'}$ in the center section Y / D = 0

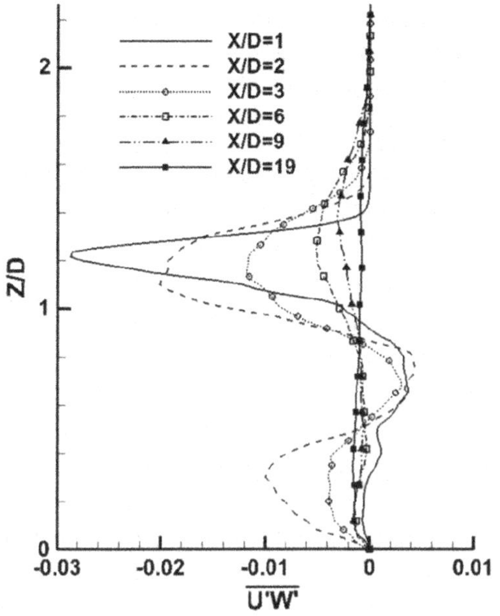

Fig. 13. Profiles of Reynolds stress $\overline{u'w'}$ in the center section Y / D = 0

The profile of the Reynolds stresses $\overline{u'v'}$ of different positions at the section Y / D = 0 (Figs.12) indicates that, peaks in different directions are alternate with each other with the increasing of the vertical distance. It is closed to zero near the outlet of the computational domain where X / D = 19. Fig. 13 shows that, at the position X / D = 1, two peaks with opposite directions are presented at Z / D = 1.2 and Z / D = 0.7, respectively. Because of the recirculation region, there is another peak exist at Z / D = 0.3 of the position X / D = 2. Similarly, at other positions in the X-direction, two peaks in the opposite directions exist. With the increase of x, peak values reduce, which is also closed to zero near the outlet of the computational domain where X / D = 19.

4 Conclusions

In this paper, the flow past a sphere placed above a ground plate is studied. The distance of gap between the sphere and the plate is 0.1D. A combined multiple-direct forcing and immersed boundary method (MDF/IBM) was used to deal with the coupling between solid and fluid with a Reynolds number of 4171 based on the sphere diameter.

Hairpin vortex is formed and sheds from the sphere, and the forest vortices are formed. In the upstream of the sphere there is a typical laminar flow velocity distribution. While near the outlet of the domain, the velocity distribution has turned to a typical turbulent layer velocity profile. The flow is separated when past the sphere. The under part forms a high velocity area and the upper part climbs upward, extending to the recirculation region behind the sphere. A recirculation region with the length of which is twice as much as the sphere diameter is formed because of the strong downwash behind the sphere.

With increasing streamwise distance, the influence of the sphere on the boundary layer decreases. The thickness of boundary layer increases, the center peak defect and the turbulence intensity decreases, and the Reynolds stresses reduction, which is close to zero near the outlet of the computational domain. With increasing vertical distance, the peak velocity defect decreases and its position is close to the center when Z/D <0.6, and moves away from the center when Z/D >0.6. The position of peak of turbulence intensity moves in a manner similar to peak velocity defect.

Acknowledgements. This work is supported by the National Natural Science Foundation of China (no. 51136006). We are grateful to that.

References

1. Schlichting, H.: Experimentelle Untersuchungen Zum Ranhigkeitsproblem. Ing. Arch. 7, 1–34 (1936)
2. Klemin, A., Schaefer, E.B., Beerer, J.G.: Aerodynamics of the Perisphere and Trylon at World's Fair. Trans. Am. Soc. Civ. Eng. 2042, 1449–1472 (1939)
3. Okamoto, S.: Turbulent Shear Flow Behind a Sphere Placed On a Plane Boundary. Turbulent Shear Flows 2, 246–256 (1980)
4. Takayuki, T.: Flow Around a Sphere in a Plane Turbulent Boundary Layer. Journal of Wind Engineering and Industrial Aerodynamics 96, 779–792 (2008)
5. Zeng, L., Balachandar, S., Fischer, P.: Interactions of a Stationary Finite-Sized Particle with Wall Turbulence. Journal of Fluid Mechanics 594, 271–305 (2008)
6. Charles, S.: Peskin: Flow Patterns Around Heart Valves: A Numerical Method. Journal of Computational Physics 10, 252–271 (1972)
7. Goldstein, D., Handler, R., Sirovich, L.: Modeling a No-Slip Flow Boundary with an External Force Field. Journal of Computational Physics 105, 354–366 (1993)
8. Le, D.V., Khoo, B.C., Lim, K.M.: An Implicit-Forcing Immersed Boundary Method for Simulating Viscous Flows in Irregular Domains. Computer Methods in Applied Mechanics and Engineering 197, 2119–2130 (2008)
9. Mittal, R., Iaccarino, G.: Immersed Boundary Methods. Annu. Rev. Fluid Mech. 37, 239–261 (2005)
10. Mohd-Yusof, J.: Combined Immersed Boundaries/B-Splines Methods for Simulations of Flows in Complex Geometries. CTR Annual Research Briefs, 317–327 (1997)
11. Xu, S., Wang, Z.J.: A 3D Immersed Interface Method for Fluid–Solid Interaction. Computer Methods in Applied Mechanics and Engineering 197, 2068–2086 (2008)
12. Le, D.V., Khoo, B.C., Peraire, J.: An Immersed Interface Method for Viscous Incompressible Flows Involving Rigid and Flexible Boundaries. Journal of Computational Physics 220, 109–138 (2006)
13. Wang, Z., Fan, J., Luo, K.: Combined Multi-Direct Forcing and Immersed Boundary Method for Simulating Flows with Moving Particles. International Journal of Multiphase Flow 34, 283–302 (2008)
14. Luo, K., Jin, J., Zheng, Y.: Direct Numerical Simulation of Particle Dispersion in Gas-Solid Compressible Turbulent Jets. Chinese Journal of Chemical Engineering 13, 161–166 (2005)
15. Zhou, Z., Wang, Z.L., Fan, J.R.: Direct Numerical Simulation of the Transitional Bounda-ry-Layer Flow Induced by an Isolated Hemispherical Roughness Element. Comput. Methods Appl. Mech. Engrg. 199, 1573–1582 (2010)

16. Luo, K., Wang, Z., Fan, J.: Full-Scale Solutions to Particle-Laden Flows: Multidirect Forcing and Immersed Boundary Method. Physical Review E 76, 066709 (2007)
17. Wang, Z., Fan, J., Cen, K.: Immersed Boundary Method for the Simulation of 2D Viscous Flow Based On Vorticity–Velocity Formulations. Journal of Computational Physics 228, 1504–1520 (2009)
18. Wang, Z.L., Fan, J.R., Luo, K.: Immersed Boundary Method for the Simulation of Flows with Heat Transfer. International Journal of Heat and Mass Transfer 52, 4510–4518 (2009)
19. Luo, K., Zheng, Y.Q., Fan, J.R.: Interaction Between Large-Scale Vortex Structure and Dispersed Particles in a Three Dimensional Mixing Layer. Chinese Journal of Chemical Engineering 11, 377–382 (2003)
20. Wang, Z., Fan, J., Luo, K.: Numerical Study of Solid Particle Erosion On the Tubes Near the Side Walls in a Duct with Flow Past an Aligned Tube Bank. AIChE J. 56, 66–78 (2010)
21. Wang, Z.L., Fan, J.R., Luo, K.: Parallel Computing Strategy for the Simulation of Particulate Flows with Immersed Boundary Method. Science in China Series E: Technological Sciences 51, 1169–1176 (2008)
22. Peskin, C.S.: The Immersed Boundary Method. Acta Numerica 11, 479–517 (2002)
23. Griffith, B.E., Peskin, C.S.: On the Order of Accuracy of the Immersed Boundary Method: Higher Order Convergence Rates for Sufficiently Smooth Problems. Journal of Computational Physics 208, 75–105 (2005)
24. Uhlmann: An Immersed Boundary Method with Direct Forcing for the Simulation of Particulate Flows. Journal of Computational Physics 209, 448–476 (2005)
25. Musker, A.J.: Explicit Expression for the Smooth Wall Velocity Distribution in a Turbulent Boundary Layer. AIAA Journal 17, 655–657 (1979)
26. Orlanski, A.: Simple Boundary Condition for Unbounded Hyperbolic Flows. Journal of Computational Physics 21, 251–269 (1976)

Airflow and Particle Deposition in a Dry Powder Inhaler: An Integrated CFD Approach

Jovana Milenkovic[1,2], Alexandros H. Alexopoulos[2], and Costas Kiparissides[1,2,3]

[1] Department of Chemical Engineering, Aristotle University of Thessaloniki, P.O. Box 472, 54124, Thessaloniki, Greece
[2] CPERI/CERTH, 6th km Harilaou-Thermi rd., 57001, Thessaloniki, Greece
[3] Department of Chemical Engineering, The Petroleum Institute, Abu Dhabi, U.A.E.
{jovbor,aleck,cypress}@cperi.certh.gr

Abstract. An integrated computational model of a commercial Dry Powder Inhaler, DPI, device (i.e., Turbuhaler) is developed. The steady-state flow in a DPI is determined by solving the Navier-Stokes equations using FLUENT (v6.3) considering different flow models, e.g., laminar, k-ε, k-ω SST. Particle motion and deposition are described using an Eulerian-fluid/Lagrangian-particle approach. Particle/wall collisions are taken to result in deposition when the normal collision velocity is less than a size-dependent critical value. The flow rate and particle deposition are determined for a range of pressure drops (i.e., 800-8800Pa), as well as particle sizes corresponding to single particles and aggregates (i.e., 0.5-20μm). Overall, the simulation results are found to agree well with available experimental data for the volumetric outflow rate as well as the local and total particle deposition.

Keywords: Dry Powder Inhaler, Turbuhaler, CFD, Particle, Deposition.

1 Introduction

Dry Powder Inhalers, DPI, are one of the principle means of delivering pharmaceuticals due to their ease of use and cost-effectiveness [16]. The main function of a DPI device is the adequate dispersion and delivery of particles. Initially particles are in the form of a loose powder which, under the action of airflow, is broken up and dispersed as particle aggregates, which are then further broken up into fine particles [4, 18, 20, 15, 3]. Powder properties, e.g., cohesion, charge, size, and size distribution, influence powder dispersion and the breakage of particle aggregates [13, 24, 12, 7].

One of the common problems with DPIs is the loss of powder/drug due to deposition within the device. In order to provide the maximum drug dose per inhalation and to ensure minimal dose-to-dose variation it is necessary to minimize drug losses due to internal deposition. Moreover, it is desired to have good control over the dispersibility of the powder, release of drug (when attached to powder particles), and breakup of aggregates in order to achieve the desired particle/aggregate size distributions at the DPI mouthpiece outflow [3]. Consequently, if the underlying

M.S. Obaidat et al. (eds.), *Simulation and Modeling Methodologies, Technologies and Applications*, Advances in Intelligent Systems and Computing 256,
DOI: 10.1007/978-3-319-03581-9_9,

processes are better understood one can achieve the desired outflow particle distribution which will conceivably minimize oropharyngeal losses and also permit better targeting for drug delivery in the respiratory tract.

Due to the complex and transient flow structures as well as the dynamic powder breakup and dispersion processes observed in most commercial DPIs only a small number of Computational Fluid Dynamics, CFD, investigations have been conducted [19, 17, 16]. Systematic computational studies have led to a better understanding of the function of DPI devices. For example, Coates et al. [8, 9, 10] studied the Aerolizer DPI in detail including the effects of air-intake, mouthpiece, and internal grid which led to improvements in the design and function of the DPI. Recently, the discrete element method, DEM, coupled to continuous phase-models has been implemented to describe the powder dispersion process within the inhaler [21, 6]. From the current state-of-the-art it is clear that the proper description of the aggregate strength as well as the particle/aggregate interaction with the inhaler walls are key processes that determine the final dispersion and size distribution of pharmaceutical powders [2].

The Turbuhaler (AstraZeneca) is a multidose DPI widely used to deliver a number of drugs (typically for asthma), e.g., terbutaline sulphate, (as Bricanyl), or budesonide (as Pulmicort), to the upper respiratory tract [23]. Each dose is initially in the form of loosely packed particle aggregates, ~1-20μm in size, which are released into a mixing/dispersion chamber, where they are partially broken up into particles which are then directed to the inhalation channel of the device [22, 23]. The proper function of the Turbuhaler is dependent on the dynamic volumetric flow as well as the peak inspiratory flow rate, PIFR, attained during inhalation, the amount of particles lost due to deposition within the device, and the adequate dispersion and breakup of the powder aggregates in the airflow exiting the mouthpiece. Recent experimental investigations have provided detailed information on particle deposition as well as the fine particle fraction and particle size distribution, PSD, of escaped particles in the outlet flow [11, 14, 1].

In this work the steady airflow in a Turbuhaler DPI is determined by CFD simulations and particle motion as well as deposition is determined by Eulerian-Fluid/Lagrangian-Particle, EFLP, simulations. In what follows the DPI geometry, the discretization procedure, and the CFD simulations are described in detail. Next the results for steady-state airflow are presented follow by the results for particle deposition. Finally, the computational results are compared to available experimental data.

2 Integrated Computational Model

The computational modeling of a DPI device is a challenging problem involving airflow, powder dispersion, aggregate breakup, and particle deposition. These are different processes operating at different spatial and temporal scales which require specific computational treatments (Figure 1). Fluid flow is typically described by CFD and can be treated separately from particle motion for small particle volume fractions (i.e., <10%). For larger volume fractions other approaches, e.g., two-phase or granular flow models, are necessary. Recently the DEM has been utilized to track the motions and interactions of individual particles in aggregates during the initial

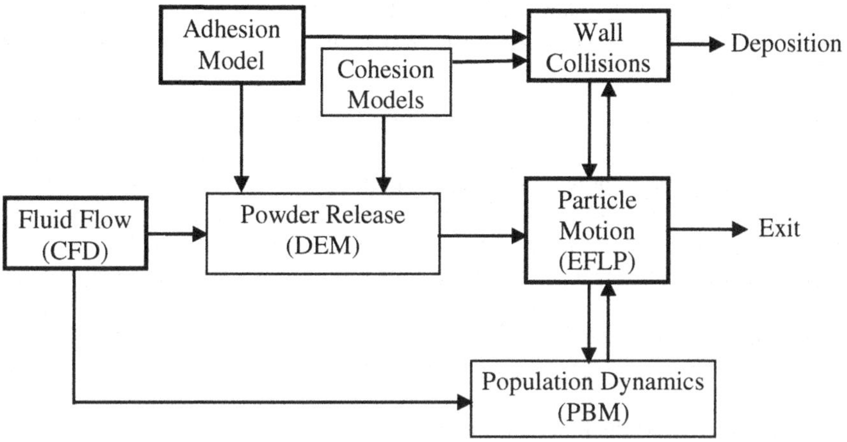

Fig. 1. Integrated computational model of a DPI. Limiting case indicated by bold boxes.

powder dispersion [6]. Particle-particle cohesion forces dominate the initial powder release and breakage of particle aggregates. The EFLP approach can be employed to follow the motion of individual particles and aggregates in the DPI. Collisions of particles with the DPI walls occur predominantly due to inertial impaction. The particle-wall collision frequency and capture efficiency (controlled by adhesion forces) determine the rate of deposition while particle cohesion forces control the rate of collision-induced breakage. Finally, the population dynamics of aggregates/ particles can be described by population balance models, PBM, if the controlling driving functions (e.g., breakage, aggregation, deposition and redispersion) can be described.

This work focuses on the limiting case of weak cohesion forces and rapid powder dispersion. In this case the powder is assumed to instantaneously break-up into a population of particles and aggregates identical to that of the free-flowing powder after which no more breakage or aggregation occurs. Consequently, the model consists of a CFD module, an EFLP module, a collision model, and an adhesion-controlled capture–efficiency model (Figure 1). Although this approach represents a limiting case it provides a means for evaluating the effects of flow rate, particle size, and adhesion forces on the local and total particle deposition in the DPI device.

The CFD approach consists of solving the continuity and momentum equations in each cell of a discretized computational domain that represents the air passage in the Turbuhaler DPI. The mass and momentum conservation equations can be expressed in scalar form as

$$\frac{\partial \rho \varphi}{\partial t} + \nabla \cdot (\rho \varphi \mathbf{u}) = \nabla \cdot \Gamma \nabla \varphi + S_\varphi \tag{1}$$

where ϕ is a scalar quantity (e.g., velocity component, temperature, concentration), Γ is the diffusivity coefficient, and S_ϕ is the source term.

Particle pathlines can be determined by the EFLP approach, that is, by solving the following force balance equation for each particle assuming an unperturbed airflow solution.

$$\frac{du_p}{dt} = F_D\left(u - u_p\right) + g\left(1 - \rho/\rho_p\right) + F_B + F \tag{2}$$

where the terms on the right hand side represent the drag force per unit particle mass, the gravitational acceleration per unit mass, the Brownian forces, and additional acceleration terms (e.g., Saffman lift force).

When particle trajectories intercept the DPI walls a particle-wall collision takes place which can result in deposition or particle rebound. In this work the capture efficiency, σ, is related to the normal particle collision velocity, v_n, according to

$$\sigma = 1 - H(v_n / v_c) \tag{3}$$

where H is the Heaviside function and v_c is the critical collision velocity which, according to the developments of Brach and Dunn [5], is given by:

$$v_c = \left(\frac{2E}{D}\right)^{10/7} \tag{4}$$

where D is the particle diameter and the effective stiffness parameter E is given by

$$E = 0.51\left(\frac{5\pi^2\left(k_s + k_p\right)}{4\rho^{3/2}}\right)^{2/5} \tag{5}$$

and k_s and k_p are determined by:

$$k_s = \frac{1 - v_s^2}{\pi E_s} \text{ and } k_p = \frac{1 - v_p^2}{\pi E_p} \tag{6}$$

where v_s and v_p and E_s and E_p are the Poisson's ratio and Young's modulus of the surface and particle, respectively. In the case of lactose particles (v_p = 0.4, E_p = 1.0GPa) colliding with polystyrene surfaces (v_s = 0.35, E_s = 4.1GPa) the critical velocity was determined to be v_c = 2.7m/s.

3 Results

The Turbuhaler DPI geometry was constructed in a CAD/CAM environment (i.e., CATIA v5 R19) and then imported into GAMBIT (v2.1) where the airflow domain was defined and a series of computational grids were constructed consisting of 2 10^5 – 2 10^6 tetrahedral cells with a maximum skewness of 0.85 (Figure 2). The computational grids were originally refined in regions where large gradients of flow were expected. Further refinements were conducted within FLUENT based on actual velocity gradients observed in initial solutions. It should be noted that the computational domain was extended from the mouthpiece outlet by 20mm in order to minimize recirculation effects at the outflow surface and to improve convergence behavior.

The Navier-Stokes equations for flow were solved using the commercial CFD software (i.e., FLUENT v6.3). The SIMPLEC scheme was employed to describe pressure-velocity coupling. Second order discretization was used for pressure and third order MUSCL for momentum and turbulent variables. Convergence of CFD simulations was assumed when the residuals were < 10^{-4}. Zero gauge pressure boundary conditions were employed at all the inflows, i.e., two powder loaded cylinders (see bottom of Figure 2) and four extra air inlets in the DPI circulation

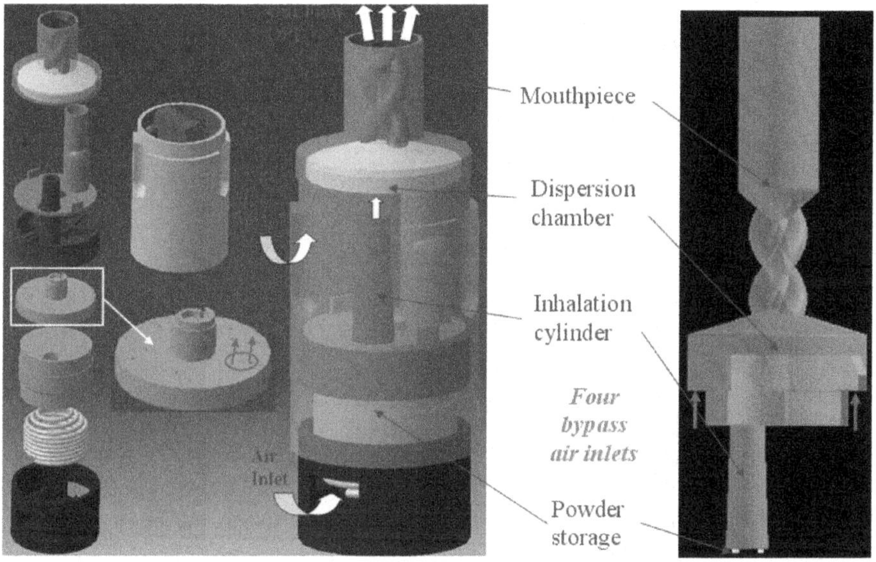

Fig. 2. Turbuhaler Dry Powder Inhaler CAD model and Computational Grid

chamber. Different steady state airflows (i.e., 20 to 70 l/min) were simulated by imposing a wide range of pressure drops at the mouthpiece outflow ranging from 800 to 8800Pa.

EFLP simulations of particle motion and deposition were conducted for particles between 0.5-20μm in size encompassing the range of single particle and particle aggregate sizes of typical pharmaceutical powders employed in the Turbuhaler DPI. Particles were assumed to be released instantaneously at t = 0 and uniformly from circular surfaces located immediately upstream (i.e., 2mm) from the powder storage sites. Powder dispersion was assumed to occur instantaneously after which no further breakage occurred. This assumption corresponds to the limit of very weak particle cohesion forces. After the initial powder release and aggregate breakage, particles in motion were taken to be constant in size.

3.1 Simulations of Airflow in the Turbuhaler DPI

According to the range of volumetric airflows examined in this work, e.g. Q = 20 – 70 l/min, the local Reynolds numbers, $Re = Q \rho / \mu A^{1/2}$, where ρ and μ are the density and the viscosity of air and A is the cross-sectional area, ranged from 130-16,000.

Consequently the transitional SST k-ω model was employed to describe the turbulent flows encountered in the DPI. Grid convergence was verified for the SST k-ω model with the 1 10^6 grid providing essentially identical results as the 2 10^6 grid and was used for the results presented in this paper.

In Figure 3 the velocity magnitudes as well as the tangential and radial velocities are displayed along an axial (i.e., zx) plane. As can be observed, the airflow in the DPI device is found to be laminar in the inhalation channel with two jet flows

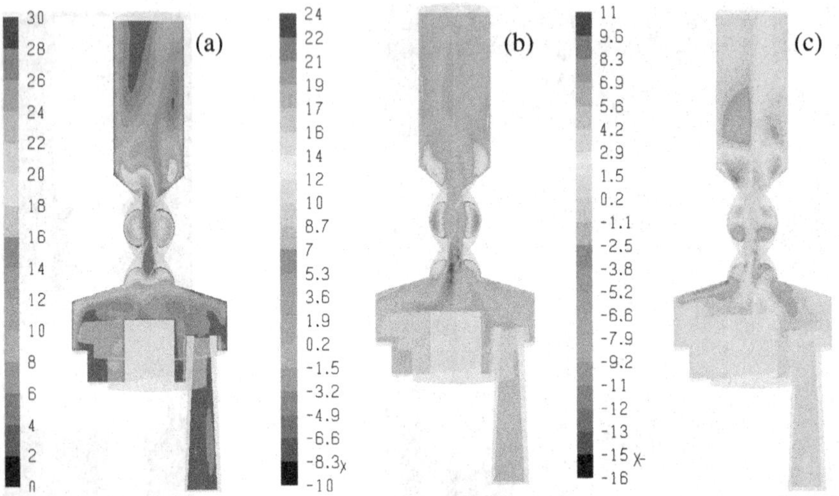

Fig. 3. Velocity contours in the Turbuhaler DPI: k-ω SST (ΔP = 800Pa). (a) velocity magntude, (b) tangential velocity, (c) radial velocity

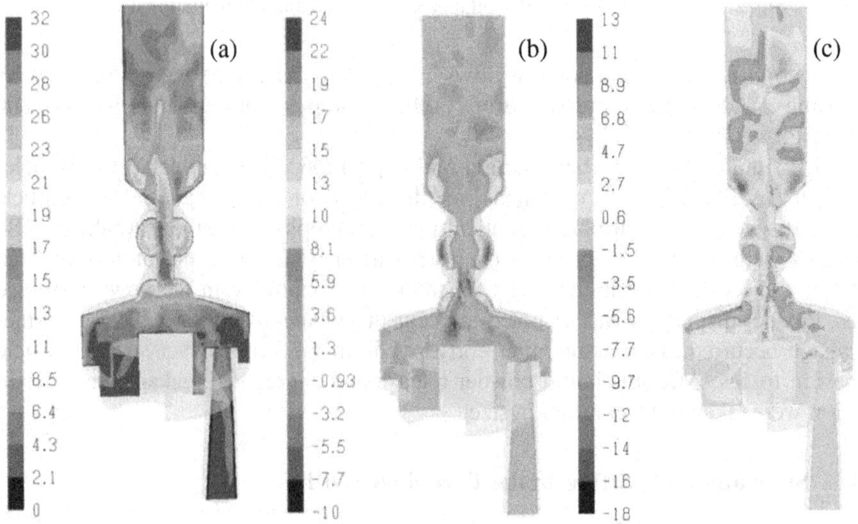

Fig. 4. Velocity contours in the Turbuhaler DPI: LES (ΔP = 800Pa). (a) velocity magnitude, (b) tangential velocity, (c) radial velocity

emanating from the powder storage cylinders. In the circulation chamber the flow is characterized by large eddies and secondary flows. In the helical region significant tangential flows develop (reaching 83% of the maximum velocity magnitude) and persist about halfway up the mouthpiece extension. It should be noted that the velocity profiles observed for larger flow rates, e.g., 60 l/min, are qualitatively similar.

Large Eddy Simulations, LES, fully resolve the large scale motion of turbulent flows thus providing more information and accurate results compared to Reynolds Averaged Navier-Stokes, RANS, approaches, e.g., k-ε, k-ω. The computational burden of LES is significant (e.g., at least an order of magnitude more than with RANS models). Consequently, only a single case (i.e., ΔP = 800Pa) of steady-state flow in the Turbuhaler DPI was simulated with LES using FLUENT. In Figure 4 the results for the mean velocity magnitude, as well as the radial and tangential components obtained with LES are shown. The main flow structures are similar with the k-ω SST results in Figure 3 but differences can be observed in the details of the flow, e.g., the secondary flows in the mouthpiece extension. The intensity of fluctuations (e.g., RMS velocity / velocity magnitude) varied within the device up to a value of ~50% indicating significant local fluctuations around the mean for the length scales resolved within the LES. The RMS range from 1-8m/s for the axial velocity component and 1-4m/s for the other components with different spatial variations within the device.

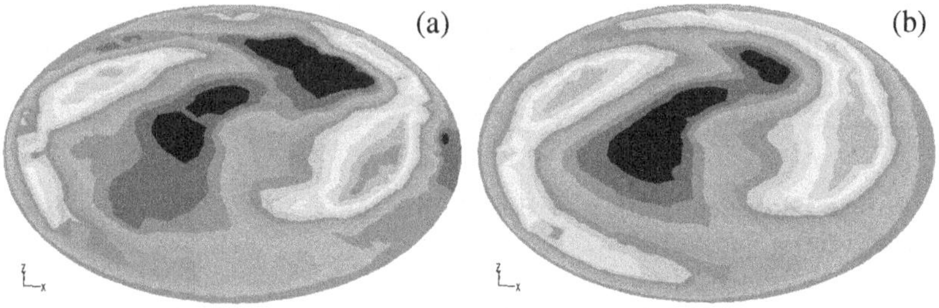

Fig. 5. Tangential velocity component at the mouthpiece exit (ΔP = 1400Pa).(a) LES (b) k-ω SST

In Figure 5 the tangential velocities at the outlet surface for ΔP = 1400Pa are shown. It is clear that the tangential velocities predicted by the k-ω SST and LES turbulence models are very similar. In fact the k-ω SST turbulence model provided the most similar to the LES results compared to the other RANS turbulence models (e.g., standard k-ε, RNG k-ε). Consequently, despite the observed differences in secondary flows, the k-ω SST model was employed for all the simulations of this work.

3.2 Simulation of Particle Motion and Deposition in the Turbuhaler DPI

Particle simulations were performed for all the flows examined in section 3.1. For effective powder dispersion the solids volume ratio in the DPI device is approximately 10^{-2}-10^{-4} depending on the location and the flow rate. Consequently, the particle phase will not influence the airflow solution and the EPLF approach can be employed. A wide range of injected particle numbers was used (i.e., 2,000-40,000). It was found that a minimum of 5,000 particles were necessary to obtain number-independent deposition results. The capture efficiency was implemented internally within FLUENT using a user-defined function.

Simulations with injections of single-sized particles as well as with particle size distributions ranging from 0.5-20μm were performed. The spatial distribution of particles deposited on the DPI walls was visualized using Tecplot. In Figure 6 the effect of particle size on the distribution of deposited particles in the DPI device is shown. Comparing particle sizes of 1, 2 and 5μm significant differences in the total

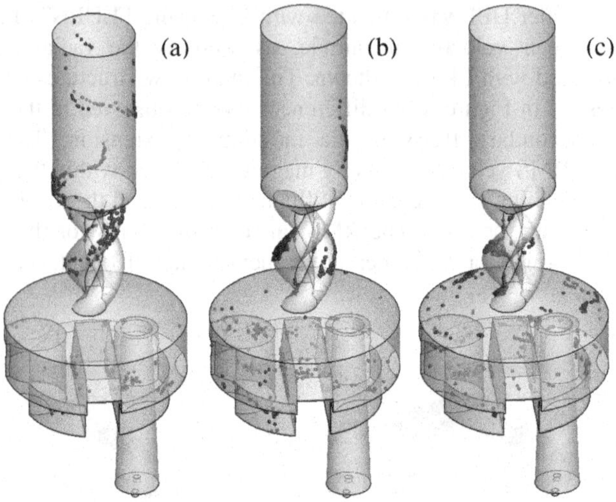

Fig. 6. Particle Deposition – Effect of Particle Size. ΔP = 800Pa, σ = 1, (a), D = 1μm, (b) D = 2μm, (c) D = 5μm.

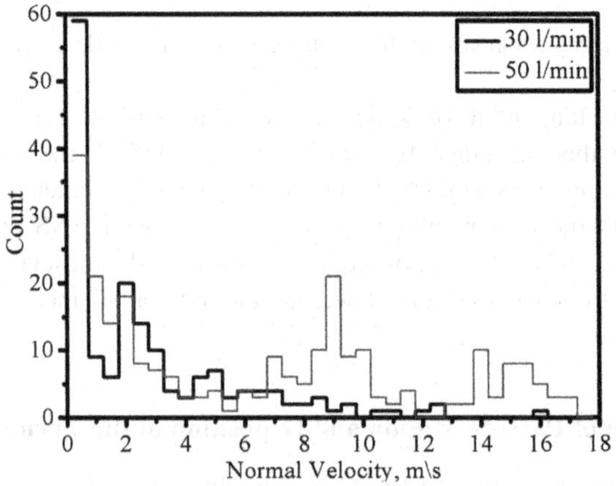

Fig. 7. Distribution of normal collision velocities (σ = 1, D = 2μm, ΔP = 1400Pa)

deposition as well as the deposition distribution are observed. The significant particle deposition that occurs in the mouthpiece region (including the helical region) is actually a common problem in many commercial DPI devices where about half the internal deposition occurs [11].

The distribution of particle collision characteristics, i.e., normal velocities and collision angles were examined for two volumetric flow rates (i.e., Q = 30 and 50 l/min). In Figure 7 the distributions of normal collision velocities are shown. For small flow rates (i.e., Q = 30 l/min) most of the particle collisions occur with normal velocities $v_n < 1$ m/s. resulting in particle capture. At larger flow rates (i.e., Q = 50 l/min) the number of particle collisions increases and shifts to larger values of normal velocity with a significant proportion between 8-10 m/s which result in rebound. It should be noted that most of the particle collisions occurred with a collision angle between 5-10° and 10-18° for Q = 30 and 50 l/min, respectively.

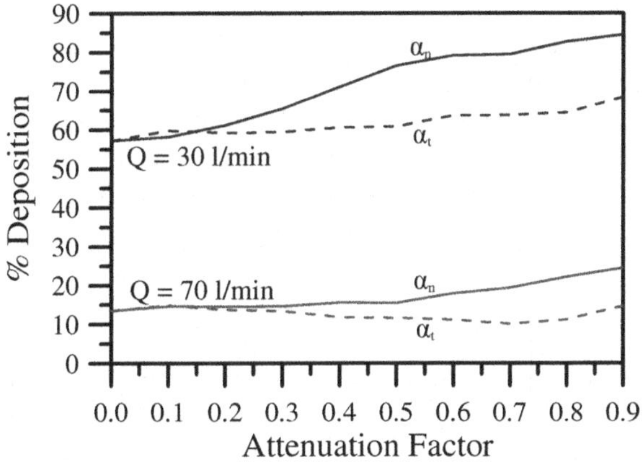

Fig. 8. Effect of attenuation factors (D = 2μm, ΔP = 1400Pa)

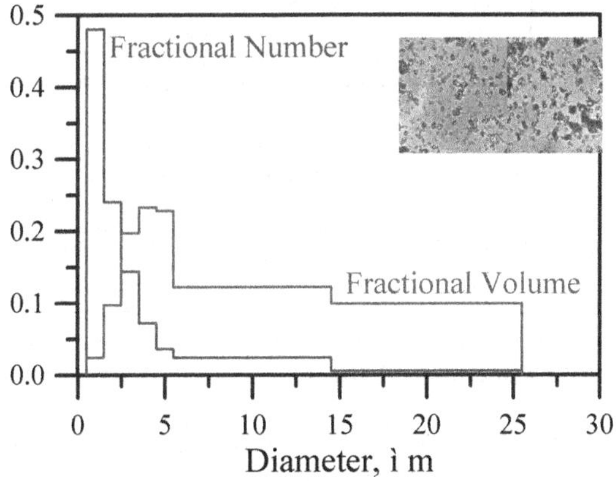

Fig. 9. Fractional particle number and volume distribution

The effect of momentum loss of rebounding particles after collision with the DPI walls was also examined. In Figure 8 the effect of a wide range of normal and tangential attenuations factors, α_n and α_t, on the total deposition of 2μm particles for Q = 30 and 70 l/min is shown. It is clear that attenuation has a more significant effect on the total deposition for the low volumetric flow case (i.e., Q = 30 l/min). Furthermore, particle deposition is more sensitive to normal attenuation than the corresponding tangential term.

Figure 9 displays the PSD of freely flowing powder containing Budesonide (Pulmicort). The peak in the number distribution is at D_0 = 2.2μm while for the volume distribution it is at 4.5μm. It was found that a Rosin-Rammler distribution, f(D), with a shape parameter value of n = 1 and a mean diameter of D_0 = 2.2μm, i.e., is a good approximation to the distribution depicted in Figure 9.

$$f(D) = (1/D_0)\, e^{-D/D_0} \tag{7}$$

The injected, escaped and deposited fractional volume distributions for ΔP= 800Pa are provided in Figure 10. It is observed that, there are very large differences in both the amounts and size distributions of escaped and deposited particles depending on the value of the capture efficiency, i.e., σ = 1 or σ evaluated by eqs. 3-6. It should be noted that small particles (e.g., 1-5μm) exhibit fewer collisions but have a larger capture efficiency than large particles (e.g., 5-10μm).

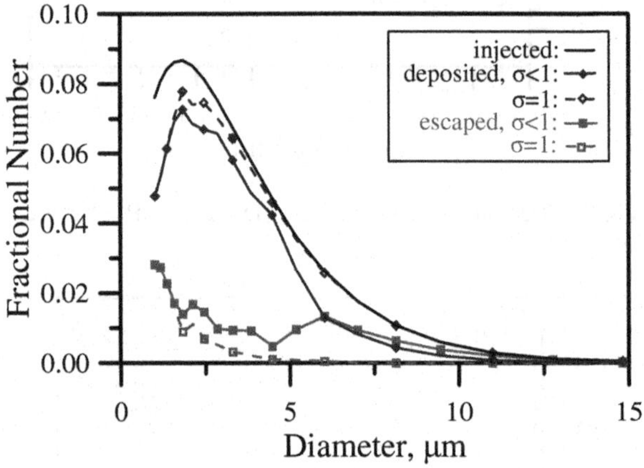

Fig. 10. Particle Deposition. (a) Fractional Cumulative Deposition, (b) Local.Deposition (ΔP = 800Pa).

3.3 Comparison to Experimental Data

The computational results of this work were compared to the experimental results of de Koning et al [11] and Abdelrahim [1] for the Turbuhaler in terms of flow and particle deposition. In Figure 11 the predicted steady-state volumetric flows are plotted against the outlet pressure drop applied at the mouthpiece. Both laminar and k-ω SST models for flow are examined. It is clear that both models agree very well with the experimental data for all flow rates with the k-ω SST model being slightly more accurate.

In Figure 12 the total, circulation chamber, and mouthpiece particle depositions for 1400Pa (corresponding to Q = 30 l/min) are compared to the experimental data of de Koning et al. [11]. The mouthpiece, circulation chamber, and total particle deposition results for Q = 30 l/min are in good agreement to the experimental data.

In Figure 13 the predicted total particle deposition are compared to the experimental data of de Koning et al [11] and Abdelrahim [1] for flowrates Q = 30, 40, 50, 60 and 70 l/min and for two different inspired volumes, i.e., 2 and 4l [1].

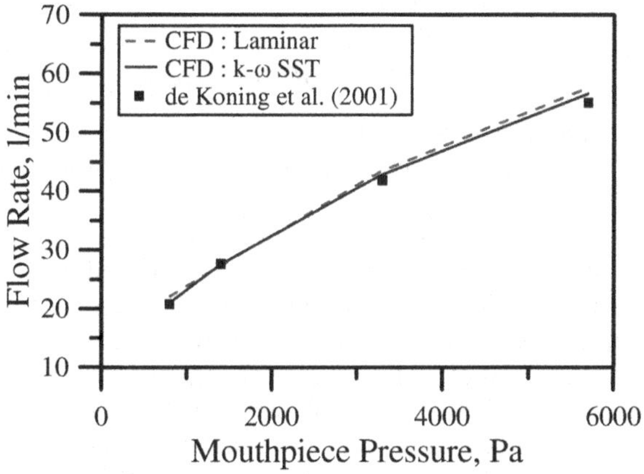

Fig. 11. Volumetric flow in the Turbuhaler

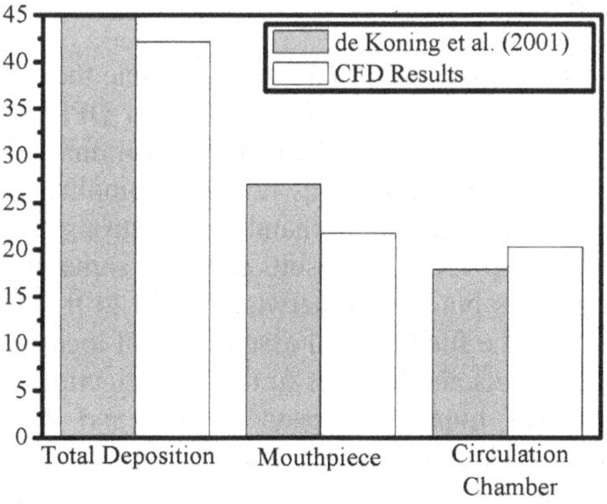

Fig. 12. Regional particle deposition in the Turbuhaler. Q = 30 l/min.

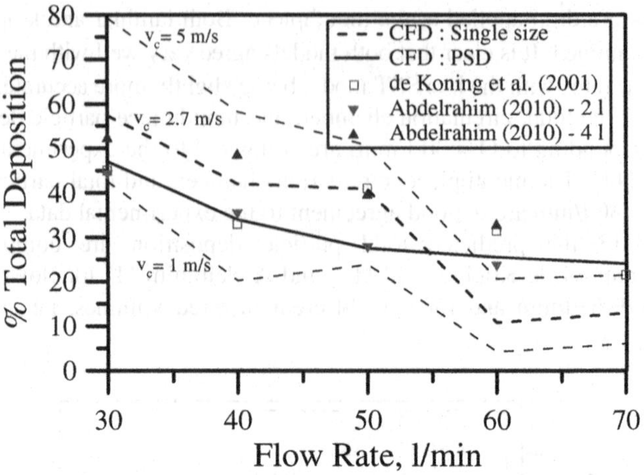

Fig. 13. Total particle deposition in the Turbuhaler. D = 2 μm. V_c = 2.7 m/s. Comparison between experimental results of de Koning et al. [11], Abdelrahim [1] and computational CFD results.

Simulations with injections of particles following a PSD characterized by eq. (6) were performed. The computational results were found to agree well with the experimental data. Simulations with injections of single-sized particles were also performed. For a particle diameter of D = 2μm and a critical velocity of v_c = 2.7m/s the agreement with the experimental data is good considering the different experimental conditions (e.g., dynamic inhalation vs. steady state simulations) and the model simplifications and assumptions. Different values of v_c are also shown in Figure 13 to provide an indication of the sensitivity of the computed particle deposition on the value of v_c.

4 Conclusions

This work has demonstrated the use of CFD to determine the complicated airflow as well as particle motion and deposition in the Turbuhaler DPI. As the flow was either locally laminar or transitionally turbulent the transitional SST k-ω model for turbulence was employed. The simulations revealed complicated flows with intense recirculation patterns in the circulation chamber and strong tangential flows in the helical region of the mouthpiece. LES results revealed some differences in the large eddies and secondary flows but were otherwise closest to the k-ω SST results. LES results also indicated that the fundamental assumption of local turbulence isotropy of the RANS models is incorrect and requires further investigation.

Particle deposition was found to depend on size and flow rate and occurred predominantly in the circulation chamber and the mouthpiece. The computational solutions were compared to experimental data for volumetric flow and regional deposition of de Koning et al. [11] and good agreement was observed. A simple collision model by Brach and Dunn [5] was employed to determine the critical velocity for particle capture, i.e., v_c, which was found to produce total particle depositions similar to the experimental values of de Koning et al [11] and Abdelrahim [1].

Future work will involve the simulation of dynamic inhalations, collision-induced breakage and will elaborate on the particle collision model which can be extended by including the effects of particle properties (e.g., size, shape, and charge), surface properties (e.g., roughness, charge), as well as humidity.

Acknowledgements. The research leading to these results has received funding from the European Union Seventh Framework Programme [FP7/2007-2013] under grant agreement n° 238013.

References

1. Abdelrahim, M.E.: Emitted dose and lung deposition of inhaled terbutaline from Turbuhaler at different conditions. Respiratory Medicine 104, 682–689 (2010)
2. Adi, S., Adi, H., Chan, H.-K., Finlay, W.H., Tong, Z., Yang, R., Yu, A.: Agglomerate strength and dispersion of pharmaceutical powders. J. Aerosol Science. 42, 285–294 (2011)
3. Alagusundaram, M., Deepthi, N., Ramkanth, S., Angalaparameswari, S., Saleem, T.S.M., Gnanaprakash, K., Thiruvengadarajan, V.S., Madhusudhana, C., Alagusundaram, C.M., et al.: Dry Powder Inhalers - An Overview. Int. J. Res. Pharm. Sci. 1(1), 34–42 (2010)
4. Ashurst, I., Malton, A., Prime, D., Sumby, B.: Latest advances in the development of dry powder inhalers. PSTT 3(7), 246–256 (2000)
5. Brach, R.M., Dunn, P.F.: A Mathematical Model of Impact and Adhesion of Microspheres. Aerosol Sci. Technol. 23, 51–71 (1992)
6. Calvert, G., Hassanpour, A., Ghadiri, M.: Mechanistic analysis and computer simulation of the aerodynamic dispersion of loose aggregates. Chemical Engineering Research and Design 89, 519–525 (2011)
7. Chan, H.-K.: Dry powder aerosol drug delivery – Opportunities for colloid and surface scientists, Colloids and Surfaces A: Physicochem. Eng. Aspects 284-285, 50–55 (2006)
8. Coates, M.S., Fletcher, D.F., Chan, H.-K., Raper, J.A.: Effect of Design on the Performance of a Dry Powder Inhaler Using Computational Fluid Dynamics. Part 1: Grid Structure and Mouthpiece Length. J. of Pharmaceutical Sciences 93, 2863–2876 (2004)
9. Coates, M.S., Chan, H.-K., Fletcher, D.F., Raper, J.A.: Influence of Air Flow on the Performance of a Dry Powder Inhaler Using Computational and Experimental Analyses. Pharmaceutical Research 22(9), 923–932 (2005)
10. Coates, M.S., Chan, H.-K., Fletcher, D.F., Raper, J.A.: Effect of Design on the Performance of a Dry Powder Inhaler Using Computational Fluid Dynamics. Part 2: Air Inlet Size. J. of Pharmaceutical Sciences 95(6), 1382–1392 (2006)
11. de Koning, J.P., Visser, M.R., Oelen, G.A., de Boer, A.H., van der Mark, T.W., Coenegracht, P.M.J., Tromp, T.F.J., Frijlink, H.W.: Effect of Peak Inspiratory Flow and Flow Increase Rate on In Vitro Drug Deposition from Four Dry Powder Inhaler Devices. In: Dry Powder Inhalation: Technical and Physiological Aspects, Prescribing and Use, Thesis, Rijksuniversiteit Groningen, ch. 6, pp. 83–94 (2001)
12. Finlay, W.: The Mechanics of Inhaled Pharmaceutical Aerosols. An Introduction. Academic Press, London (2001)
13. French, D.L., Edwards, D.A., Niven, R.W.: The Influence of Formulation on Emission Deaggregation and Deposition of Dry Powders for Inhalation. J. Aerosol Sci. 27(5), 769–783 (1996)

14. Hoe, S., Traini, D., Chan, H.-K., Young, P.M.: Measuring charge and mass distributions in dry powder inhalers using the electrical Next Generation Impactor (eNGI). European J. of Pharmaceutical Science 38, 88–94 (2009)
15. Islam, N., Gladki, E.: Dry powder inhalers (DPIs) - A review of device reliability and innovation. Int. J. of Pharmaceutics 360, 1–11 (2008)
16. Islam, N., Clearly, M.J.: Developing an efficient and reliable dry powder inhaler for pulmonary drug delivery – A review for multidisciplinary researchers. Medical Engineering & Physics 34, 409–427 (2012)
17. Ligotke, M.W.: Development and characterization of a dry powder inhaler. In: Dalby, R.N., Byron, P.R., Peart, J., Farr, S.J. (eds.) Respiratory Drug Delivery VIII, vol. I, pp. 419–422. Serentec Press Inc., Tucson (2002)
18. Newman, S.P., Busse, W.W.: Evolution of dry powder inhaler design, formulation, and performance. Respir. Med. 96(5), 293–304 (2002)
19. Schuler, C., Bakshi, A., Tuttle, D., Smith, A., Paboojian, S., Snyder, H., Rasmussen, D., Clark, A.: Inhale's dry-powder pulmonary drug delivery system: Challenges to current modeling of gas-solid flows. In: Proceedings of FEDSM 1999: 3rd ASME/JSME Joint Fluids Engineering Conference and 1999 ASME Fluids Engineering Summer Meeting, FEDSM 1999-7895 (1999)
20. Tobyn, M., Staniforth, J.N., Morton, D., Harmer, Q., Newton, M.E.: Active and intelligent inhaler device development. Int. J. of Pharmaceutics 277, 31–37 (2004)
21. Tong, Z.B., Yang, R.Y., Chu, K.W., Yu, A.B., Adi, S., Chan, H.-K.: Numerical study of the effects of particle size and polydispersity on the agglomerate dispersion in a cyclonic flow. Chemical Engineering Journal 164, 432–441 (2010)
22. Tsima, M.P., Martin, G.P., Marriott, C., Gardenton, D., Yianneskis, M.: Drug delivery to the respiratory tract using dry powder inhaler. Int. J. of Pharmaceutics 101, 1–13 (1994)
23. Wetterlin, K.: Turbuhaler: A New Powder Inhaler for Administration of Drugs to the Airways. Pharmaceutical Research 5(8), 506–508 (1988)
24. Zeng, X.-M., Martin, G.P., Marriott, C., Pritchard, J.: The influence of carrier morphology on drug delivery by dry powder inhalers. Int. J. of Pharmaceutics 200, 93–106 (2000)

Solar Soldier: Virtual Reality Simulations and Guidelines for the Integration of Photovoltaic Technology on the Modern Infantry Soldier

Ioannis Paraskevopoulos and Emmanuel Tsekleves

School of Engineering and Design, Brunel University, U.K.
{Ioannis.paraskevopoulos,emmanuel.tsekleves}@brunel.ac.uk

Abstract. Following recent advances in the field of thin and flexible materials, the use of product integrated photovoltaics (PIPV) for light harvesting and electric power generation has received increased attention today. PIPV is one of the most promising portable renewable energy technologies of today, especially for the defense industry and the modern infantry soldier. Nevertheless, there is limited work on light harvesting analysis and power generation assessment for its use in various military scenarios including on how to best integrate the technology on the infantry soldier. This study aims to fill this gap by accurately analyzing the light harvesting through virtual reality simulations. Following the virtual light analysis, an assessment of the power generation potential per scenario and investigation of the optimum integration areas of flexible PV devices on the infantryman are presented. Finally, there is a discussion of the key results, providing the reader with a set of guidelines for the positioning and integration of such renewable energy technology on the modern infantry soldier.

Keywords: Renewable Energy, 3D Simulation, Virtual Reality, Photovoltaic, Solar Energy Harvesting, Computer Simulation, Daylight Simulation, Infantry Soldier, Design, Human Factors, Usability, Military Environment, Product Integrated Photovoltaics (PIPV), Wearable Photovoltaics.

1 Introduction

Despite modern advances in military technology, the infantry soldier continues to play a significant role in defense. In the age of stealth jets, nuclear munitions and guided weapons, it is still the infantry soldier that examines and secures a location to ascertain whether the target area is cleared and the enemy is defeated. The modern infantry soldier utilizes the electronic technology and resources available today, in order to penetrate into hostile and difficult terrain, where armored vehicles cannot trespass and overcome the enemy. The power requirements of such electronic technology, critically essential for the modern soldier, are much higher when compared to the power requirements of a civilian counterpart. Furthermore, the environment of operation is far more hostile and challenging than those of the civilian applications and the loss of power may endanger the infantry soldier's life. That is the

M.S. Obaidat et al. (eds.), *Simulation and Modeling Methodologies, Technologies and Applications*, Advances in Intelligent Systems and Computing 256,
DOI: 10.1007/978-3-319-03581-9_10, © Springer International Publishing Switzerland 2014

main reason behind the massive overload of batteries constituting nearly 25% [1] of the overall equipment load (including lethal, survival and communication). Owing to the aforementioned fact, there is an uncontested restriction of maneuverability and operational range, as well as a significant burden, both physical and cognitive. The recent advances in the field of sustainable energy, and particularly the innovative flexible and wearable photovoltaic (PV) technologies, could offer a potential solution to this issue, by removing, or reducing to a great extent, the use of batteries. The Solar Soldier project, which is partly funded by the Defense Science and Technology Laboratory (DSTL) of the MoD in the United Kingdom (UK) and the Engineering and Physical Sciences Research Council (EPSRC), investigates this research challenge. Part of the work conducted in the boundaries of that project is presented in this article, which focuses on how one could integrate the PV technology epitomizing the Solar Soldier concept from a human interface and design perspective. The objectives of this challenge are the following:

- To assess the incorporation of the PV technology on the uniform and equipment of the infantry soldier;
- To accurately measure and evaluate its effectiveness (amount of light captured under various scenarios) taking into account usability (human comfort, intuitiveness).
- To assess the effectiveness of each area (amount of power generated under various scenarios) as well as to investigate the areas that yield the same power values all over their extent for further research on usability (human comfort, intuitiveness).

This chapter is organized in a number of sections. Section 2 offers a background literature review on related work. This is followed by a presentation of the adopted overall methodology, including the validation study of the simulation software platform employed. Next, the outcomes of the simulations are presented and discussed and finally, the inferred conclusions on the design aspects along with guidelines on the integration of PV on the infantryman are offered.

2 Related Work

The theoretical background of the study presented in this paper falls within various research areas, including Modeling and Simulation (M&S), virtual reality (VR) applications and product design aspects.

2.1 Virtual Reality and Defense Applications

The advances of VR, in recent years, have led to the development of new areas of applications beyond the entertainment industry. Research and development in interactive VR has been employed in the areas of training, education, health and simulation, with one of the major areas of interest being military and defense applications [2]. VR can be utilized for military applications to perform a wide range of simulations. These range from cognitive and behavior simulations in battle to ergonomic simulations; all serving the improvement of the welfare of the modern soldier. These simulations have to be conducted in a virtual framework often consisting of assets that offer three dimensional (3D) graphical representations of terrains, human avatars and objects, as well as weather and daylight-augmented systems. All these elements create a Virtual world on a computer-based simulated environment. This is of significant interest and importance to research, as it offers a very useful alternative reality, especially for situations (as in the case of this research works), where actual experiments are not feasible or dangerous to conduct in real life [3], [4]. More precisely, Chryssolouris et al. [5] have conducted research in the area of human ergonomics in an assembly line and Reece [6] has studied the movement behavior of soldier agents on a virtual battlefield. Furthermore, the Santos [7],[8] project offers a virtual platform for human ergonomics in military environments and Shiau and Liang [9] present a real-time network VR military simulation system comprising weather, physics and network communications. Blount et al. [10] have introduced the aspect of physical fitness into simulations for infantry soldiers and others, such as Cioppa et al. [11] and Bitinas et al. [12], have worked with agent-based simulations and their military applications, focusing on human factors in military combat and non-combat situations, respectively. Apart from these aforementioned articles, there is a recently published three-volume edition containing an extensive literature on VR and applications: The PSI handbook of virtual environment training and education: developments for the military and beyond [13]. The second volume of this text contains subjects such as 'Mixed and augmented reality for training', 'Evaluating virtual environment component technologies' and 'Enhancing virtual environments to support training'. The aforementioned literature focuses mainly on simulating human factors and ergonomics, either in the production line or in military environments. However, the applications of VR human-centered simulations are not restricted to ergonomics. The aspect of Human-centered Design (HCD) that this article examines is the integration of renewable energy devices on the human vesture, and in particular the integration of PV technology on the uniform or equipment of the modern infantry soldier in terms of light capture efficiency.

2.2 Simulation of Solar Light Harvesting

Currently, the main focus of PV technology and its corresponding simulations has been on building and infrastructure applications. The very recent developments in the area of PV devices [14], [15], along with the introduction of thin films and flexible materials for light absorption [16], have attracted the focus of harvesting renewable energy to human-centered applications as well. The study of the performance of the so-called Product Integrated Photovoltaics (PIPV) [17] is twofold: firstly, to investigate the performance and electrical characteristics of the PV device itself; secondly, to study the effectiveness of light harvesting, which is also the main focus and aim of our work. The effectiveness of light harvesting depends on the interaction of the device with the environment, as well as on the type of integration of the PV technology on the product (e.g. attached on clothing, embroidered or woven onto the fabric). The environmental conditions would require the modeling of daylight and shading in a 3D authoring and simulation tool, whilst the integration guidelines would require simulated scenarios and results that would infer the most effective method of integration.

Daylight and Shading Modeling

With regards to daylight modeling, there have been numerous studies on methods to maximize solar system outputs [18]. Furthermore, there are studies on the shading effect of the environment, which investigate the effects of random shading on PV energy production [19], [20]. Apart from research studies, there has been major development in the corresponding software industry, with very intelligent and complex packages developed for daylight simulations, including 3D Studio Max Design (3DSMD) by Autodesk, which is the software, utilized in this project. 3DSMD was chosen mainly because it comprises a toolset for animation and because it includes the feature of light analysis of a 3D scene, which is essential for a HCD project such as this. 3DSMD also offers extension capabilities through its embedded programming language, Maxscript. It can thus be used to semi-automate the procedures as described in the work of Paraskevopoulos and Tsekleves [21]. The results of the light analysis of 3DSMD have been validated by Reinhart et al [22] and Paraskevopoulos and Tsekleves [21] and the software has been used in a number of other studies regarding light harvesting for PV technology [17], [23]. Nevertheless, all of them have focused on simulations where the PIPV device was in a static position and none of these has used simulation to analyze the effects of movement on PIPV light harvesting. Furthermore, no previous work has offered any conclusions or guidelines on the design aspects of wearable PV devices in terms of the light capturing efficiency of mobile agents.

Integration of PV Technology on Commercial Products

Although the integration of PV technology on commercial products is not a new idea, the emergence of flexible and thin-film materials has extended the possibilities of integration into more products with a smaller scale factor, which can be portable. However, until recently and as stated by Mestre and Diehl [24], there have been no guidelines for the integration of PV technology on products in the context of either human comfort or efficiency of energy harvesting. The work of Reinders [25] examines in depth the options for PV systems and portable devices and presents their advantages and drawbacks. Among the drawbacks, one indicates the lack of PV technology penetration in our society and market. This is mainly due to limited knowledge of this technology by product designers and manufacturers, restricting in turn the extension of applications for this technology. The work presented in this article aims to fill in this gap by deploying design guidelines and a simulation platform on the integration of PV technology on military garments or equipment initially and commercial products in the future. As already mentioned in the introduction of this article, the use of VR simulations is a prerequisite for military applications, due to the hostile and extremely hazardous environment. Randall et al. [26] have integrated solar modules to use them as light sensors in order to collect physical measurements and not for the purposes of light analysis simulation. With regards to the design aspects of the integration of PV technology on clothing, Schubert and Werner [27] have presented an overview of flexible solar cell technologies applied on wearable renewable sources. This, however, focuses only on the material aspect of PV technology. In their paper, Schubert and Werner [27] reference Gemmer, who has performed experimental investigation on light harvesting under different daylight scenarios and has calculated energy yield for various user profiles. For example a 'regular clerk', an 'outdoor construction worker' and a 'night shift nurse'. In the system we propose, these profiles can be very easily modeled (3D avatars and motion capture) and simulated (light analysis tool, 3DSMD) for all various light conditions (daylight system, 3DSMD) and encompassing environments (3D terrain models). The outcomes of such simulations would infer the design guidelines of the most efficient manner of integration of PV on clothing in terms of light harvest.

3 Methodology

The overall methodology adopted for this project is outlined in this section. The problem stated in the Introduction of this paper requires the employment of a virtual framework able to conduct a number of experiments and collect measurements, which are impossible to collect due to the hazardous nature of the real environment. The methodology that fulfils the development of such a virtual framework is Modelling and Simulation (M&S) [28]. The application of M&S presented in this article is aimed at applying an existing feature of a 3D authoring commercial software, 3D Studio Max Design (3DSMD), by extending its capabilities and applying it to simulation of daylight for sustainable energy applications of military interest. The lighting analysis system of 3DSMD will be employed in a virtual military environment framework. Nevertheless, before the development of the simulation platform, the software plugin is validated against real sunlight measurements in order to facilitate the accuracy required for the purposes and conditions of the project. Only after the accuracy provided proof of our concept we moved on with the design of the light sensors. These are offered in the design assets palette of the software and are attached on specific areas of the soldier's uniform and equipment to assess the incorporation of PV technology. The Block Diagram of Figure 1 illustrates the adopted methodology of the Solar Soldier project.

3.1 Validation Study

As stated above, the initial step of the methodology is to validate the accuracy of the simulated daylight values against real daylight measurements on a virtual reality model that would match the actual circumstances. This took place at Brunel University campus, since light intensity data have collected for the site on a daily basis since October 2006 by the weather station of the SunnyBoy project conducted by Chowdhury et al. [29]. The next element towards the validation is the 3D model of the campus. Accurate models of this were available prior to the start of this study, which were utilized here. A more detailed presentation of the validation methodology can be found in the work of Paraskevopoulos and Tsekleves [30].

3.2 The Light Analysis Platform

Moving on to the main purpose of this study, the simulation platform is assembled according to M&S methodology very similarly to the validation study that preceded;

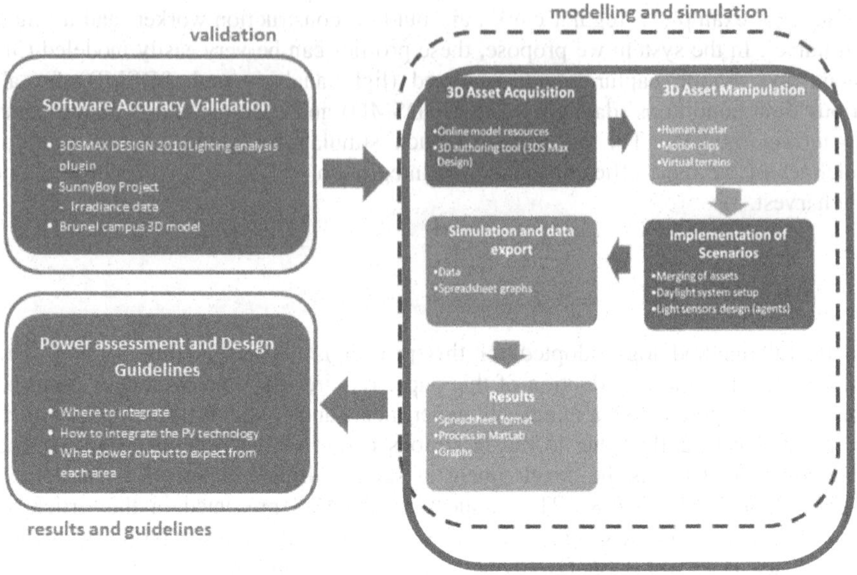

Fig. 1. Block diagram of the overall methodology

however engaging more complex set of models and daylight conditions. The software consists of a light analysis plugin, which is the tool we extended to fulfill the requirements of our study. The main purpose of the light analysis plugin of 3DSMD is for building light assessment and is mostly used by civil engineers. The aim of our study requires light analysis of non-static sensors attached on the soldier's uniform and equipment. Therefore, by the aid of the software's API we developed a script that would enable the light analysis of mobile sensors. Following is the assembly of the simulation platform. The platform consists of an animated human avatar, namely the UK infantry soldier, various terrain types and a virtual daylight system. Merging the assets together derive to a set of scenarios which are simulated under different daylight conditions. The daylight conditions vary in terms of times of the day, season and global location. The complete set of scenarios is illustrated in the tree of Figure 2. For the purposes of this project and as the VR simulations are performed for various global positions a database of irradiance values is required as an input for the VR daylight system. This data is available via the Photovoltaic Geographical Information System (PVGIS) [31]. PVGIS is an online system developed by the Joint Research centre for Energy and Renewable Energy Units. Another essential element of the simulation platform is the light sensors. These are designed and attached on the soldier, as demonstrated in Figure 3 in order to follow the motion directed on the soldier avatar by the animation clips. The set of areas examined is produced by trial simulations as well as suggestions and recommendations of the corresponding liaising expert of DSTL.

The gathering of all assets facilitates the simulations from which the results are gathered in an organized fashion and automatically stored into spreadsheet format through the developed script. Those files infer the light harvest capabilities of each of the simulated scenarios and further processed into Matlab in the following steps will

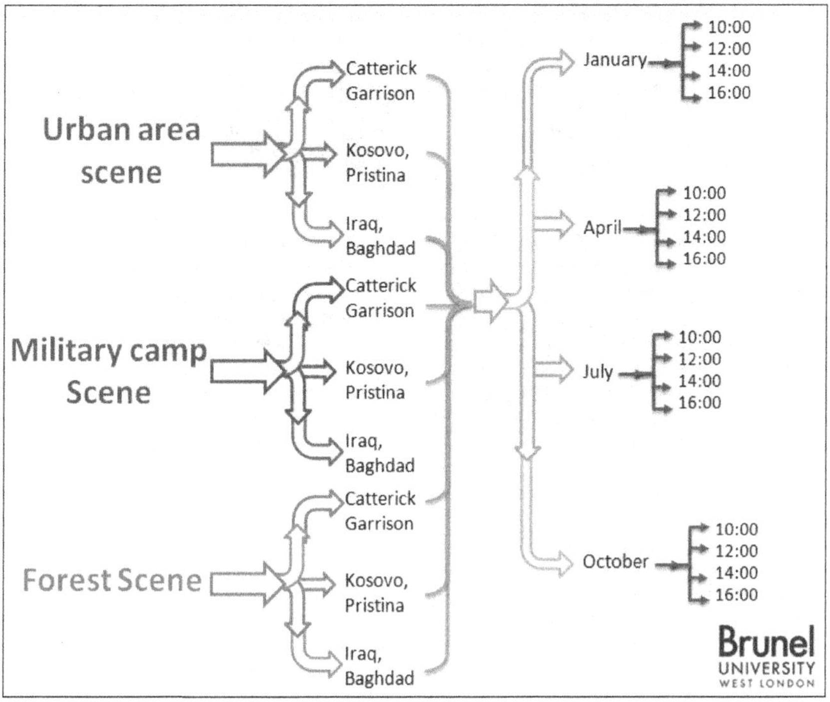

Fig. 2. Scenarios tree

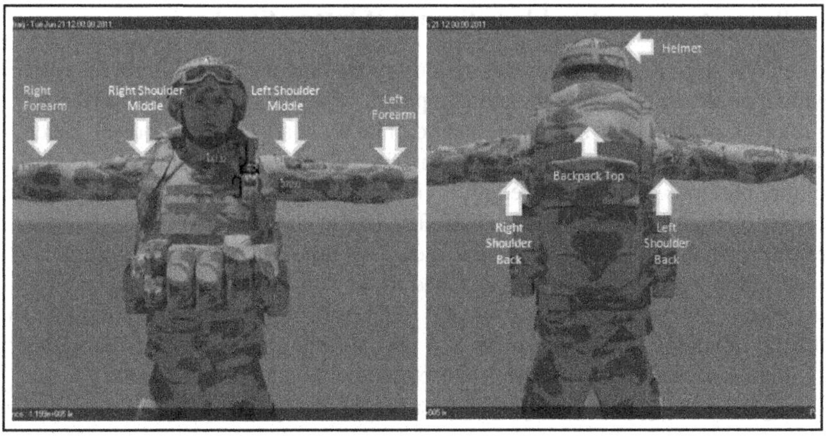

Fig. 3. Distribution of sensors. Front and Back view.

derive to the power assessment of each of the proposed areas of integration of PV technology. An extended version of that study is available in the work of Paraskevopoulos and Tsekleves [30].

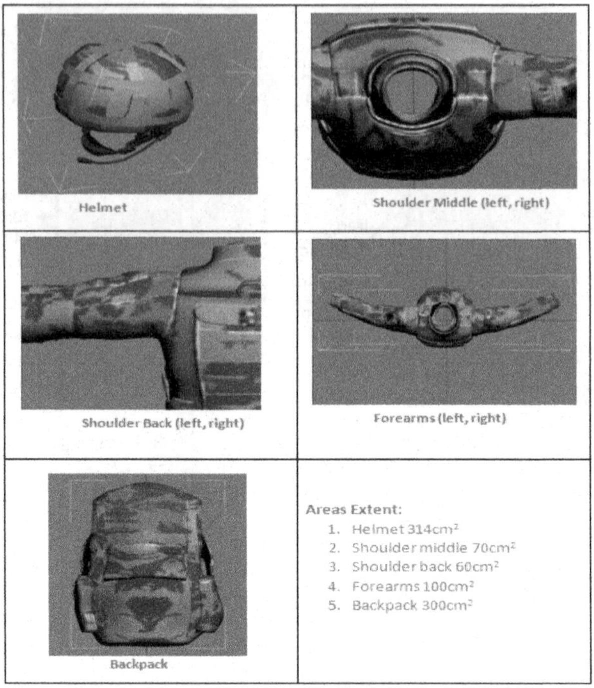

Fig. 4. Areas of integration

3.3 Power Assessment and Investigation of Integration Areas

The previous step of this project was aimed towards the analysis of light harvested on individual points on the uniform and equipment of an infantryman. Nevertheless, the requirements of the Solar Soldier project dictate an analysis of the power generation capabilities for the proposed PIPV system. Therefore a further analysis of the data from the previous section is required, in order to assess the power generation efficiency of each of the proposed areas. The extent of each area under investigation is defined as an entity by the variance it shows with the centre of each aforementioned area, on which the light sensor was attached for reasons of computational economy. In other words, the areas extend from the light sensor up to the point they show significant deviation. From that point and beyond, the area is taken as being another entity. Figure 4 depicts each of the proposed areas and their extent in cm^2:

The light intensity data is imported into Matlab using the appropriate script. Then the units are converted appropriately and further analyzed to derive to the power generation assessment of each area. Therefore, hitherto the concept of sensor is replaced by the concept of integration area. Using the value of the area (in cm^2) and the efficiency value (5%) of the prototype PV device developed for the Solar Soldier project by our corresponding project partners we are able to calculate and produce the average power graphs of each sensor (in watt) [32].

4 Results

Each step of the methodology presented above produces massive sets of results and plots, with each set illustrating and inferring specific conclusions towards the final guidelines. The size of the chapter does not facilitate the presentation of each set of results in detail. Therefore only the key results that derive to the conclusions defined by the aims and objectives of the entire Solar Soldier project will be presented here. However readers who wish to find more information on the detailed results are encouraged to look at [21], [30] and [32].

4.1 Light Intensity Assessment

As the results of the validation study [21] showed very low discrepancy between the simulated and actual light intensity values for the outdoor conditions examined, we simulated the scenarios presented on Figure 4. These simulations will manifest the classification of the various areas on the uniform and equipment in the context of higher light intensity incidents on these areas. Engineers and designers can then, by using the data of this study, have a draft blueprint of how and where to incorporate the PV device on the soldier's uniform and equipment. There are a total of 144 scenarios simulated with 8 sensors each. The stratification of areas according their light harvesting capabilities is allocated in Table 1 that follows:

Table 1. Classification of light harvesting capabilities of each candidate area of integration[30]

	Scene		
Location	Forest Scene	Military Base	Urban Area
Season:	Jan/Apr/Jul/Oct	Jan/Apr/Jul/Oct	Jan/Apr/Jul/Oct
Baghdad	1. Helmet 2. Forearms 3. Backpack 4. Shoulders Middle 5. Shoulders Back	1. Helmet 2. Forearms 3. Backpack 4. Shoulders Back 5. Shoulders Middle	1. Helmet 2. Shoulders Middle 3. Forearms 4. Backpack 5. Shoulders Back
Catterick Garrison	1. Helmet 2. Forearms 3. Backpack 4. Shoulders Middle 5. Shoulders Back	1. Helmet 2. Forearms 3. Backpack 4. Shoulders Back 5. Shoulders Middle	1. Helmet 2. Shoulders Middle 3. Forearms 4. Backpack 5. Shoulders Back
Pristina	1. Helmet 2. Backpack 3. Forearms 4. Shoulders Middle 5. Shoulders Back	1. Helmet 2. Forearms 3. Backpack 4. Shoulders Back 5. Shoulders Middle	1. Helmet 2. Shoulders Middle 3. Forearms 4. Backpack 5. Shoulders Back

The table above indicates that the change in location or season does not affect the order in the classification immensely, although the divergence between two sensors is not constant. Hence, the classification of the sensor for each scene can be simplified to the following (Table 2):

Table 2. Classification of light harvesting capabilities of integration areas [30]

Scene	Sensor Classification
Forest Scene	**1. Helmet**
	2. Forearms
	3. Backpack
	4. Shoulders Middle
	5. Shoulders Back
Military Base	**1. Helmet**
	2. Forearms
	3. Backpack
	4. Shoulders Back
	5. Shoulders Middle
Urban Area	**1. Helmet**
	2. Shoulders Middle
	3. Forearms
	4. Backpack
	5. Shoulders Back

This overall classification of the areas manifests the light harvesting capabilities of each area examined. Nevertheless, the extent of each area plays significant role in the power generation assessment. A larger area that is not very efficient might be able to provide more power due to its extent. Therefore, further analysis of the light intensity assessment is required as mentioned in the methodology. The results of the power generation assessment of each integration area that follows tackles this issue.

4.2 Power Generation Assessment

The elements that facilitate the calculations of power generation capacity for each area or integration, apart from the extent of this particular area and the light harvesting capabilities, is the efficiency of the PV device. As stated in the methodology, the efficiency of the flexible PIPV developed in the boundaries of Solar Soldier by our project partners is in the range of 5%. Therefore, the light intensity data produced by the results of the previous section appropriately process with the aforementioned setup derive to the power generation assessment of the proposed integration areas. The areas then are classified in a descending order in the context of the higher power yield. The results of this step can be used by engineers and PIPV designers as a draft blueprint of how and where to incorporate the PV devices on the infantry soldier according to each scenario. The average power values in W for each of the scenarios and for all areas are organized and presented in figures in [32]. These results were interpreted and the classification of the areas can be inferred by comparing the values of each season. Table 4 provides the overall classification in terms of power generation for the examined scenarios as well as the extent of each area in cm^2:

Table 3. Classification of power generation capabilities of integration areas

Scene	Area Classification
Forest	1. **Backpack** 300cm² 2. Helmet 314cm² 3. Forearms 100cm² 4. Shoulder Mdl 70cm² 5. Shoulder Back 60cm²
Military Base	1. **Helmet** 2. Backpack 3. Forearms 4. Shoulder Middle 5. Shoulder Back
Urban Area	1. **Helmet** 2. Backpack 3. Forearms 4. Shoulder Middle 5. Shoulder Back

5 Discussion and Guidelines

As stated in the previous sections, the usability of the PV device proposed by the Solar Soldier project is examined by liaising and interacting with DSTL. This interaction derived to a preliminary set of guidelines for the integration of PV on the uniform or equipment. The feedback we received from DSTL enabled us to shortlist the potential areas where PIPV could be integrated and thus reduced the number of light sensors to use in our simulations. For instance, it was gathered that the chest and the back of the uniform areas would not constitute good candidate areas for installation of PIPVs as they are constantly occluded by the gun and hands holding it and by the backpack respectively. The second set of guidelines derived from the case studies of the three different environments presented above. The light analysis and the power generation assessment derived the results presented in Tables 2 and 3. Those results provide a draft guideline for designers and manufactures of wearable PV devices in military environments; especially for the positioning of such devices on the uniform and equipment of the infantry soldier for the examined areas. The light harvesting assessment is slightly different to the power generation assessment as the two differ in the extent of areas examined. For instance in the light harvesting assessment, the Helmet area came first in all scenarios indicating that it is the optimum area of integration. However, owing to the actual size of this area the power generation is not the optimum in all cases according to the power generation assessment but still a very efficient area. This fact was apparent but yet not validated by any study so far. Moving on, we notice that the top of the backpack as well as the forearms as a set (right and left) qualify as important area candidates for integration even if they showed, in cases, poor light harvesting capabilities. Particularly, the backpack showed low light intensity in some scenarios but was proven to perform well in terms of power generation for the same scenarios. The rest areas qualify only as supplementary areas as they show poor performance and score low in most classifications. Combining the simulation data presented in this paper along with the feedback on the HCD ascertained from DSTL we can provide the following set of guidelines and recommendations with regards to the integration of PIPVs on the modern infantry soldier:

1. The best places on the soldier's uniform in terms of ergonomics and power generation are the helmet followed by the backpack and forearms. These three positions will provide the PV system with constant exposure to solar radiation, which can be converted to energy even when the soldier is on the move.

2. It is recommended that the entire backpack is covered with PVs as, on one hand, more PV panels can be placed and thus more energy can be harvested at all times as the soldier can easily leave the backpack in the sun whilst resting. Although the helmet yields the highest amount of light its consistent supply of power to the solar harvesting system may be stopped in certain cases. In very warm environments of operations the soldier will seek shade under natural and man-made constructions such as trees and buildings and may even take off the helmet whilst resting. This necessitates further the need to place PVs on the backpack as it can be removed and placed under the sun.

3. Integrating PV directly into the uniform is not recommended as this is washed in extremely boiling hot water. As fabric and nano-material technology evolves it may be able to interweave the solar panel nano-material onto the uniform that will withstand extremely high temperatures. Until then it is recommended that the solar panels are attached onto Velcros so that the PV can be attached and detached. This would also enable the interchange of the PV positioning on the uniform according to the environment and location of operation.

6 Conclusions

Infantry soldiers today carry a lot of electronic equipment essential for the operations, which have high power consumption requirements. This forces them to carry, in dismounted operations, several heavy and bulky batteries, which increase dramatically their total equipment load and may affect their maneuverability. Renewable energy technology such as the incorporation of PVs can substitute batteries and relieve the soldier from the physical and cognitive load or carrying and replacing batteries. This study has proposed a virtual simulation framework that mimics closely the military environment for the purposes of investigating the integration of PIPV technology on the infantry soldier. By analyzing and measuring the effectiveness of light capture on various areas of the uniform and equipment of the virtual soldier. The examined case studies covered several basic military environments as well as the several potential areas of integration of the PV device after interacting with the army personnel. After performing numerous simulations, the resulting data were organized and presented in such a manner enabling the classification of the examined areas in order of power generation efficiency. The derived overall classification infers draft yet qualitative guidelines for any designer or practitioner of wearable military applications.

References

1. British Army – Vehicles and Equipment, http://www.army.mod.uk/documents/general/285986_ARMY_VEHICLESEQUIPMENT_V12.PDF_web.pdf
2. Zyda, M.: From visual simulation to virtual reality to games. Computer 38(9), 25–32 (2005)
3. Jarvenpaa, S., Leidner, D., Teigland, R., Wasko, M.: MISQ Special Issue: New Ventures in Virtual Worlds. MIS Quarterly Call for Papers (2007)
4. Chaturvedi, A., Dolk, D.R., Drnevich, P.L.: Design Principles for Virtual Worlds. MIS Quarterly 35(3), 673–684 (2011)
5. Chryssolouris, G., Mavrikios, D., Fragos, D., Karabatsou, V.: A virtual reality- based experimentation environment for the verification of human-related factors in assembly processes. Robotics and Computer Integrated Manufacturing 16, 267–276 (2000)
6. Reece, D.: Movement Behaviour for Soldier Agents on a Virtual Battlefield. Presence 12(4), 387–410 (2003)
7. Yang, J., Rahmatalla, S., Marler, T., Abdel-Malek, K., Harrison, C.: Validation of Predicted Posture for the Virtual Human Santos^{TM}. In: Duffy, V.G. (ed.) Digital Human Modeling, HCII 2007. LNCS, vol. 4561, pp. 500–510. Springer, Heidelberg (2007)
8. Abdel-Malek, K., Yang, J., Kim, J.H., Marler, T., Beck, S., Swan, C., Frey-Law, L., Mathai, A., Murphy, C., Rahmatallah, S., Arora, J.: Development of the Virtual-Human SantosTM. In: Duffy, V.G. (ed.) Digital Human Modeling, HCII 2007. LNCS, vol. 4561, pp. 490–499. Springer, Heidelberg (2007)
9. Shiau, Y.H., Liang, S.J.: Real-time network virtual military simulation system. In: 11th International Conference Information Visualization, IV 2007, July 4-6, pp. 807–812 (2007)
10. Blount, E.M., Ringleb, S.I., Tolk, A., Bailey, M., Onate, J.A.: Incorporation of physical fitness in a tactical infantry simulation. The Journal of Defence Modelling and Simulation: Applications, Methodology, Technology (September 8, 2011)
11. Cioppa, T.M., Lucas, T.W., Sanchez, S.M.: Military applications of agent-based simulations. In: Ingalls, R.G., Rossetti, M.D., Smith, J.S., Peters, B.A. (eds.) Proceedings of the 2004 Winter Simulation Conference, pp. 171–180 (2004)
12. Bitinas, E.J., Henscheid, Z.A., Truong, L.V.: A new agent-based simulation system. Technology Review Journal, 45–58 (2003)
13. Schmorrow, D., Cohn, J., Nicholson, D. (eds.): The PSI Handbook of Virtual Environment Training and Education: Developments for the Military and Beyond. VE Components and Training Technologies, vol. 2. Praeger Security International, Westport CT (2009)
14. Parida, B., Iniyan, S., Goicm, R.: A review of solar photovoltaic technologies. Renewable and Sustainable Energy Reviews 15(3), 1625–1636 (2011)
15. Chaar, L.E., Lamont, L.A., Zein, N.E.: Review of photovoltaic technologies. Renewable and Sustainable Energy Reviews 15(5), 2165–2175 (2011)
16. Hashmi, G., Miettunen, K., Peltola, T., Halme, J., Asghar, I., Aitola, K., Toivola, M., Lund, P.: Review of materials and manufacturing options for large area flexible dye solar cells. Renewable and Sustainable Energy Reviews 15(8), 3717–3732 (2011)
17. Reich, N.H., van Sark, W.G.J.H.M., Turkenburg, W.C., Sinke, W.C.: Using CAD software to simulate PV energy yield – The case of product integrated photovoltaic operated under indoor solar irradiation. Solar Energy 84, 1526–1537 (2010)
18. Mousazadeh, H., Keyhani, A., Javadi, A., Mobli, H., Abrinia, K., Sharifi, A.: A review of prin-ciple and sun-tracking methods for maximizing solar systems output. Renewable and Sustainable Energy Reviews 13(8), 1800–1818 (2009)

19. Nguyen, D.D., Lehman, B., Kamarthi, S.: Solar Photovoltaic Array's Shadow Evaluation Using Neural Network with On-Site Measurement. In: Electrical Power Conference, EPC 2007, October 25-26, pp. 44–49. IEEE, Canada (2007)
20. Cortez, L., Cortez, J.I., Adorno, A., Muñoz-Hernandez, G.A., Cortez, E.: Study of the effects of random changes of solar radiation on energy production in a photovoltaic solar module. Canadian Journal on Electrical and Electronics Engineering 1(4) (June 2010)
21. Paraskevopoulos, I., Tsekleves, E.: Simulation of Photovoltaics for Defence and Commercial Applications by Extending Existing 3D Authoring Software - A Validation Study. In: International Conference on Simulation and Modelling Methodologies, Technologies and Applications (SIMULTECH), Netherlands, July 28-31 (2011)
22. Reinhart, C., Breton, P.F.: National Research Council Canada, Institute for Research in Construction, Ottawa, Canada (2001-2008), Harvard University, Graduate School of Design, Cambridge, MA, USA, Autodesk Canada, Media & Entertainment, Montreal, Canada, Experimental Validation of 3DS Max® Design 2009 and DAYSIM 3.0, Building Simulation (2009)
23. Reinders, A.: A design method to assess the accessibility of light on PV cells in an arbitrary geometry by means of ambient occlusion. In: Proceedings of 22nd EU Photovoltaic Solar Energy Conference and Exhibition, Milan (2007)
24. Mestre, A., Diehl, J.C.: Ecodesign and Renewable Energy: How to Integrate Renewable Energy Technologies into Consumer Products. In: Fourth International Symposium on Environmentally Conscious Design and Inverse Manufacturing, Eco Design 2005, December 12-14, pp. 282–288 (2005)
25. Reinders, A.H.M.E.: Options for photovoltaic solar energy systems in portable products. In: Proceedings of TMCE, Wuhan, China (2002)
26. Randall, J., Bharatula, N., Perera, N., Von Buren, T., Ossevoort, S., Troster, G.: Indoor Tracking using Solar Cell Powered System: Interpolation of Irradiance. In: International Conference on Ubiquitous Computing (2004)
27. Schubert, M.B., Werner, J.H.: Flexible solar cells for clothing. Materials Today 9(6) (June 2006)
28. Bruzzone, A.G.: Critical issues in advancing Modelling and Simulation. In: Keynote Speaker: International Conference on Simulation and Modelling Methodologies, Technologies and Applications (SIMULTECH), Netherlands, July 28-31 (2011)
29. Chowdhury, S., Day, P., Taylor, G.A., Chowdhury, S.P., Markvart, T., Song, Y.H.: Supervisory Data Acquisition and Performance Analysis of a PV Array Installation with Data Logger. In: IEEE Power and Energy Society General Meeting: Conversion and Delivery of Electrical Energy in the 21st Century, pp. 1–8. Institute of Electrical and Electronics Engineers, USA (2008)
30. Paraskevopoulos, I., Tsekleves, E.: Simulating the integration of photovoltaic tech-nology on the modern infantry soldier using Agent-based modelling: simulation scenarios and guidelines. Journal of Defense Modeling and Simulation. Special Issue in Simulation and Intelligent Agents to support Defense and Homeland Security (2012), doi:10.1177/1548512912458194
31. Sŭri, M.: Solar resource data and tools for an assessment of photovoltaic systems. In: Jäger-Waldau, A. (ed.) Status Report 2006, Office for Official Publications of the European Communities, Luxembourg, pp. 96–102 (2007), Data source `http://re.jrc.ec.europa.eu/`
32. Paraskevopoulos, I., Tsekleves, E.: Simulation of Photovoltaics For Defence: Applications Power Assessment and Investigation of the available integration areas of photovoltaic devices on a virtual infantry man. In: International Conference on Simulation and Modeling Methodologies, Technologies and Applications (SIMULTECH), Rome, July 28-31 (2012)

Simulation and Realistic Workloads to Support the Meta-scheduling of Scientific Workflows

Sergio Hernández, Javier Fabra, Pedro Álvarez, and Joaquín Ezpeleta

Aragón Institute of Engineering Research (I3A),
Department of Computer Science and Systems Engineering,
University of Zaragoza, Spain
{shernandez,jfabra,alvaper,ezpeleta}@unizar.es

Abstract. When heterogeneous computing resources are integrated to create more powerful execution environments, new scheduling strategies are necessary to allocate work units to available resources. In this paper we apply simulation results to schedule the execution of scientific workflows in a resource integration platform. A simulator built upon Alea and GridSim has been implemented to simulate the behaviour of the grid and cluster computing resources integrated in the platform. Simulations are generated using realistic workloads and then analysed by a meta-scheduler to decide the most suitable resource for each workflow task execution. To improve simulation results synthetic workloads are dynamically created considering the current resources state and a set of log-recorded historical executions. The paper also reports the impact of the proposed techniques when experimentally applied to the execution of the Inspiral analysis workflow.

Keywords: Grid Modelling And Simulation, Scientific Workflow, Workloads, Performance Analysis.

1 Introduction

Grid computing emerged as a paradigm for the development of computing infrastructures able to share heterogeneous and geographically distributed resources [1]. Due to their computational and networking capabilities, this type of infrastructure has turned into execution environments suitable for scientific workflows, which require intensive computations as well as complex data management. The comparison of existing grid workflow systems has shown relevant differences in the building and execution of workflows that causes experiments programmed by scientists and engineers to be strongly coupled to the underlying system responsible for their execution [2, 3]. Therefore, two of the most interesting open challenges in the field of scientific computing are the ability to program scientific workflows independently of the execution environment and the flexible integration of heterogeneous execution environments to create more powerful computing infrastructures for their execution. Both points are considered in this work.

This new generation of computing infrastructures requires new strategies of resource brokering and scheduling to facilitate the utilization of multiple-domain resources and

M.S. Obaidat et al. (eds.), *Simulation and Modeling Methodologies, Technologies and Applications*, Advances in Intelligent Systems and Computing 256,
DOI: 10.1007/978-3-319-03581-9_11,

the allocation and binding of workflow activities to them. An emerging topic in this discipline is the use of simulation environments to help in the scheduling process, improving the overall execution requirements in terms of resource usage, time and costs. Some approaches propose the use of simulation tools to evaluate different scheduling policies and then select the most suitable meta-scheduling strategy, such as GMBS [4] or P-GRADE [5]. This simulation-based evaluation is carried out previously to the execution of workflows. On the other hand, another research topic focus on the development of a novel scheduling algorithm and its execution over a simulated environment. Results are then compared with other similar approaches in order to classify the algorithm with respect to some predefined criteria. Strategies are normally compared in terms of makespan [6–8], simulation times [9] or queue times [8, 9].

Regardless of the problem to be solved, a solution based on the use of simulation techniques requires execution environment models and workloads. The first one provides a complete specification of architectures and configurations of the execution environment. Meanwhile, workloads are logs of job sets based on historical data or statistical models representing jobs to be executed in the environment. The relation between workloads and scheduling policies turns around the necessity of using a workload fitting the characteristics of jobs executed in the infrastructure in order to evaluate the suitability of a concrete scheduling algorithm in real terms. In [10], the benefits of using workloads as well as how to use them to evaluate a system are discussed. However, their use is still rather limited, due mainly to the complexity of its creation, being the process automation a difficult task. Therefore, workloads are mainly used just for the analysis of grid systems [11, 12]. Understanding these real workloads is a must for the tuning of existing grids and also for the design of future grids and cloud infrastructures.

In [13], a framework for the deployment and execution of scientific workflows whose main features are described in Section 2 was presented. This framework facilitates the flexible integration of heterogeneous computing environments, addressing the challenge of creating more powerful infrastructures. Besides, its architectural design guarantees that workflow programmers do not need to be aware of this heterogeneity. In this paper, we integrate new components into the proposed framework (a meta-scheduler and a set of simulators, more specifically) for enhancing its scheduling capabilities. Unlike other approaches, in our proposal simulation results are internally used to make scheduling decisions transparently to programmers and their workflows. Obviously, the complexity of this simulation-based scheduling is increased by the evolving nature of the underlying computing infrastructure. Internally, the new meta-scheduler can be configured to use different existing scheduling algorithms in order to carry out some optimization process depending on the parameters to be optimized. For instance, a better-execution-time algorithm or a cost reduction strategy could be used with the information provided by simulators.

The remainder of this paper is organized as follows. The main features of the developed framework in which the presented simulation approach is integrated are described in Section 2. The design and implementation of the simulator is sketched by means of the application to a cluster managed by HTCondor in Section 3. The flexibility and reuse capabilities of the component are then depicted in Section 4 by means of the integration of another grid managed by gLite. Then, the simulation approach integration is applied

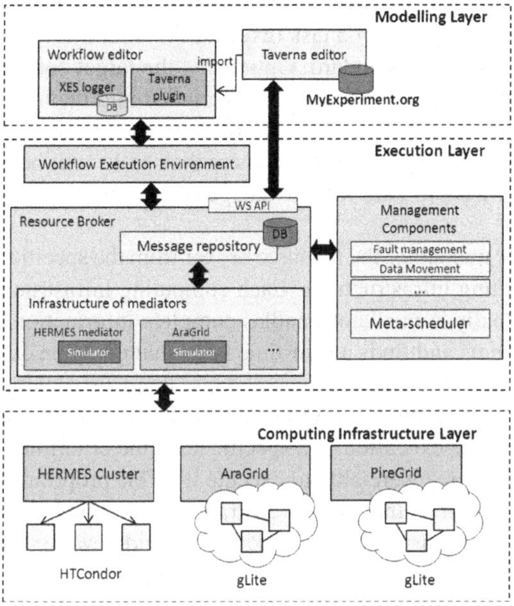

Fig. 1. Architecture of the execution platform

to the development of a real case study, the LIGO Inspiral analysis workflow in Section 5. Finally, Section 6 concludes the paper and addresses future research directions.

2 Evolving towards the Adaptable Deployment of Scientific Workflows

The proposed framework for scientific workflow deployment is able to tackle some of the open challenges in the field of grid computing. From the programmer's point of view, workflows can be programmed independently of the execution environment where the related tasks will be executed. On the other hand, the proposed framework is open and flexible from the computing resource integration point of view. First, and in accordance with this feature, it is able to simultaneously work with different distributed computing infrastructures (cluster, grid or cloud, for instance) and different middleware technologies. And, secondly, heterogeneous execution environments and framework components can be added, modified or even removed without previous announcement in a flexible and transparent way. Therefore, the combination of these features turns the proposed solution into a novel and suitable proposal in the field of scientific workflows [3].

Figure 1 shows the high-level architecture of the proposed framework. A more detailed description can be found in [13]. Let us concentrate on the process of executing workflow tasks and the architectural components involved in it.

Once a workflow has been deployed, the *workflow execution environment* is responsible for controlling its execution and submitting tasks to the *resource broker*. Submitted tasks are then stored into the *message repository* as messages that encapsulate the information needed for the execution of a task (executable, input files, etc.) These messages are described using the JSDL standard. Optionally, the target computing environment responsible for the task execution can be also included into the message. This type of tasks is called *concrete tasks*. Nevertheless, workflows will be usually programmed independently of the execution environment where their tasks will be executed (*abstract tasks*). This decision tries to take full advantage of the integration capabilities of the framework.

A set of *mediators* uncouples the resource broker from the specific and technological details of the computing infrastructures. Each computing infrastructure is represented by a mediator. Internally, a mediator handles complete information about the grid infrastructure it represents and finds at run-time tasks that could be executed by its middleware. Therefore, mediators are responsible for making dispatching decisions related to the execution of tasks. Obviously, in this dispatching model more than one mediator could compete for the execution of a specific task (the criterion would be that their corresponding middlewares were able to execute it). This proposal is an alternative to traditional solutions based on the use of a centralized task scheduler.

Finally, each mediator dispatches its tasks to the middleware that manages the infrastructure it represents, monitors the execution of these tasks and stores the results of the executed tasks into the message repository. These results will be subsequently recovered by the workflow execution environment for controlling the execution of the deployed workflow.

2.1 Enhancing the Scheduling Capabilities of the Framework

The dispatching strategy of our proposal presents a set of drawbacks: 1) performance issues related to the execution of tasks are not considered by the mediators (therefore, a task could be executed by an inappropriate infrastructure degrading the performance of the whole workflow); 2) dispatching decisions are locally adopted by each mediator and, consequently, one of them could monopolize the execution of pending tasks (this could cause unnecessary overloads on its corresponding infrastructure); and, finally, 3) the real behaviour of the computing environments and the state of their resources is also ignored by the mediators.

In order to solve the previous drawbacks and also to enhance the proposed infrastructure, a *meta-scheduler* based on simulation techniques will be integrated into the framework. Figure 2 represents the alternative process of executing workflow tasks using the meta-scheduler. Initially, abstract tasks are stored into the message repository. The meta-scheduler retrieves this type of tasks for determining where they will be finally executed. Scheduling decisions are made by simulating the execution of each task in the available computing environments and analyzing the simulation results. With these results, the task is made concrete and then submitted to the message repository, allowing the task to be executed by the selected mediator.

The interface of mediators has been extended to support this process. Now, each mediator exposes a set of operations able to simulate the execution of a task. Internally, a simulator has been integrated into each mediator for providing the required functionality. More specifically, the simulator is able to: 1) model the corresponding computing

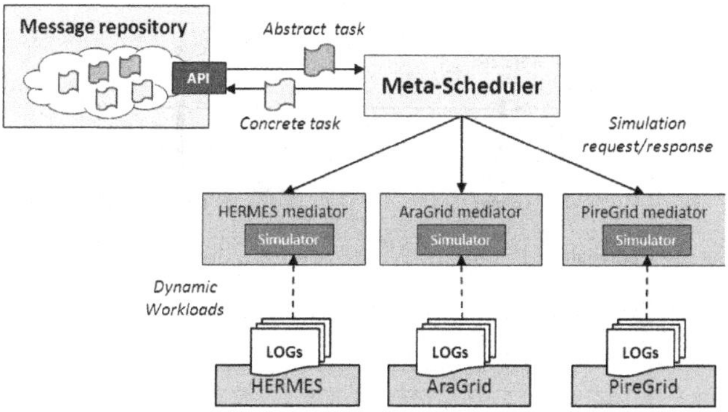

Fig. 2. Architectural components for the simulation-based scheduling

environment managed by the mediator (computing resources, memory, network bandwidth, user and scheduling internal policies, etc.); 2) select the most suitable workload for representing the real behaviour of the computing environment and the state of its resources (execution logs are used for creating these workloads); and, finally, 3) simulate the execution of tasks measuring parameters such as execution time, queuing time, etc.

In the following, the design and implementation of the simulator component is depicted. As it will be shown, it is flexible enough to allow an easy adaptation for different computing infrastructures with different scheduling policies.

3 Simulating Workflow's Execution

As stated, the simulator component has been integrated as an internal component in each mediator. Therefore, each computing infrastructure can handle different and customized simulation capabilities. Anyway, simulators are accessed through a well defined API, so adding new simulators to the framework is a guided and easy process. Also, coupling simulation components with mediators allows developers to introduce new computing infrastructures without implementing them, considering only the corresponding scheduling policy.

In the following, the design and implementation of the simulation component is sketched by means of the description of two real use cases: the HERMES cluster (in this Section) and the AraGrid multi-cluster grid (in Section 4).

3.1 Overview of the HERMES Cluster

HERMES is a cluster hosted by the Aragón Institute of Engineering Research (I3A) (http://i3a.unizar.es) managed by the HTCondor [14] middleware. In general terms, HERMES consists of 126 heterogeneous computing nodes with a total amount of 1308 cores and 2.56 TB of RAM. Also, computing nodes in HERMES are connected by Gigabit links, allowing high-speed data transfers.

The cluster is used by a vast variety of researchers. Because of this diversity, jobs submitted to HERMES have very different characteristics, in terms of execution time,

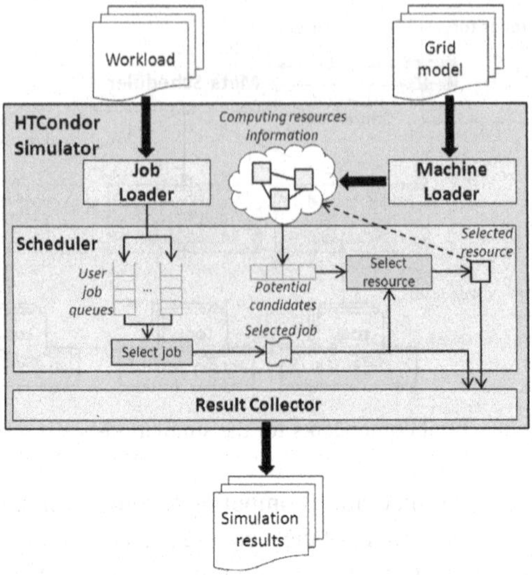

Fig. 3. Architecture of the HTCondor simulator based on Alea

size of the input data, memory consumption, etc. The analysis of workloads shows that users are unaware of load peaks or advanced configuration issues, which normally produces that experiments last extremely long, require oversized resources or are queued for long times. In this scenario, our proposal for a framework which would optimize such situations is extremely useful from both the researcher and also the system usage perspectives.

3.2 Implementation Details of the HERMES Simulator

Alea [15] has been used to implement the internal simulator in the HERMES mediator component. Alea is an event-based simulator built upon the GridSim toolkit [16]. It extends GridSim and provides a central scheduler, extending some functionalities and improving scalability and simulation speed. Alea has been designed to allow an easy incorporation of new scheduling policies and to easily extend its functionalities. Also, it provides an experimentation environment easy to configure and use, which helps in the quick and rapid development of simulators when a new infrastructure is going to be added to the system.

The original implementation of Alea has been extended to allow some HTCondor features such as user priorities, RAM requirements and preemptions. Figure 3 depicts the structure of the simulator. As shown, it consists of two input files, the *workload* and the *Grid model*, and four main internal components, the *Job Loader*, the *Machine Loader*, the *Scheduler* and the *Result Collector*, respectively. Now we will discuss the most important aspects of each component. More details of the internal components can be found on [17].

The simulator takes as input the workload containing tasks to be simulated and the model of the computing infrastructures. On the one hand, multiple *workloads* have been composed using the cluster execution logs from the last year and identifying common

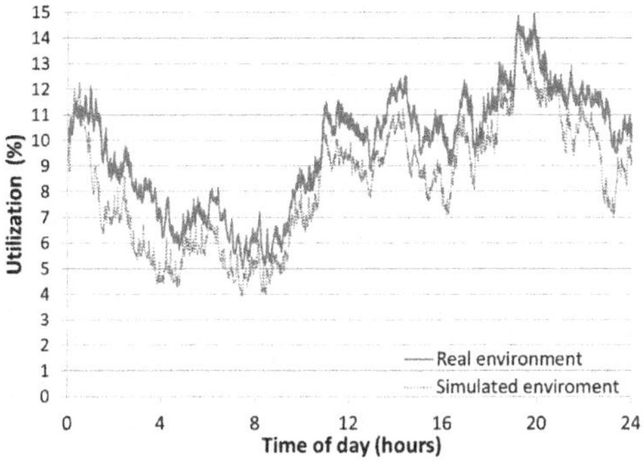

Fig. 4. HTCondor cluster utilization for the real and simulated environment

situations of resource utilization and job submission. More details on the creation of workloads are provided in subsection 5.1. On the other hand, the *Grid model* contains the hardware and network details of each computing node and includes a failure model.

Internally, the *Job Loader* reads the job descriptions and sends them to the scheduler. The *Machine Loader* is responsible for reading the resource description from a file containing the *Grid model*. The *Scheduler* has been extended to support the HTCondor scheduling policy. It works as follow: when a job reaches the scheduler, it is queued in the right user queue. This queue is ordered by the job user priority and the job arrival time. When the scheduler requests a new job to be executed, the job with the highest priority is chosen. Then, the machines with available resources (CPUs and RAM) and also the machines that could have available resources (if some running jobs are evicted) are selected as potential candidates to execute the job. The list of all potential candidates is ordered by multiple criteria (job preferences, machine preferences, etc.) to get the most suitable resource. If there is no resource available to execute the job, this is queued again and the scheduler looks for the next job. Finally, when a job and a resource have been chosen, the job is sent to the resource and its state is updated. In addition, some of the current running jobs are evicted from the selected resource if necessary to execute the new job. These evicted jobs are requeued and will be reexecuted later. Finally, the *Result Collector* stores simulations results and provide them as output.

3.3 Validation of the HERMES Simulator

The aim of the developed simulator is to be used as a decision tool at meta-scheduling level. In terms of simulation accuracy, its validation is a key issue to verify its feasibility and usefulness for this purpose [18]. Figure 4 shows a comparison of the actual cluster utilization, extracted from the logs, and the simulated utilization, obtained from the simulation of the tasks described in the workload. The comparison is presented as a daily cycle in which the horizontal axis indicates the time (in hours) and the vertical axis shows the CPU utilization rate (in percentage). As it can be observed, the simulation results are very similar to real results. Both plots follow the same trend, being the simulation utilization slightly lower. In terms of the deviation of the simulation results, an average error of 15.09% and a standard deviation of 8.03% is observed.

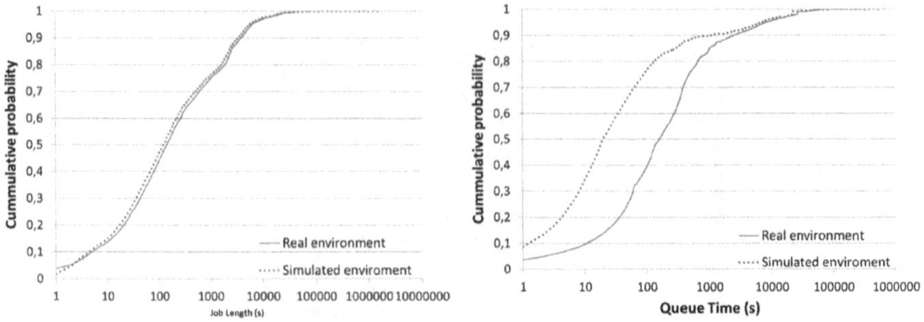

(a) Job execution time (b) Job queue time

Fig. 5. Job performance comparison between real data and simulation results in terms of: (a) job execution time, (b) job queue time

In order to validate the job performance indicator, two metrics are provided: the execution time and the queue time. Figure 5 shows the cumulative distribution function for the execution time (Figure 5(a)) and the queue time (Figure 5(b)). For the sake of clarity, the horizontal axis is shown on a log scale. Figure 5-a illustrates that job execution time is almost the same in the simulation and the real environment. In contrast, there is an important difference between queue time in both environments, which can be explained because the simulator is able to schedule a job without delay when there are available resources to execute a job. However, HTCondor middleware suffers for several delays due to different reasons such as delay notifications between distributed components, scheduling cycle duration or status update. To fix this error and reduce its influence on the results, two techniques are proposed: the first one adds a synthetic delay to the job execution time, whereas the second one adds the synthetic delay to the job queue time results. Also, how this feature can be incorporated in the simulator to get more accurate simulations is being studied for the meantime.

4 Experience Reuse for the Simulation of a gLite Grid

In this section, how a simulator for a multi-cluster grid can be easily implemented replacing some parts of the previously developed simulator is shown. Also, we illustrate the usefulness of the methodology presented to validate the simulator results.

4.1 Overview of the AraGrid Grid

AraGrid (http://www.araGrid.es/) is a research and production grid hosted by the Institute for Biocomputation and Physics of Complex Systems (BIFI) (http://bifi.es/). AraGrid consists of four homogeneous sites located at four different faculties in different geolocated cities. In total, AraGrid consists of 1728 cores and 3.375 TB of RAM. Both sites and nodes are interconnected by Gigabit links. The grid is managed by the gLite [19] middleware.

AraGrid infrastructure is oriented to long-term experiment in the fields of physics, biochemistry, social behaviour analysis, astronomy, etc. Users are more conscious of

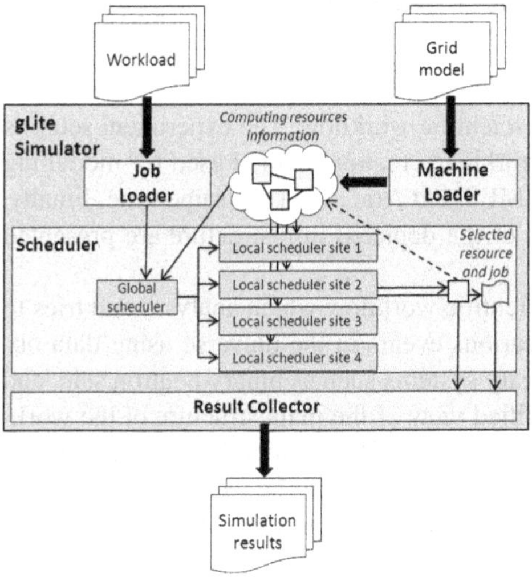

Fig. 6. Architecture of the gLite simulator with detail of a local scheduler

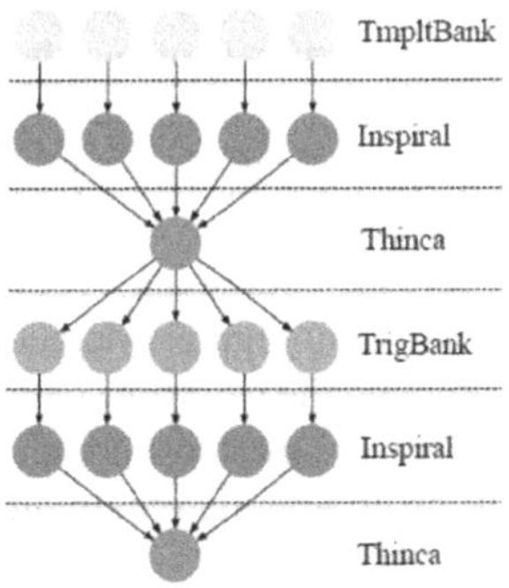

Fig. 7. Workflow of the LIGO Inspiral analysis scientific workflow

loads and resource usage, although they deploy experiments similarly to the HERMES case, getting long waiting times.

4.2 Implementation and Validation of the AraGrid Simulator

Starting from the simulator structure, the HTCondor simulator has been reused to develop a gLite simulator valid for AraGrid. This is an easy and quick implementation process, and the resulting simulator can be easily adapted to another gLite infrastructure. The reasons to implement these two simulators are twofold. On the one hand, HERMES (HTCondor) and AraGrid (gLite) are connected using high speed Gigabit links, which enhances data movement performance (which is left out of the scope of this paper). On the other hand, HTCondor and gLite are well known and widespread grid middlewares in the research community.

The only component that needed a custom adaptation to fit the behaviour of AraGrid with respect to the HERMES simulator component is the scheduler. The scheduler's policy follows a hierarchical approach, as shown in Figure 6. Jobs sent by the Job Loader are managed by the global scheduler component that sends them to the right local scheduler considering job requirements, job rank and site occupation. Meanwhile, every local scheduler uses a custom First Come First Serve (FCFS) policy. Also, it is important to consider a special case. As some sites are shared with EGI, the scheduler checks that some jobs can only be executed in the shared sites.

The resulting simulator has been integrated into the AraGrid gLite mediator. The validation of the component has been carried out following the same approach depicted in subsection 3.3. In this case, the results are more accurate than in the HERMES case. That is because AraGrid scheduling policy is easier to replicate. The average error is of 1.19% with a standard deviation of 0.85%.

5 A Case Study: Inspiral Analysis Workflow

In this section, the proposed simulation-based approach is applied in order to improve the performances of the Inspiral analysis scientific workflow. The experiment setup is detailed, with particular attention to the workload creation method used for modelling other users jobs that are executed in HERMES and AraGrid at the same time. Finally, performance results showing the benefits of the depicted infrastructure are presented and discussed.

The Inspiral Analysis Workflow is a scientific workflow which analyses and tries to detect gravitational waves produced by various events in the universe using data obtained from the coalescing of compact binary systems such as binary neutron stars and black holes [20]. Figure 7 depicts a simplified view of the main structure of the workflow. Although the workflow has a simple structure, it allows a high level of parallelism. Inspiral jobs are the most computationally intensive tasks in the workflow, generating most of the computing requirements. The remaining tasks perform checking and filter operations mostly.

Several scientific workflows management systems could be used to develop the workflow. In this case, a high level Petri nets implementation (Reference nets) has been developed using the workflow editor provided by the framework depicted in Section 2. However, the workflow implementation details are out of the scope of this paper.

5.1 Experiment Setup

The experiment setup is not specific for this experiment or case study, but it is a general setup automatically generated by the components of the framework. This design simplifies the use of the infrastructure, making the simulation-based meta-scheduling completely transparent to the user.

The process is as follows: first, when a mediator receives a simulation request, it builds a workload describing the tasks to be simulated. Next, it gets information about the state of the computing infrastructure it represents. These data are used to adapt the predefined grid model to its current situation (introducing resource failures) and to build a second workload representing the infrastructure state during the simulation. Details about the creation of this second workload are shown below. Once both workloads have been created, they are combined into one that is used as the simulation input. Then, the simulation starts its execution. Once it has finished, the simulation results are analysed by the mediator and only the information concerning the target tasks is provided to the meta-scheduler. Finally, the meta-scheduler chooses the best computing infrastructure based on data obtained from several simulations. For that purpose, the meta-scheduling policy uses a better-execution-time algorithm. Nevertheless, more complex policies involving the information obtained in previous simulations could be easily used.

The creation of the workload used to represent the state of the computing infrastructure is a key step in the simulation process. The importance of using an appropriate workload has been identified as a crucial input in some previous work [10, 12]. Using a wrong workload can cause the simulation results not to correspond to the actual behaviour of the involved grids. These research papers propose the generation of a single workload based on historical information from a long period of time and only considering representative periods (e.g. the peak hours during weekdays in job-intensive months). It is assumed that the longer the observation period is, the more representative is the workload, which allows tuning the grid in extreme situations [10]. Nevertheless, for simulation purposes these approaches are not valid because the state of the resources must be considered as the simulation starts. If an average or extreme workload is used, it is very likely to get very inaccurate results that lead to wrong scheduling decisions. Our proposal is to build several representative workloads with different situations depending on the state of the infrastructure (e.g. low load, average load and high load) and date. Therefore, the current computing infrastructure state is obtained before starting a simulation and used to select the most suitable workload. Also, the recovered infrastructure information, including currently running jobs and queued jobs, is added at the beginning of the workload, obtaining this way a workload describing the current infrastructure state and its evolution.

The model proposed in [21] has been used for workload creation. This model incorporates the notions of different users and jobs in *Bag-of-Tasks* (BoTs) to the Lublin-Feitelson model [22]. Due to the fact that the HERMES and AraGrid analysis has shown that more than 90% of jobs belongs a BoT and a few users are responsible for the entire load, this model is suitable for modelling jobs in the available infrastructures.

5.2 Analysis of the Results

To prove the usefulness of the proposed approach, the workflow has been executed for a whole day (24 hours). Figure 8 depicts the CPU load observed in HERMES and AraGrid during the experiment. Note that HERMES load is different from the one sketched in figure 4. That is because the load in Figure 4 is an average load extracted from the execution log corresponding to the whole last year, whereas Figure 8 shows the cluster load on a particular day. As it can be observed, both computing infrastructures have different load models. Throughout the day there are better periods of time for submitting jobs to HERMES (mostly at early morning and night), and times more appropriate to submit jobs to AraGrid (in the afternoon). However, this is not the only criterion to be considered as the performance of a grid infrastructure depends on many factors.

The use of simulation as a decision tool for meta-scheduling deals with this complexity and improves the performance obtained in the workflow execution as shown in Figure 9. The figure shows the total execution time for each stage of the Inspiral workflow entirely executed in each computing infrastructure (HERMES on the left bar and AraGrid on the right bar) and using the framework with the simulation-based meta-scheduling strategy (center bar) depicted previously. The results show that the use of the proposed approach leads to an improvement of 59% in HERMES execution time and a 111% in AraGrid execution time.

Regarding the simulation overhead in terms of execution time, the simulation process for HERMES is more complex (more iterative structures) and can take up to 3-4 minutes for a bag of 10000 tasks, whereas for gLite it takes one minute approximately. Therefore, simulation times are insignificant in comparison to execution times of each stage. Also, data movement has been measured. For the sake of clarity, as HERMES

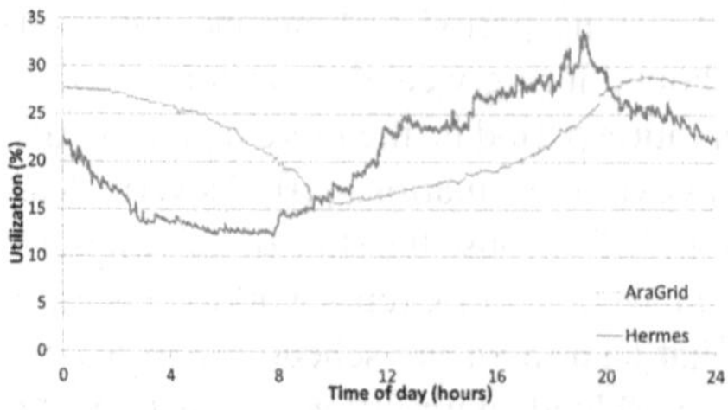

Fig. 8. HERMES and AraGrid utilization (in percentage) observed during workflow execution

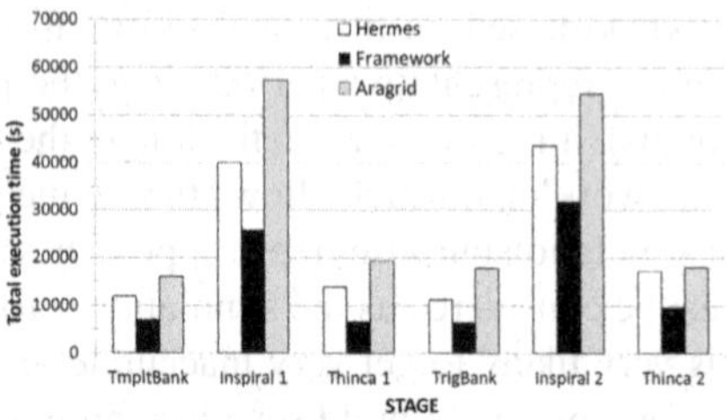

Fig. 9. Experimental results for LIGO Inspiral analysis workflow

and AraGrid are connected by a Gigabit link, these times are small and can be avoided in the calculation of the overall execution time.

6 Conclusions

In this paper, we have proposed the use of simulation techniques to implement a meta-scheduler, which has been integrated into a framework able to execute scientific workflows in multiple heterogeneous computing infrastructures. The key of our proposal is using realistic workloads describing the current state of the computing infrastructures and its likely future evolution based on analysis of historical data. Using these workload, the developed simulator provides accurate information about expected execution time or queue times, for instance. This information is used by the meta-scheduler to improve decision making getting a performance increase in the execution of workflows, as shown in the Inspiral analysis workflow.

Currently, the proposed simulation component is being extended to support the dynamic building of workloads. Also, the addition of new features in the simulator is being addressed in order to get more accurate queue times in simulations. Finally, the incorporation of complex meta-scheduling approaches that can use the information provided by the simulation process will be studied.

Acknowledgements. This work has been supported by the research project TIN2010-17905, granted by the Spanish Ministry of Science and Innovation.

References

1. Foster, I., Kesselman, C.: The Grid 2: Blueprint for a New Computing Infrastructure. Morgan Kaufmann Publishers Inc., San Francisco (2003)
2. Rahman, M., Ranjan, R., Buyya, R., Benatallah, B.: A taxonomy and survey on autonomic management of applications in grid computing environments. Concurrency and Computation: Practice and Experience 23, 1990–2019 (2011)
3. Yu, J., Buyya, R.: A taxonomy of scientific workflow systems for grid computing. SIGMOD Record 34, 44–49 (2005)

4. Kertész, A., Kacsuk, P.: GMBS: A new middleware service for making grids interoperable. Future Generation Computer Systems 26, 542–553 (2010)
5. Kacsuk, P., Kiss, T., Sipos, G.: Solving the grid interoperability problem by P-GRADE portal at workflow level. Futur. Gener. Comp. Syst. 24, 744–751 (2008)
6. Hamscher, V., Schwiegelshohn, U., Streit, A., Yahyapour, R.: Evaluation of Job-Scheduling Strategies for Grid Computing. In: Buyya, R., Baker, M. (eds.) GRID 2000. LNCS, vol. 1971, pp. 191–202. Springer, Heidelberg (2000)
7. Abraham, A., Liu, H., Zhang, W., Chang, T.G.: Scheduling Jobs on Computational Grids Using Fuzzy Particle Swarm Algorithm. In: Gabrys, B., Howlett, R.J., Jain, L.C. (eds.) KES 2006, Part II. LNCS (LNAI), vol. 4252, pp. 500–507. Springer, Heidelberg (2006)
8. Yu, Z., Shi, W.: An Adaptive Rescheduling Strategy for Grid Workflow Applications. In: IEEE International Parallel and Distributed Processing Symposium, IPDPS 2007, pp. 1–8 (2007)
9. Ludwig, S.A., Moallem, A.: Swarm Intelligence Approaches for Grid Load Balancing. Journal of Grid Computing 9, 279–301 (2011)
10. Feitelson, D.G.: Workload Modeling for Performance Evaluation. In: Calzarossa, M.C., Tucci, S. (eds.) Performance 2002. LNCS, vol. 2459, pp. 114–141. Springer, Heidelberg (2002)
11. Iosup, A., Epema, D.H.J.: Grid Computing Workloads. IEEE Internet Computing 15, 19–26 (2011)
12. Li, H., Groep, D., Wolters, L.: Workload characteristics of a multi-cluster supercomputer. In: Feitelson, D.G., Rudolph, L., Schwiegelshohn, U. (eds.) JSSPP 2004. LNCS, vol. 3277, pp. 176–193. Springer, Heidelberg (2005)
13. Fabra, J., Hernández, S., Álvarez, P., Ezpeleta, J.: A framework for the flexible deployment of scientific workflows in grid environments. In: Proceedings of the Third International Conference on Cloud Computing, GRIDs, and Virtualization, Cloud Computing 2012, pp. 1–8 (2012)
14. HTCondor Middleware, `http://research.cs.wisc.edu/htcondor/` (accessed March 5, 2013)
15. Klusáček, D., Rudová, H.: Alea 2 – Job Scheduling Simulator. In: Proceedings of the 3rd International ICST Conference on Simulation Tools and Techniques, SIMUTools 2010 (2010)
16. Sulistio, A., Cibej, U., Venugopal, S., Robic, B., Buyya, R.: A toolkit for modelling and simulating data Grids: an extension to GridSim. Concurrency and Computation: Practice and Experience 20, 1591–1609 (2008)
17. Hernández, S., Fabra, J., Álvarez, P., Ezpeleta, J.: A Simulation-based Scheduling Strategy for Scientific Workflows. In: Proceedings of the 2nd International Conference on Simulation and Modeling Methodologies, Technologies and Applications. SIMULTECH 2012, pp. 61–70 (2012)
18. Sargent, R.G.: Verification and validation of simulation models. In: Proceedings of the 2010 Winter Simulation Conference, WSC 2010, pp. 166–183 (2010)
19. gLite Middleware, `http://glite.cern.ch/` (accessed March 5, 2013)
20. Taylor, I.J., Deelman, E., Gannon, D.B., Shields, M.: Workflows for e-Science: Scientific Workflows for Grids. Springer-Verlag New York, Inc., Secaucus (2006)
21. Iosup, A., Sonmez, O., Anoep, S., Epema, D.: The performance of bags-of-tasks in large-scale distributed systems. In: Proceedings of the 17th International Symposium on High Performance Distributed Computing, HPDC 2008, pp. 97–108 (2008)
22. Lublin, U., Feitelson, D.G.: The workload on parallel supercomputers: modeling the characteristics of rigid jobs. Journal of Parallel and Distributed Computing 63, 1105–1122 (2003)

Dynamic Simulation of the Effect of Tamper Resistance on Opioid Misuse Outcomes

Alexandra Nielsen and Wayne Wakeland

Systems Science Graduate Program, Portland State University,
1604 S.W. 10th Av., Portland, OR 97201, U.S.A.
{alexan3,wakeland}@pdx.edu

Abstract. The objective of the study was to develop a system dynamics model of the medical use of pharmaceutical opioids, and the associated diversion and nonmedical use of these drugs. The model was used to test the impact of the a tamper resistance intervention in this complex system. The study relied on secondary data obtained from the literature and from other public sources for the period 1995 to 2008. In addition, an expert panel provided recommendations regarding model parameters and model structure. The behavior of the resulting systems-level model compared favorably with reference behavior data. After the base model was tested, logic to simulate the replacement of all opioids with tamper resistant formulations was added and the impact on overdose deaths was evaluated over a seven-year period, 2008-2015. Principal findings were that the introduction of tamper resistant formulations unexpectedly increased total overdose deaths. This was due to increased prescribing which counteracted the drop in the death *rate*. We conclude that it is important to choose metrics carefully, and that the system dynamics modelling approach can help to evaluate interventions intended to ameliorate the adverse outcomes in the complex system associated with treating pain with opioids.

Keywords: Prescription Drug Abuse, System Dynamics Modeling, Opioid Analgesics, Public Health.

1 Introduction

A dramatic rise in the nonmedical use of pharmaceutical opioid pain medicine has presented the United States with a substantial public health problem [6]. Despite the increasing prevalence of negative outcomes, such as nonfatal and fatal overdoses, nonmedical use of pharmaceutical opioids remains largely unabated by current policies and regulations (see [8]. Resistance to policy interventions likely stems from the complexity of medical and nonmedical use of pharmaceutical opioids, as evidenced by the confluence of the many factors that play a role in medical treatment, diversion, and abuse of these products in the US.

Complex social systems are well known to resist to policy interventions, often resulting in unintended consequences or unanticipated sources of impedance [24]. These undesirable outcomes can result from our inability to simultaneously consider a

M.S. Obaidat et al. (eds.), *Simulation and Modeling Methodologies, Technologies and Applications*,
Advances in Intelligent Systems and Computing 256,
DOI: 10.1007/978-3-319-03581-9_12,

large number of interconnected variables, feedback mechanisms, and complex chains of causation [10]. Prescription opioid use, diversion, and nonmedical use constitute a complex system with many interconnected components, including prescribers, pharmacists, persons obtaining opioids from prescribers for medical use, persons obtaining drugs from illicit sources, and people giving away or selling drugs. Interactions among these actors result in chains of causal relationships and feedback loops in the system. For example, prescribing behaviors affect patients' utilization of opioids; adverse consequences of medical and nonmedical use influence physicians' perceptions of the risks associated with prescribing opioids; and physicians' perception of risk affects subsequent prescribing behaviors [19,14].

This paper presents a system dynamics model which simulates the system described above. The model is designed to provide a more complete understanding of how medical use, nonmedical use, and trafficking are interrelated, and to identify points of high leverage for policy interventions to reduce the adverse consequences associated with the epidemic of nonmedical use. An intervention corresponding to the introduction of relatively less-abusable, tamper-resistant formulation is simulated, and possible downstream effects are highlighted.

Policymakers striving to ameliorate the adverse outcomes associated with opioids could benefit from a systems-level model that reflects the complexity of the system and incorporates the full range of available data. Such a model could be used to study the possible effectiveness of a tamper resistant drug.

2 Background

Between 1999 and 2006, the number of U. S. overdose deaths attributed to opioids tripled–increasing more than five-fold among youth aged 15 to 24 [26]–signaling the onset of a major public health concern. Overdose deaths involving opioid analgesics have outnumbered cocaine and heroin overdoses since 2001 [3], and estimates from the 2009 National Survey on Drug Use and Health (NSDUH) suggest that 5.3 million individuals (2.1% of the U.S. population aged 12 and older) used opioids for nonmedical purposes within the previous month [23]. Earlier data from NSDUH suggest that the rate of initiating nonmedical usage increased drastically from 1994 to 1999 [21], and has continued at high rates, with over 2 million individuals reporting the initiation of nonmedical use of pain relievers in 2009 [23]. Recent increases in prescribing opioids stem in part from increases in chronic pain diagnosis and the development of highly effective long-acting pharmaceutical opioid analgesics.

One problem that arose with these new long-acting formulations was the ease with which they could be tampered with to enhance the effects when used non-medically [15,20,28]. To combat this trend in abuse, many manufacturers are developing or have already developed opioid formulations that use a physical barrier to resist tampering, or a mix of pharmacologically active ingredients that deter abuse [16]. Post-marketing studies have been conducted on tamper resistant opioids currently on the market that imply lower abuse rates (e.g. [1] for OxyContin), but the long term effects of large scale adoption of tamper resistant opioids on the treatment of pain and opioid abuse are still unknown.

3 System Dynamics Simulation Model

The current work features a system dynamics simulation model that represents the fundamental dynamics of opioids as they are prescribed, trafficked, used medically and nonmedically, and involved in overdose mortality. The model was developed over a two-year period through collaborative efforts of a modeling team and a panel of pain care and policy experts. The SD modeling approach uses a set of differential equations to simulate system behavior over time. SD models are well suited to health policy analyses involving complex chains of influence and feedback loops that are beyond the capabilities of statistical models [25], and have been successfully applied to the evaluation of policy alternatives for a variety of public health problems [2,11,13,18]. The SD approach can help identify points of high leverage for interventions, as well as possible unanticipated negative consequences of those interventions. This provides policymakers with information that is not available from research focused on individual aspects of a system [25].

The model was developed iteratively, starting with a brainstorming session that included both subject matter experts (SMEs) and computer simulation team members and encompasses the dynamics of the medical treatment of pain with opioids, the initiation and prevalence of nonmedical usage; the diversion of pharmaceutical opioids from medical to nonmedical usage; and overdose fatalities. Discussion of each sector includes a description of empirical support, a narrative on model behavior, and a stock and flow diagram showing model structure. Bracketed numbers in the text correspond to specific points in the diagrams. The model contains 40 parameters, 41 auxiliary variables, and 7 state variables, as well as their associated equations and graphical functions.

3.1 Nonmedical Use Sector

12%-14% of individuals who use opioids nonmedically meet the criteria for opioid abuse or dependence [5], either of which is associated with a high frequency of nonmedical use. Extrapolation from heroin findings indicates that higher frequency opioid use is associated with a significantly higher mortality rate (WHO; see [7]; and [12]) and supports a distinction between two subpopulations of nonmedical users (low- and high-frequency) in this model sector.

As illustrated in Figure 1, a percentage of the US population {1} is assumed to initiate nonmedical use each year {2}, all of whom start out in a stock of 'low-frequency nonmedical users,' and a small percentage of whom advance to a stock of 'high-frequency nonmedical users' {3} during each subsequent year. The total number of individuals using opioids nonmedically {4} is divided by the current number of individuals in the US who are using other drugs nonmedically {5} to calculate the relative popularity of opioids for nonmedical use {6}. As the popularity of using opioids nonmedically increases, the rate of initiation increases, creating a positive feedback loop that *ceteris paribus* would result in an exponential increase in the rate of initiation.

Nonmedically used opioids are obtained through many routes, but of key interest for the current research is opioid 'trafficking' (i.e., buying or selling) via persons who

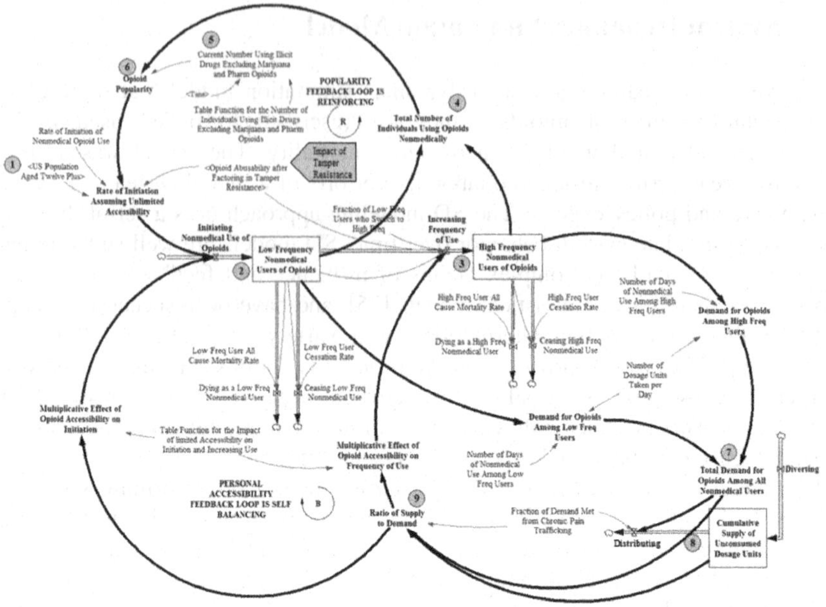

Fig. 1. Stock and flow diagram of the Nonmedical Use Sector. Circled numbers correspond to bracketed notations in the text.

are receiving these products ostensibly for treatment. Extrapolation of results from the 2006 NSDUH survey [22] suggests around 25% of the nonmedical demand for opioids is met via trafficking.

In the model, demand for opioids is calculated from the number of individuals in low- and high-frequency populations {7}. As noted above, 25% of demand is assumed to be met by trafficking {8}, with the rest coming from sources not modeled explicitly (mostly interpersonal sharing among friends and relatives, per [22]). When the trafficking supply is ample relative to demand, the rate of initiation {2} and the rate of advancement from low-frequency to high-frequency use {3} are assumed to be somewhat enhanced. When the trafficking supply is limited, however, rates of initiation and advancement are assumed to decrease dramatically. The ratio of supply to demand {9} indicates the degree to which opioids are accessible for nonmedical use. As the populations of nonmedical users increase beyond what trafficking can support, accessibility becomes limited, decreasing initiation and advancement. This creates a negative feedback loop that eventually equilibrates the otherwise exponential increase in nonmedical use driven by the popularity feedback loop.

3.2 Medical Use Sector

Increases in opioid abuse and addiction have led to regulatory policies which have lead many physicians to avoid prescribing opioids to patients out of fear of overzealous regulatory scrutiny [14]. Or, they may decrease the amount of opioids they prescribe, and shift their prescribing towards short-acting opioid products because long-acting

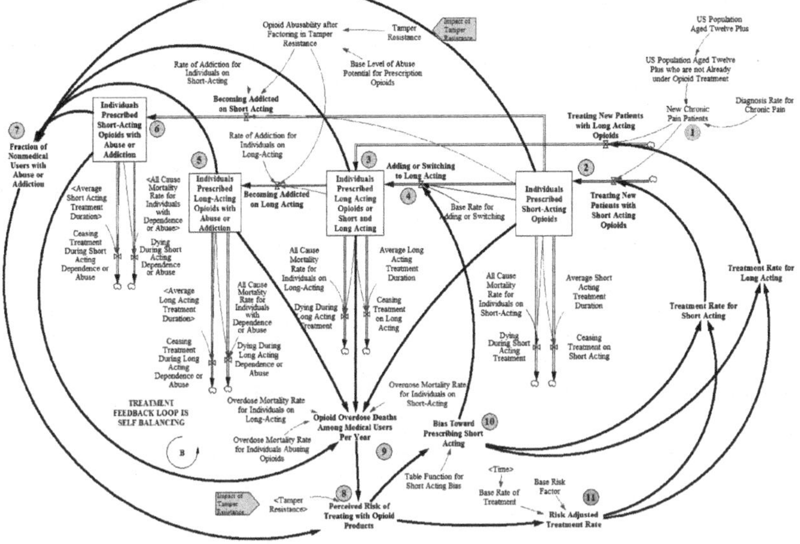

Fig. 2. Stock and flow diagram of the Medical Use Sector. Circled numbers correspond to bracketed notations in the text.

opioids have been shown to have a higher rate of abuse than immediate-release opioid analgesics when abuse rates are normalized for the number of individuals exposed [4]. Thus, physicians exhibit more caution in prescribing long-acting opioids [19].

As illustrated in Figure 2, the system model assumes that a proportion of the US population is diagnosed with a chronic pain condition each year {1}. A fraction of these people are subsequently treated with either short-acting {2} or long-acting {3} opioid formulations, and become members of one of the stocks (populations) of patients under opioid treatment ostensibly for chronic pain. Patients who begin treatment with short-acting formulations may cease treatment if their condition improves, or they may switch to long-acting formulations if their pain conditions appear to worsen {4}.

Each year some individuals move from the stocks of 'individuals receiving opioids' {2-3} to the stocks of 'individuals receiving opioids with abuse or addiction' {5-6}. The fraction of opioid-prescribed individuals with abuse or addiction {7} influences physicians' perception of the risk involved in opioid prescribing {8}, as does the total number of overdose deaths among medical users each year {9}. As physicians perceive higher levels of risk {8} they become increasingly biased toward prescribing short-acting formulations {10}, and their overall rates of opioid prescribing decrease {11}. Because of these balancing feedback loops, the increase in the amount of abuse and addiction {7} is slowed. Physicians' responses to increasing rates of abuse, addiction, and overdose effectively move the model towards a state of dynamic equilibrium.

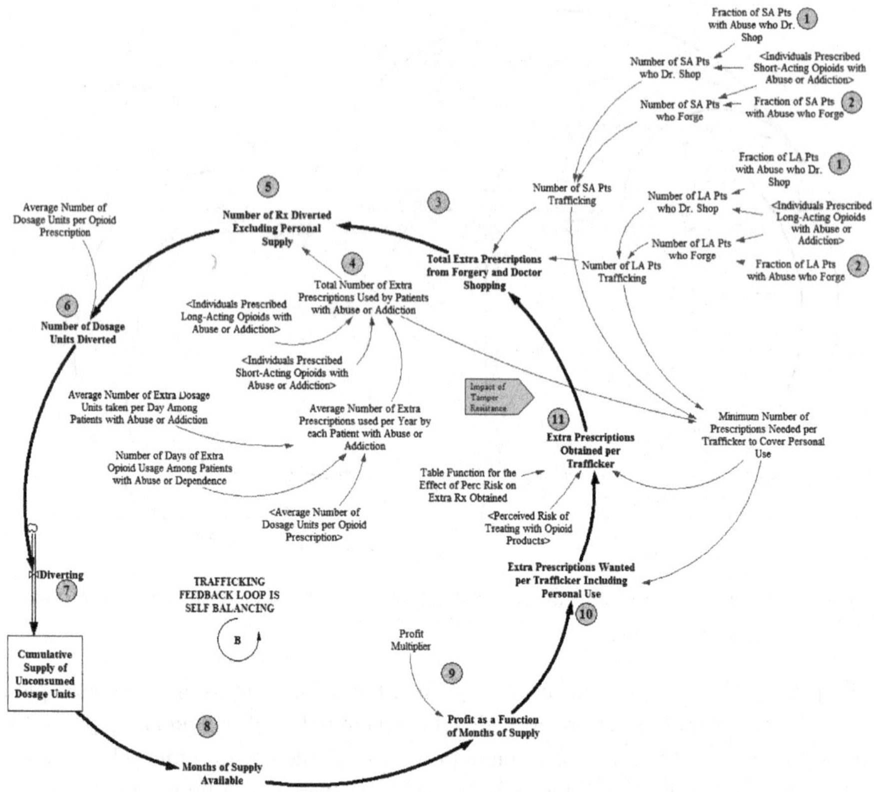

Fig. 3. Stock and flow diagram of the Trafficking Sector

3.3 Trafficking Sector

Findings from Manchikanti et al. [17] indicate that 5% of chronic pain patients engage in doctor shopping and around 4% engage in forgery. In the model, forgery and doctor shopping by persons interacting with prescribers are assumed to be exhibited entirely by those with abuse or addiction, which constitute around 7% of individuals receiving opioid prescriptions for chronic pain. This would imply that about 70% of persons with abuse or addiction (5 out of 7) engage in doctor shopping and over half (4 out of 7) engage in forgery. More research is needed to support these parameters and the associated logic.

As shown in Figure 3, a fixed fraction of persons with abuse or addiction are assumed to engage in trafficking each year, including doctor shopping {1} and forgery {2}. The number of extra prescriptions acquired {3} is calculated as a product of (a) the total number of individuals engaging in trafficking and (b) the number of extra prescriptions obtained per trafficker {11}. Some proportion of these excess prescriptions is assumed to be used by the traffickers themselves, rather than diverted to other nonmedical users {4}. This number is calculated as a product of (a) the number of individuals with abuse or addiction and (b) the average number of extra prescriptions used per year by such individuals. The number of prescriptions that are used "in excess" by medical users is subtracted from the number of extra prescriptions acquired. The remainder is converted to dosage units {5} and assumed to be diverted to nonmedical users {6}.

Trafficked opioids accumulate in a stock of dosage units {7} that are consumed according to demand from the nonmedical use sector. Supply can also be expressed as 'months of supply available' {8}, which indicates the extent to which the trafficked supply is able to meet the demand at any given time. When the supply of opioids becomes limited, a profit motive emerges {9} and motivation to forge and doctor shop increases. When supply is large compared to demand, motivation to commit fraud for the purpose of sale is small. As this motivation fluctuates, the number of extra prescriptions each trafficker would like to obtain {10} also changes. But the number of prescriptions that can be successfully trafficked is attenuated by cautious dispensing when perceived risk is high among physicians and pharmacies {11}, which creates a balancing feedback loop that stabilizes the amount of trafficking.

4 Model Testing

The model was tested in detail to determine its robustness and to gain an overall sense of its validity. As is often the case with system dynamics models, the empirical support for some of the parameters was limited, as indicated in Tables 1-3 in the Appendix. System Dynamics models are generally more credible when their behavior is not overly sensitive to changes in the parameters that have limited empirical support. Therefore, to determine sensitivity of primary outcomes to changes in parameter values, each parameter in turn was increased by 30% and then decreased by 30%, and the outcome was recorded in terms of cumulative overdose deaths. One parameter with limited empirical support which has a substantial influence on model behavior is the impact of limited accessibility on the initiation rate. Another parameter, the rate of initiation of nonmedical use, also strongly influenced model behavior but is less worrisome because it *does* have sufficient empirical support. Because model testing revealed a high degree of sensitivity to certain parameters for which empirical support is limited, study results should be considered exploratory and viewed with caution.

When empirical support was available, model outputs were validated against reference data for the historical period. While this reference period is relatively short, the model does fit the data well, as shown in Figure 4, 5, and 6.

Figure 4 shows the number of prescription opioid overdose deaths from a baseline model run for the historical period overlaid on a plot of the reported number of overdose deaths obtained from the CDC multiple cause of death database.

The total opioid-related deaths resulting from all types of medical and nonmedical use has been reported to be 13,755 in 2006 and approximately 14,000 in 2007. Both the model and the data exhibit sigmoidal growth and with a Mean Average Percent Error of 22%, suggesting a moderate fit for this metric.

Figure 5 shows the total number of individuals using prescription opioids non-medically overlaid on reference data for the historical time period. The graph of historical data is not smooth, but again, the general pattern of growth is S-shaped. The graphical output from a baseline model run is a smooth S-shaped curve that is a good fit for the limited time series data available (MAPE = 9.1%).

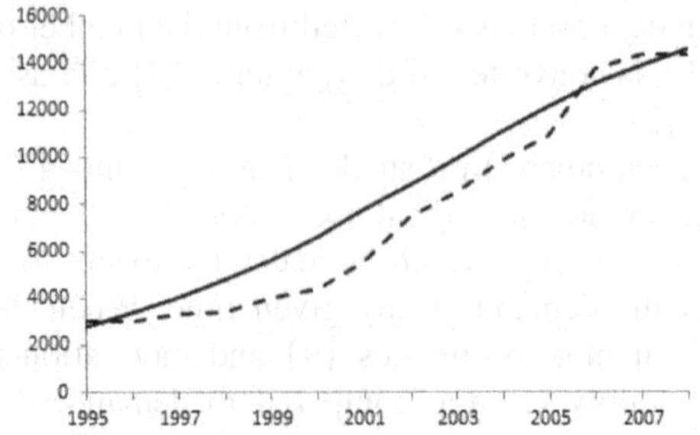

Fig. 4. Model output (solid) versus reference behavior (dashed) of the total prescription opioid overdose deaths per year (MAPE 22%)

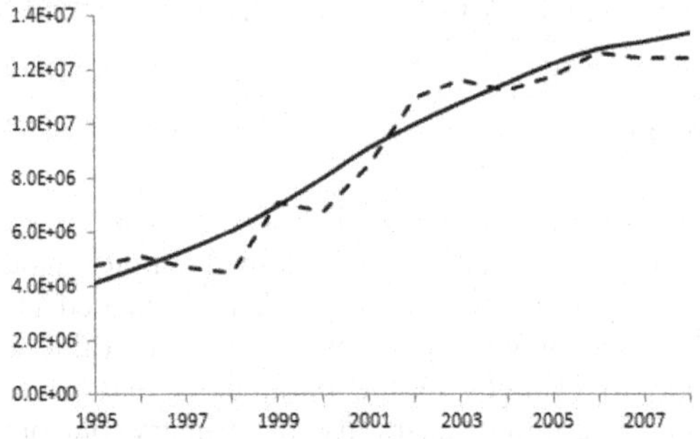

Fig. 5. Model output (solid) versus reference behavior (dashed) of the total number of nonmedical users of prescription opioids (MAPE 9.9%)

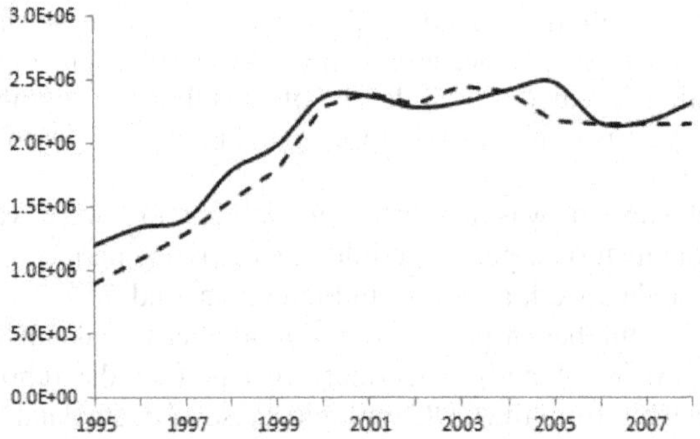

Fig. 6. Model output (solid) versus reference behavior (dashed) of the total number of individuals initiating nonmedical opioid use per year (MAPE 9.9%)

Figure 6 gives model output and reference data for the number of individuals initiating nonmedical use of prescription opioids. The reference behavior pattern here is highly non-linear and the baseline model run matches the reference behavior pattern well (MAPE = 9.9%).

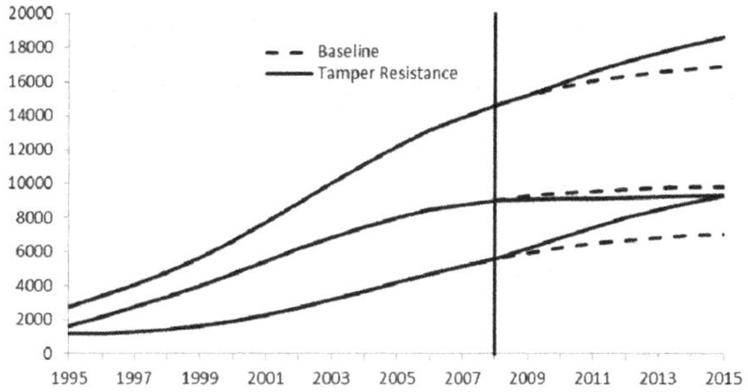

Fig. 7. Baseline model run (dashed) and the effect of the simulated tamper resistance intervention (solid) on opioid overdose deaths

Overall, model results closely track the complex patterns graphs of empirical data despite exhibited. Thus, baseline results were deemed sufficiently plausible to proceed with intervention analysis.

5 Results

To test the intervention, the model time horizon was extended to 2015 and a baseline run was made. The intervention was then formulated and tested.

5.1 Tamper Resistant Formulation

Logic representing the introduction of a tamper resistant drug formulation was added to the model. The model was run over a time period of twenty years, which was divided into an historical period from 1995 to 2008, and an evaluation period from 2008 to 2015. The intervention was represented as simple toggle switch that doubled beneficial parameters and/or halved harmful parameters. The response of the model to this simulated intervention is shown in Figure 5.

This intervention of a new drug formulation being introduced in 2008 was implemented as a 50% decrease in: 1) the rate of abuse or addiction among opioid-treated persons, 2) the fraction of low-frequency nonmedical opioid users who become high-frequency users per year, 3) the rate of initiation of nonmedical opioid use, and 4) the perceived risk of opioid abuse amongst prescribers (this increased the prescribing rates for all opioids). Figure 7 shows that this change caused an increase in the total number of overdose deaths in the model, due to a sizable increase in deaths among medical users and a small decrease in deaths among nonmedical users. This was not expected.

Figure 8 and 9 explain why this happened. Figure 8 shows that the number of individuals receiving treatment increased sharply, and this increase, even when coupled with lower death rates, led to the net increase in the total number of overdose deaths compared to baseline (Figure 7).

Figure 9 shows the number of deaths divided by number of individuals receiving opioid treatment (and then divided by 10,000 to yield an indicator in the 0 to 10 range). This indicator, which was beginning to increase as of 2008, declined as a result of the intervention, especially in the nonmedical sector. So, although the fraction of deaths among patients did decrease as anticipated, the rise in patient populations (due to the lower risk perception associated with tamper resistant formulations amongst prescribers) obscured the benefits of the lower death fraction.

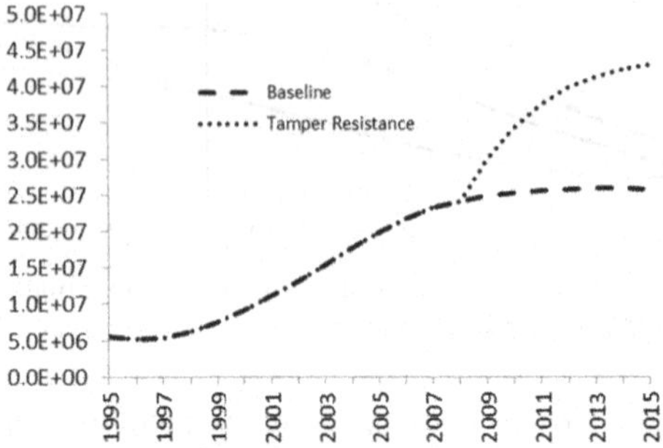

Fig. 8. Tamper resistance results in a dramatic increase in the number of patient who receive opioid therapy

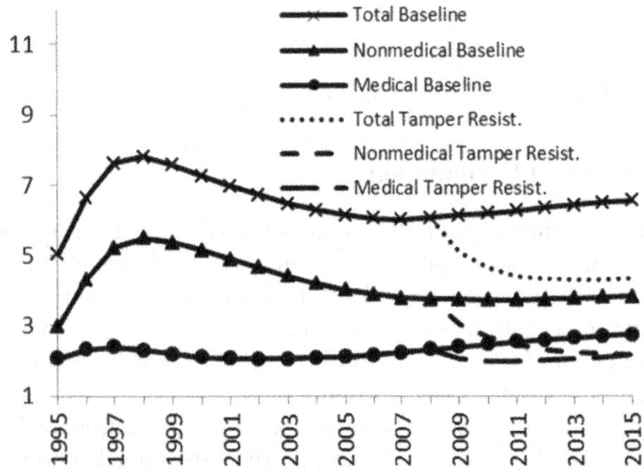

Fig. 9. The dramatic increase in the number of treated patients results in a smaller ratio of the overdose deaths divided by the number of patients receiving opioid therapy/10000

6 Discussion

Results from the model indicate that SD modeling holds promise as a tool for understanding the complex challenges associated with the epidemic of nonmedical use of opioids, and for evaluating the potential impact (on overdose deaths) of interventions to minimize the risks of opioid analgesics. By deliberately exaggerating the direct effects, downstream effects were also accentuated to make as obvious as possible any unintended consequences or counterintuitive results.

Since previous research has indicated that over half of opioid overdose deaths are individuals who have never been prescribed opioids directly (Hall et al., 2008), it is important to consider distal effects of medical sector-related interventions on nonmedical use and overdose deaths. Results of the intervention that simulated the introduction of tamper-resistant formulations also show that it is important to be aware of the metrics used to judge effectiveness. When using the metric deaths per 10,000 treated patients, tamper resistance appears to be effective at reducing the rate of overdose deaths as proportion of the medical users.

6.1 Limitations

Tamper resistance is only one possible intervention in the system of opioid misuse and since it is a pharmaceutical intervention, we do not take into account the social forces that influence health behavior or drug use. Additionally, though there are tamper resistant formulations currently available in the United States, a complete replacement of all opioids with tamper resistant formulations is unlikely even if congress passes the Stop Tampering Prescription Pills Act of 2013, which aims to stimulate competition in tamper resistant technologies by limiting competition from medicines without tamper resistant or abuse deterrent technologies.

Furthermore, despite great efforts to find empirical support for all model parameters, parameter validity remains a primary limitation in the study (see [27]. Several parameters have weak empirical support, and a number of potentially important factors have been excluded. For example, the model is limited because it focuses on chronic pain, and ignores the vastly-larger number of persons who receive opioids to treat acute pain. The prescribing of opioids to treat acute pain accounts for a much larger fraction of the opioids dispensed annually, so it is likely to contribute the supply of opioids for the nonmedical use sector, as well as to physician's perception of risk in the medical use sector.

The model may also be exaggerating the notion of profit as a motive for trafficking. Since the fraction of demand met by interpersonal sharing is large, it may be necessary to model this mechanism in a more detailed fashion.

Additionally, poly-drug use and abuse, opioid treatment programs, alternative treatments, and institutional factors that impact opioid use, such as payer policies and formularies, can all influence rates of medical and nonmedical use of opioids and the outcomes associated with such use. The exclusion of these factors imposes limitations on the model's ability to provide conclusive inferences.

7 Conclusions

The principal strength of this study is its system-level perspective and deliberate recognition of the complex interconnections and feedback loops associated with the use of opioids to treat pain and the associated adverse outcomes. From a systems perspective it is clear that interventions focused on prescribing behavior can have implications beyond the medical aspects of the system, and that a multifaceted approach which also addresses illicit use is warranted. The present study serves well to demonstrate how a systems-level model may help to evaluate the relative potential efficacy of interventions to reduce opioid-related overdose deaths.

Acknowledgements. Funding Was Provided NIDA Grant Number 1R21DA031361-01A1. the Authors Also Gratefully Acknowledge Support from J. David Haddox, John Fitzgerald, Jack Homer, Lewis Lee, Louis Macovsky, Dennis Mccarty, Lynn R. Webster, and Aaron Gilson.

References

1. Butler, S.F., Cassidy, T.A., Chilcoat, H., Black, R.A., Landau, C., Budman, S.H., Coplan, P.M.: Abuse Rates and Routes of Administration of Reformulated Extended-Release Oxycodone: Initial Findings From a Sentinel Surveillance Sample of Individuals Assessed for Substance Abuse Treatment. The Journal of Pain: Official Journal of the American Pain Society (2012), doi:10.1016/j.jpain.2012.08.008
2. Cavana, R.Y., Tobias, M.: Integrative System Dynamics: Analysis of Policy Options for Tobacco Control in New Zealand. Systems Research and Behavioral Science 25, 675–694 (2008)
3. Centers for Disease Control and Prevention. Unintentional Drug Poisonings in the United States. CDC Issue Brief (2010), `http://www.cdc.gov/HomeandRecreationalSafety/pdf/poison-issue-brief.pdf` (retrieved on September 14, 2010)
4. Cicero, T.J., Surratt, H., Inciardi, J.A., Munoz, A.: Relationship between Therapeutic Use and Abuse of Opioid Analgesics in Rural, Suburban, and Urban Locations in the United States. Pharmacoepidemiol Drug Safety 16(8), 827–840 (2007)
5. Colliver, J.D., Kroutil, L.A., Dai, L., Gfroerer, J.C.: Misuse of Prescription Drugs: Data from the 2002, 2003, and 2004 National Surveys on Drug Use and Health. DHHS Publication No. SMA 06-4192, Analytic Series A-28. Rockville, MD: Substance Abuse and Mental Health Services Administration, Office of Applied Studies (2006)
6. Compton, W.M., Volkow, N.D.: Major Increases in Opioid Analgesic Abuse in the United States: Concerns and Strategies. Drug and Alcohol Dependence 81(2), 103–107 (2006)
7. Degenhardt, L., Hall, W., Warner-Smith, M., Lynskey, M.: Ezzati, M., Lopez, A.D., Rodgers, A., Murray, C.J.L. (eds.): Comparative Quantification of Health Risks: Global and Regional Burden of Disease Attributable to Selected Major Risk Factors, Geneva, Switzerland, vol. 1, pp. 1109–1175 (2004)
8. Fishman, S.M., et al.: Regulating opioid prescribing through prescription monitoring programs: Balancing drug diversion and treatment of pain. Pain Medicine 5(3), 309–324 (2004)
9. Hall, A.J., et al.: Patterns of Abuse among Unintentional Pharmaceutical Overdose Fatalities. Journal of the American Medical Association 300(22), 2613–2620 (2008)
10. Hogarth, R.: Judgment and Choice, 2nd edn. John Wiley and Sons, Chichester (1987)
11. Homer, J.B.: Projecting the Impact of Law Enforcement on Cocaine Prevalence: A System Dynamics Approach. Journal of Drug Issues 23(2), 281–295 (1993)
12. Hser, Y., Hoffman, V., Grella, C.E., Anglis, M.D.: A 33-year follow-up of narcotics addicts. Archives of General Psychiatry 58, 503–508 (2001)
13. Jones, A., Homer, J., Murphy, D., Essein, J., Milstein, B., Seville, D.: Understanding Diabetes Population Dynamics through Simulation Modeling and Experimentation. American Journal of Public Health 96, 488–494 (2006)
14. Joranson, D.E., Gilson, A.M., Dahl, J.L., Haddox, J.D.: Pain Management, Controlled Substances, and State Medical Board Policy: A Decade of Change. Journal of Pain Symptom Management 23, 138–147 (2002)

15. Katz, N., Dart, R.C., Bailey, E., Trudeau, J., Osgood, E., Paillard, F.: Tampering with prescription opioids: nature and extent of the problem, health consequences, and solutions. The American Journal of Drug and Alcohol Abuse 37(4), 205–217 (2011)
16. Lourencco, L.M., Matthews, M., Jamison, R.N.: Abuse-deterrent and tamper-resistant opioids: how valuable are novel formulations in thwarting non-medical use? Expert Opinion on Drug Delivery, 1–12 (2012)
17. Manchikanti, L., Cash, K., Damron, K.S., Manchukonda, R., Pampati, V., McManus, C.D.: Controlled Substance Abuse and Illicit Drug Use in Chronic Pain Patients: An Evaluation of Multiple Variables. Pain Physician 9, 215–226 (2006)
18. Milstein, B., Homer, J., Hirsch, G.: Analyzing National Health Reform Strategies with a Dynamic Simulation Model. American Journal of Public Health 100(5), 811–819 (2010)
19. Potter, M., et al.: Opioids for Chronic Nonmalignant Pain: Attitudes and Practices of Primary Care Physicians in the UCSF/Stanford Collaborative Research Network. Journal of Family Practice 50(2), 145–151 (2001)
20. Raffa, R.B., Pergolizzi Jr, J.V.: Opioid Formulations Designed to Resist/Deter Abuse. Drugs 70(13), 1657–1675 (2010)
21. SAMHSA. Overview of findings from the 2005 National Survey on Drug Use and Health. Office of Applied Studies, NSDUH Series H-30, DHSS Publication No. SMA 06-4194. Rockville, MD (2006)
22. SAMHSA. Results from the 2006 National Survey on Drug Use and Health: National findings (2007), `http://www.oas.samhsa.gov/nsduh/2k6nsduh/2k6results.pdf` (retrieved on June 12, 2010)
23. SAMHSA. Results from the 2009 National Survey on Drug Use and Health: Volume I. Summary of National Findings. Office of Applied Studies, NSDUH Series H-38A, HHS Publication No. SMA 10-4586. Rockville, MD (2010)
24. Sterman, J.D.: Business Dynamics: Systems Thinking and Modeling for a Complex World. McGraw-Hill, Boston (2000)
25. Sterman, J.D.: Learning from Evidence in a Complex World. American Journal of Public Health 96, 505–514 (2006)
26. Warner, M., Chen, L.H., and Makuc, D.M.: Increase in Fatal Poisonings Involving Opioid Analgesics in the United States, 1999-2006. NCHS Data Brief, No. 22 (2009), `http://www.cdc.gov/nchs/data/databriefs/db22.pdf` (retrieved on August 20, 2010)
27. Wakeland, W., et al.: Key Data Gaps for Understanding Trends in Prescription Opioid Analgesic Abuse and Diversion among Chronic Pain Patients and Nonmedical Users. In: College on Problems of Drug Dependence, 72nd Annual Scientific Meeting, Scottsdale, AZ (June 2010)
28. Webster, L.: Update on Abuse-Resistant and Abuse-Deterrent Approaches to Opioid Formulations. Pain Medicine 10(S2), S124–S133 (2009)

A Multi-GPU Approach to Fast Wildfire Hazard Mapping

Donato D'Ambrosio[1], Salvatore Di Gregorio[1], Giuseppe Filippone[1], Rocco Rongo[1], William Spataro[1], and Giuseppe A. Trunfio[2]

[1] Department of Mathematics and Computer Science, University of Calabria, 87036 Rende (CS), Italy
[2] DADU, University of Sassari, 07041 Alghero (SS), Italy

Abstract. Burn probability maps (BPMs) are among the most effective tools to support strategic wildfire and fuels management. In such maps, an estimate of the probability to be burned by a wildfire is assigned to each point of a raster landscape. A typical approach to build BPMs is based on the explicit propagation of thousands of fires using accurate simulation models. However, given the high number of required simulations, for a large area such a processing usually requires high performance computing. In this paper, we propose a multi-GPU approach for accelerating the process of BPM building. The paper illustrates some alternative implementation strategies and discusses the achieved speedups on a real landscape.

Keywords: GPGPU, Cellular Automata, Wildfire Simulation, Wildfire Susceptibility, Hazard Maps.

1 Introduction

Among the several tools recently developed to support fire hazard management, there are the so-called *burn probability maps* (BPMs), which attempt to provide an estimate of the probability of a point in a landscape to be burned under certain environmental conditions. To cope with the nonlinear interactions between the many factors that determine the fire behaviour, models for simulating wildfire spread are increasingly being used to build BPMs [1, 2]. In particular, the typical approach is based on carrying out a high number of simulations (e.g. many thousands), under different weather scenarios and ignition locations [1].

In order to obtain reliable results in reasonable time, such an approach must be based on fast and accurate simulation models operating on high-quality high-resolution remote sensing data (e.g., Digital Elevation Models, vegetation description, etc). Among the different wildfire simulation techniques [3], those based on Cellular Automata (CA) [4–7] represent an ideal approach to build a BPM. This is because they provide accurate results and can often perform the same simulations in a fraction of the run time taken by different methods [6].

However, because of the required high number of explicit fire propagations, even using the most optimized algorithms, the building of simulation-based BPMs for a large area often results in a highly intensive computational process. For example, building a

DOI: 10.1007/978-3-319-03581-9_13,

high-resolution BPM for a regional territory typically requires the use of high performance computing (HPC).

Among the different HPC alternatives is the recently emerging General-Purpose computing on Graphics Processing Units (GPGPU), in which multicore Graphics Processing Units (GPU) perform computations traditionally carried out by the CPU. In this paper, we apply GPGPU, in conjunction with a wildfire simulation model, to the process of BPM computation.

In particular, the proposed approach is based on a CA simulation model which represents a suitable trade-off between accuracy and speed of execution. The adopted parallel computation consists of the iterative simultaneous simulation of a number of wildfires with GPGPU, in order to cover the whole area under study. In the paper we illustrate two different implementation strategies together with a multi-GPU approach. In addition, we discuss some numerical results obtained on a real Mediterranean landscape, which is historically characterized by a high incidence of wildfires.

The paper is organized as follows. In the next section we outline the main characteristics of the adopted CA simulation model and illustrate some details of the typical approach for BPM computation. Then, in section 3 we present some introductory elements of the adopted GPGPU approach. In section 4 we outline the proposed parallel approaches and in section 5 we investigate some of their computational characteristics. The paper ends with section 6 in which we draw some conclusions and outline possible future work.

2 Simulation-Based Burn Probability Mapping

2.1 The Wildfire Simulation Model

As mentioned above, CA methods for simulating wildfires can be highly optimized from the computational point of view. For this reason they are well suited for the process of building BPMs. In particular, in this study we use the CA-based wildfire simulation model described in [8] (to which the reader is referred for the details).

The adopted simulator is based on the Rothermel fire model [9], which provides the heading rate and direction of spread given the local landscape and wind characteristics. An additional constituent is the commonly assumed elliptical description of the spread under homogeneous conditions (i.e. spatially and temporally constant fuels, wind and topography) [10]. Under the above hypothesis, given the assumption of homogeneity at the cell level, the CA transition function uses the elliptical model for producing the complex patterns that correspond to the fire spread in heterogeneous conditions.

In brief, the two-dimensional fire propagation is locally obtained by a growing ellipse (see Figure 1) having the semi-major axis along the direction of maximum spread, the eccentricity related to the intensity of the so-called *effective wind* and one focus acting as a 'fire source' [5, 6]. At each CA step the ellipse's size is increased according to both the duration of the time step and maximum rate of spread. Afterwards, a neighbouring cell invaded by the growing ellipse is considered as a candidate to be ignited by the spreading fire. In case of ignition, a new ellipse is generated according to the amount of overlapping between the invading ellipse and the ignited cell.

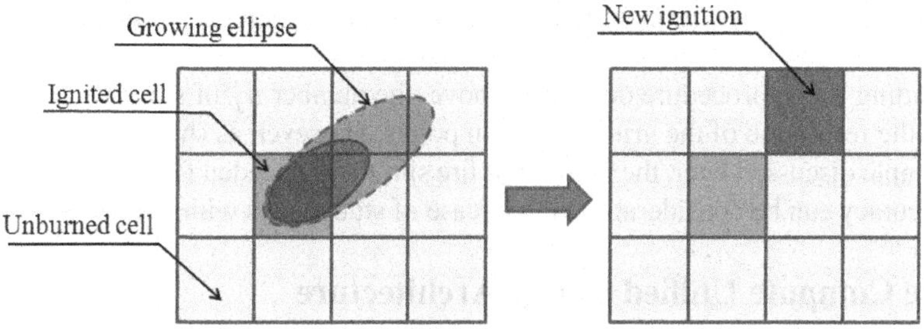

Fig. 1. Growth of the ellipse locally representing the fire front. The forward spread is incremented at each CA step according to the local spread rate and to the size of the time step.

A relevant feature of the model consists of the dynamic adaptation of the duration of each time step. In practice, the step size is computed on the basis of the minimum amount of time that elapses before the fire may have traveled from a cell on the current fire front to a neighbouring cell [4, 6, 7].

2.2 A Simulation-Based Approach for Building BPMs

In the latest years, the use of hazard maps based on the explicit simulation of natural phenomena has been increasingly investigated as an effective and reliable tool for supporting risk management [11, 1, 2, 12].

In the case of wildfire, the most general approach for computing a BPM on a landscape [1, 2] consists of a Monte Carlo approach in which a high number of different fire spread simulations are carried out, sampling from suitable statistical distributions the random variables relevant to the fire behaviour. For example, the wind direction for each simulated fire can be sampled in a range corresponding to the typical directions of severe wind for the area. At the end of the process, the local risk is computed on the basis of the frequency of burning.

The technique for computing the BPMs adopted in this study is based on a prefixed number n_f of simulation runs, where each run represents a single simulated fire. The adopted weather scenario (i.e. wind and fuel moisture content) is stationary and corresponds to extreme conditions for the area with regards to relevant historical fires. A regular grid of ignition locations is adopted, which corresponds to the assumption of a uniform ignition probability for each point of the landscape. Also, all the simulated fires have the same duration. The latter is selected considering the duration of historical fires in the regions under study. All the other relevant characteristics are kept constant during the simulations.

Once these have been carried out, the resulting n_f maps of burned areas are overlaid and cells' fire frequency are used for the computation of the fire risk. In particular, a *burn probability* $p_b(\mathbf{c})$ for each cell $\mathbf{c}$ is computed as:

$$p_b(\mathbf{c}) = \frac{f(\mathbf{c})}{n_f}; \tag{1}$$

where $f(\mathbf{c})$ is the number of times the cell $\mathbf{c}$ is ignited during the n_f simulated fires. The burn probability for a given cell is an estimate of the likelihood that a cell will burn given a single random ignition within the study area and given the assumed conditions in terms of fire duration fuel moisture and weather.

According to the procedure described above, the number n_f of simulation runs depend on the resolution of the grid of ignition points. However, as shown in the application example discussed later, the number of fire simulation needed for achieving a good BPM accuracy can be considerably high in case of study areas with great extensions.

3 The Compute Unified Device Architecture

A natural approach to deal with the high computational effort related to construction of the BPMs is the use of parallel computing. Among the different parallel architectures and computational paradigms, the recently emerging GPGPU is particularly suited for accelerating CA-based simulations [13, 14].

Modern GPUs are multiprocessors with a highly efficient hardware-coded multi-threading support. The key capability of a GPU unit is thus to execute thousands of threads running the same function concurrently on different data. Hence, the computational power provided by such an architecture can be fully exploited through a fine grained data-parallel approach when the same computation can be independently carried out on different elements of a dataset.

The GPGPU platform investigated in this paper is the one provided by nVidia, which consist of a group of Streaming Multiprocessors (SMs) able to support a limited number of co-resident concurrent threads, which share the SM's limited memory resources. Furthermore, each SM consists of multiple Scalar Processor (SP) cores. In order to program the GPU, the C-language Compute Unified Device Architecture (CUDA) [15] is used. In a typical CUDA program, sequential host instructions are combined with parallel GPU code. The idea underlying this approach is that the CPU organizes the computation (e.g. in terms of data pre-processing), sends the data from the computer main memory to the GPU global memory and invokes the parallel computation on the GPU. After, and/or during the latter, the CPU invokes the copying of the computed results into the main memory for post-processing and output purposes.

In CUDA, the GPU activation is obtained by writing device functions in C language, which are called *kernels*. When a kernel is invoked by the CPU, a number of threads (e.g. typically several thousands) execute their code in parallel on different data. According to the nVidia approach to GPGPU, threads are grouped into blocks and executed on the SMs.

The GPU can access different types of memory. For example, to each thread block can be assigned a certain amount of fast shared memory (which can be used for some limited intra-block communication between threads). Also, all threads can access a slower but larger global memory which is on the device board but outside the computing chip. The device global memory is slower if compared with the shared memory but it can deliver significantly higher bandwidth than the main computer memory. The latter is typically linked to the GPU card through a relatively slow bus. As a results, the parallel computation should be organised in such a way to minimize data transfers between the host and the device.

Even though the GPUs global memory offers a considerably higher memory bandwidth compared to CPUs host memory, it still requires hundreds of clock cycles to start the fetch of a single element. However, the massively threaded architecture of the GPU is used to hide such memory latencies by rapid switching between threads: when a thread stalls on a memory fetch, the GPU switches to a different thread. To fully exploit such latency hiding strategy, it is usually beneficial to use a large number of blocks.

In the following, the concepts outlined above are applied to the GPGPU-based parallelization of the procedure for building BPMs previously described.

4 GPGPU-Based Wildfire Hazard Mapping

The procedure described in the following is based on accelerating through GPGPU the execution of the many CA simulations required by the BPM. As in the sequential version, the CA simulation model requires two memory regions, which will be called CA_{cur} and CA_{next}, representing the *current* and *next* states for the cells respectively. For each CA step, the neighbouring values from CA_{cur} are read by the local transition function, which performs its computation and writes the new state value into the appropriate element of CA_{next}.

More in particular, accordingly to the recent literature in the field, in our implementation most of the automaton data (i.e. both the current and next memory areas mentioned above) is stored in the GPU global memory. This involves: *(i)* initialising the current state through a CPU-GPU memory copy operation (i.e. from host to device global memory) before the beginning of the simulation and *(ii)* retrieving the final state of the automaton at the end of the simulation through a GPU-CPU copy operation (i.e. from device global memory to host memory). Also, at the end of each CA step a device-to-device memory copy operation is used to re-initialise CA_{cur} with CA_{next}.

In order to speed up the access to memory, an array with the size corresponding to the total number of cells was allocated in the CPU memory for each of the CA substates. All of such arrays were then mirrored in the GPU together with some additional auxiliary arrays (e.g. for storing the neighbourhood structure and the model parameters).

A key step in the parallelization of a sequential code for the GPU architecture according to the CUDA approach, consists of identifying all the sets of instructions that can be grouped in CUDA kernels. In particular, in the CA model for wildfire simulation adopted in this study (see section 2), two main CUDA kernels have been developed:

- the kernel implementing the fire propagation logic (i.e. the cell-level mechanism of fire contagion);
- the kernel for dynamically adapting the time-step duration. Since this involves finding the minimum of all allowed time-step sizes among the cells on the current fire front, such kernel simply implements a standard parallel reduction (PR) algorithm.

The above kernels can be executed independently of each other on the different cells of the automaton. However, the GPGPU parallelization object of this study raises several issues.

First, in the whole automaton, only the cells belonging to the current fire front perform actual computation. Hence, launching one thread for each of the automaton cells would result in a certain amount of dissipation of the GPU computational power. In other words, a high percentage of threads would be uselessly scheduled. In addition,

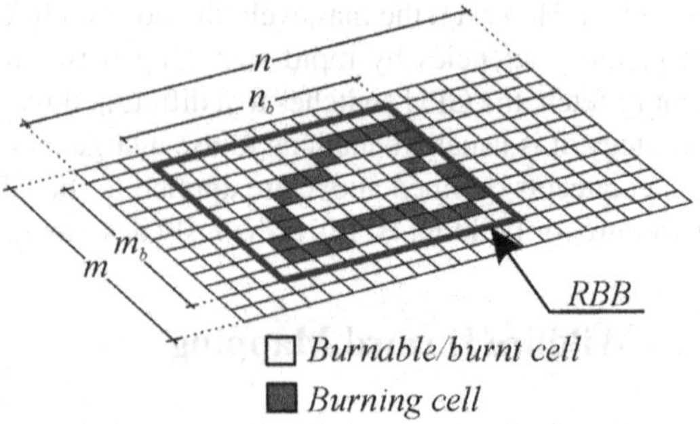

Fig. 2. The RBB of a fire in a $n \times m$ raster. Only the cells inside the RBB, which is recomputed at each CA step, are mapped into the CUDA grid of threads.

given the small size of most fires, the number of active threads generated by the simulation of a fire is usually too low to allow the GPU to effectively activate the latency hiding mechanism.

For the above reasons, two main strategies were adopted. In particular, besides the most straightforward parallel implementation, labelled as *WCAM*, in which the CUDA kernels operate on the whole automaton, we have developed an additional implementation in which the grid of threads is dynamically computed during the simulation in order to keep low the number of computationally irrelevant threads.

In such an approach, labelled as *RBBM* and represented in Figure 2, the smallest rectangular bounding box (RBB) that includes any cells on the current fire front is computed at each CA step using the efficient CUDA atomic instructions. Then, all kernels required by the CA step (i.e. the PR for the time step adaptation and the transition function) are mapped on such RBB.

Another measure adopted in this study consists of handling many fires simultaneously. In particular, clusters of fires are simulated up to covering the entire area under study (i.e. until all the required n_f fires are propagated). Each cluster is composed of a block of fires originated by spatially-contiguous ignition points taken from the regular grid of fire ignitions.

Obviously, such an approach requires: (*i*) the use of an additional array for storing an independent combustion state of each cell for each simultaneous fire; (*ii*) the computation of a RBB that is common to all the simultaneous fires.

It is worth noting that since the efficiency of the *RBBM* approach depends on the actual distribution of the simultaneous fires over the automaton, we choose from contiguous ignition points the fires that are simulated together.

In order to carry out many simulations simultaneously, a particular grid of threads is used is such a way that for each cell a separate thread handles the different simultaneous fires (see Figure 3). This corresponds to a three-dimensional grid in which the vertical dimension is associated with the simultaneous fires. It is also worth noting that the alternative approach, in which a single thread per cell iterates on the different simultaneous fires, would generate a frequent divergence between the threads of the same warp, with a consequent decline of efficiency.

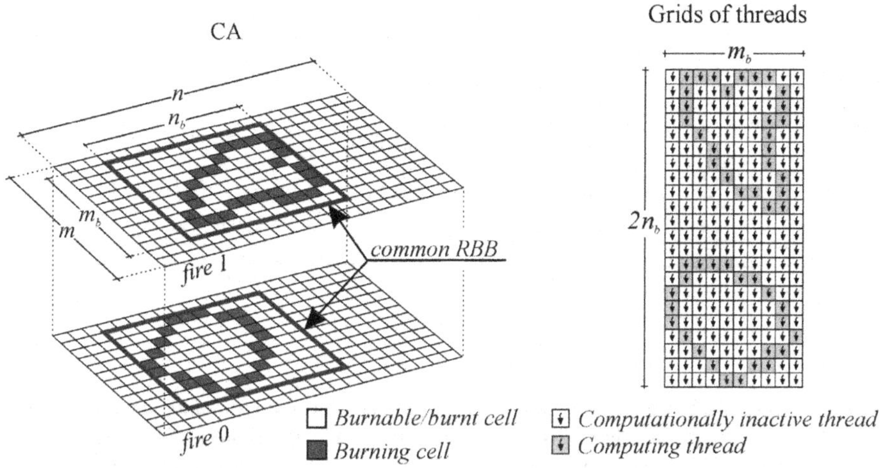

Fig. 3. The *RBBM* mapping of the CA transition function into a CUDA grid of threads in case of the two simultaneous fires shown on the left. The kernel is mapped on a grid in which a separate copy of the RBB is associated with each of the different simultaneous fires. This corresponds to a three-dimensional grid, with the base represented by the RBB and the vertical dimension corresponding to the fires.

Before starting the BPM construction, a pre-processing sequential phase takes place in which:

- for each cell the maximum rate of spread, its direction and the local ellipse eccentricity (see Figure 1) are computed using the proper model equations [9, 16]. Such pre-computed quantities determine, together with the landscape topography, the wildfire spread at the cell-level.
- the maximum time-step size for each cell is computed and stored in an array in order to speed-up the time-step adaptation during the CA iterations.
- a data structure $\mathcal{C}$ is built, which contains all the clusters of contiguous ignition points corresponding to fires that must be simulated together (see Figure 4).

It is worth noting that the first two steps above make sense because the weather conditions are considered stationary, which is a common assumption for computing BPMs.

As explained above, the required CA simulations are carried out operating on clusters of fires iteratively extracted from the data structure $\mathcal{C}$. In particular, in the GPGPU wildfire simulation, each CA step essentially consists of:

- executing the CUDA kernel that finds the current time step size for each simultaneous fire;
- executing the CUDA kernel that implements the propagation mechanism (and updates the RBB in the RBBM approach);
- activating a device-to-device memory copy operation to re-initialise CA_{cur} with CA_{next};
- incrementing the current time for each simultaneous fire by the corresponding time step size.

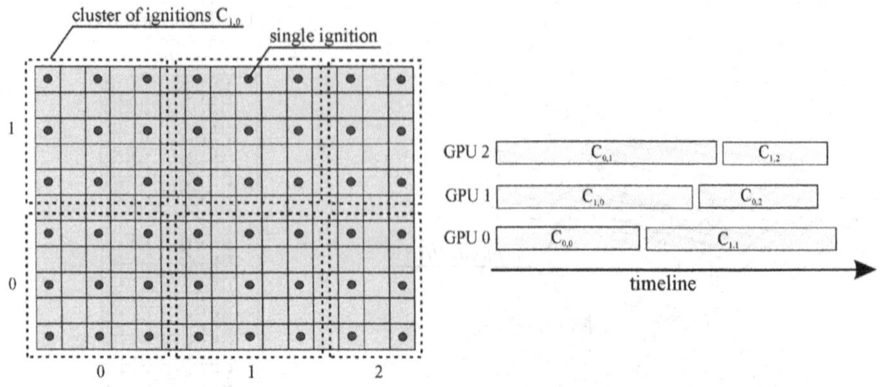

Fig. 4. An hypothetical example of regular grid of ignition points used for building a BPM. Contiguous ignitions are grouped into clusters and the corresponding fires are simulated simultaneously. In the multi-GPU implementation each cluster is assigned to the first available GPU.

The above steps are iterated until the current time reaches the desired final time for each fire of the current cluster.

4.1 The Multi-GPU Implementation

In the multi-GPU implementation, which is based on the *Pthreads* library, to each available GPU in the system we permanently associate a separate CPU thread using the specific CUDA function *cudaSetDevice*. In addition, a separate copy of the automaton is mirrored in each GPU.

After such initialization phase, the process for the BPM computation is concurrently carried out by each CPU thread as follows:

1. extraction, using the mutual exclusion mechanism provided by *PThreads*, of the first available cluster C of fire ignitions from the data structure $\mathcal{C}$ (see Figure 4) mentioned above.
2. execution of a CUDA kernel to set the ignition points included in its current cluster C into the automaton cells stored in its GPU global memory;
3. execution, in its associated GPU, of the CA steps described above, until the current time reaches the desired final time;
4. at the end of the simulations, execution of a further CUDA kernel on the whole automaton to update an array $\mathbf{f}_d$ in which each element represents the number of times that a cell has been burned since the beginning of the process;
5. if there are still elements in $\mathcal{C}$, extraction of a new cluster of ignitions and iteration of the above steps $1-4$; otherwise, the CPU thread ends.

When all the CPU threads have completed their cycles of simulations, all the arrays $\mathbf{f}_d$ are moved to the host memory and collected into a single array $\mathbf{f}_n$. The latter, divided by the total number of simulations n_f, gives an estimate of the burn probability for each cell of the automaton.

Given that the CPU threads are essentially independent on each others, the above procedure can achieve a satisfactory level of efficiency. However, a limiting factor of the scalability originates from the exchange of data between GPUs and CPU at each CA step (for example the duration of the steps or the RBB).

It is worth noting that, in a general case of heterogeneous landscape, the time required for the simulation of each cluster $C \in \mathcal{C}$ cannot be predicted in advance. In fact, it depends on the specific conditions of the area covered by C (e.g. fuel and terrain characteristics or local wind vector). However, according to the above scheduling procedure, as soon as a GPU finishes its current cluster computation, it fetches the next available cluster of fires. This can be seen in Figure 4, where an hypothetical example of scheduling is represented. Therefore, if $\mathcal{C}$ contains enough clusters (which is always the case for large areas) the workload of each GPU is automatically balanced. In other words the BPM computation runs flawlessly even in case of GPUs with different computational power.

5 Results on a Real Landscape

The application presented here concerns an area of the Ligury region, in Italy. The landscape, shown in Figure 5 was modelled through a Digital Elevation Model composed of 461×445 square cells with side of $40\,m$. In the area, the terrain is relatively complex with an altitude above sea level ranging from 0 to 250 m. The fuel bed, depicted in Figure 5, was based on the land cover map from the CORINE EU-project. In particular, the CORINE land-cover codes were mapped on the standard fuel models used by the CA. Plausible values of fuel moisture content were obtained from literature data. Also, a North direction open-wind vector, having an intensity of $20\,km\,h^{-1}$, was used for producing the wind field. A duration of 10 hours was adopted for all simulated fires. Over the area, a regular grid of 92×89 ignition points was superimposed, leading to 7391 fires to simulate and to the BPM shown in Figure 6.

For the numerical experiments, we used a workstation based on an Intel Xeon X5660 (2.80 GHz) 6-Core CPU and equipped with two GPUs, namely a nVidia Tesla C2075 and a nVidia Geforce GTX 680 graphic card. The latter card belongs to the new nVidia's Kepler GPU architecture while the former is endowed with the older Fermi GPUs. The Tesla C2075 was used in single precision floating point and with the ECC disabled. In Table 1 we report some of the relevant characteristics of the two GPU devices. To quantify the speedup of the parallel implementations, we also included in the same CUDA program the C-language sequential versions of the *WCAM* and *RBBM* approaches described above. In particular, the program allows the user to select the desired version of the algorithm, as well as whether to use the available GPUs or only the CPU.

It is worth noting that, in the sequential case only one fire at a time was propagated since the advantages of simulating multiple fires are not significant. Specifically, using only the CPU for simulating the 7391 independent fires required by the case-study BPM, the *WCAM* approach took $14167.4\,s$ while the *RBBM* strategy took $2082.6\,s$.

Using the adopted GPU devices in both single and multi-GPU mode, the BPM was built with the *WCAM* approach and a variable number of simultaneous fires. According to the results shown in Figure 7, the two GPUs working together achieved the lowest elapsed time of $197.8\,s$, simultaneously simulating 81 fires. As can be seen, the simultaneous simulation of many fires was beneficial since it allowed a significant computing

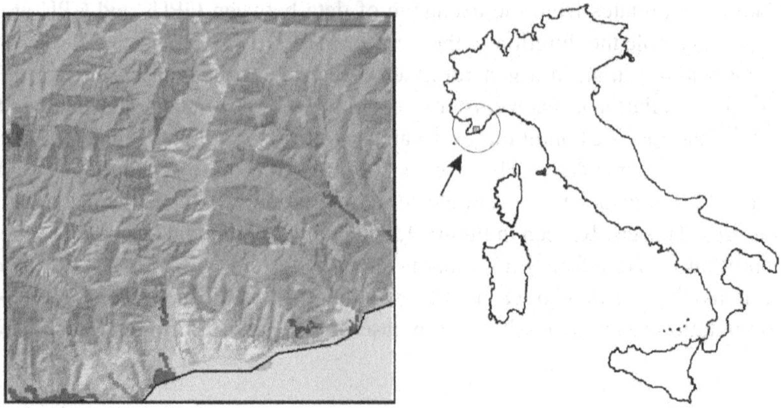

Fig. 5. The landscape under study: a $18km \times 18km$ area in Ligury, Italy. Colors refer to the standard CORINE land-cover data.

Table 1. Some relevant characteristics of the adopted GPGPU hardware for all carried out experiments. Note that GFLOPs refers to the theoretical peak performance in single-precision floating point operations.

	GTX 680	Tesla C2075
SM count	8	14
CUDA cores	1536	448
Clock rate [MHz]	1006	1150
Bandwidth [GB/s]	192.3	144.0
GFLOPs	3090.4	1030.4

time decrement. However, for all the GPU configurations, incrementing the size of the clusters in C above a certain threshold did not lead to significant advantages. This is due to the fact that, once the number of computationally-relevant threads is sufficient for enabling latency hiding, a further increase of simultaneous fires is not necessary.

Interestingly, using the two GPUs together lead to a speedup ranging between 1.82 and 1.90 over the faster GPU, namely the GTX 680. Particularly relevant is the speedup over the sequential run, which attained the value of 71.6.

Figure 7 also shows the times taken by the parallel *RBBM* approach as a function of the number of simultaneous fires. According to the graph, in this case the joint computational effort of the two GPUs gave the lowest elapsed time of 33.8 s, corresponding to a parallel speedup of 61.6.

In the *RBBM* runs, using the two GPUs together lead to a speedup over the faster GPU (i.e. the GTX 680) ranging between 1.72 (196 simultaneous fires) and 1.89 (16 simultaneous fires). Not surprisingly, given the multi-GPU scheduling scheme described in section 4.1, in case of GPUs with different computational power the use of small clusters of fires (i.e. the use of more clusters) gives the opportunity to better exploit the faster GPU.

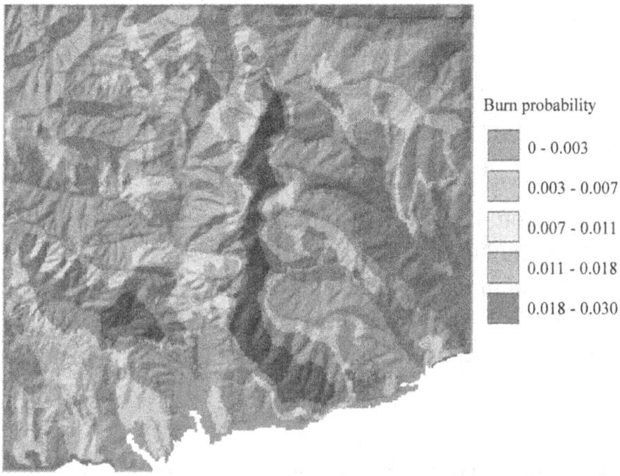

Fig. 6. The BPM obtained for the landscape under study

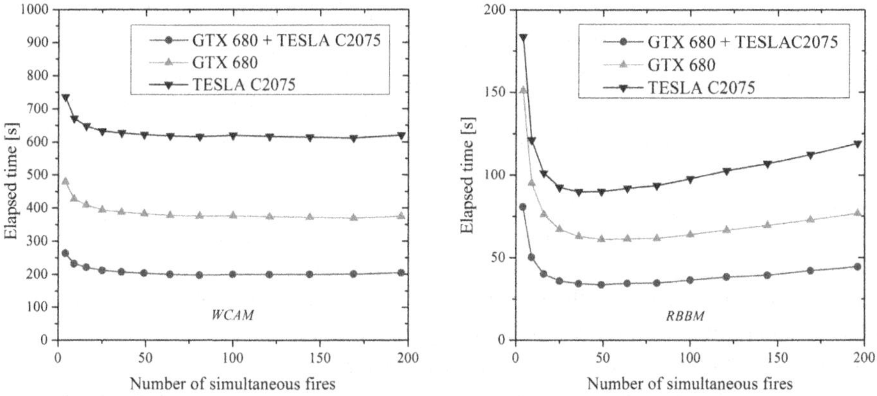

Fig. 7. Time taken by the *WCAM* and *RBBM* approaches for simulating the 7391 independent fires required by the BPM shown in Figure 6. The elapsed time is plotted as a function of the number of fires that are simultaneously simulated by the GPU.

As expected, the simultaneous simulation of many fires was much more beneficial than in the $WCAM$ case. The reason is that, having available a relatively limited number of simultaneous fires, the percentage of active cells is higher in a small RBB

Table 2. Best elapsed times (in seconds) for the computation of the BPM shown in Figure 6. In brackets is the optimum number of simultaneous fires.

	WCAM	RMBB
CPU	14167.4	2082.6
GTX 680	369.5 (169)	61.1 (49)
Tesla C2075	612.2 (169)	89.8 (36)
GTX 680 + Tesla C2075	197.8 (81)	33.8 (49)

Table 3. Best parallel speedup achieved with the used GPGPU devices

	WCAM	RMBB
GTX 680	38.3	34.1
Tesla C2075	23.1	23.2
GTX 680 + Tesla C2075	71.6	61.6

than in the entire automaton. However, also in this case it is not advantageous to simulate more than a certain number of simultaneous fires.

For clarity reasons, a summary of the elapsed times achieved for solving the test problem are shown in Table 2. Moreover, the corresponding parallel speedups are shown in Table 3. As can be seen, even using a single GPU lead to a significant acceleration of the BPM computation.

6 Conclusions and Future Work

The parallel speedups of the multi-GPU based BPM computation procedure presented in this paper are very satisfactory. The main advantage of such a parallelization lies in enabling the building of BPMs for very large areas (e.g. at a regional level), which otherwise may not be possible by adopting a standard computation. However, starting from the application described above, several research directions can be explored.

For example, the multi-GPU approach may also be adopted for the automatic planning of risk-mitigation interventions on the landscape (i.e. the so-called *fuel treatments*) [17]. Moreover, the fast simulation of a number of wildfires can make easier and more accurate the assimilation of real-time sensor data [18]. However, in this case a new approach with non-stationary weather conditions should be developed. This would require computing with the GPU also the fire characteristics at the cell level at the price of an increased computing time, though obtaining higher speedups.

Also, the GPGPU parallelization strategies investigated in this paper can be exploited, with the required adjustments, to other research and application areas. In fact, the algorithm for building BPMs is somewhat similar to those adopted for dealing with other natural hazards, such as the risk induced by debris or lava flows (e.g. [11, 19]).

Eventually, another possible direction of research consists of making the GPGPU approach available in more general libraries for supporting CA modelling and simulation, such as the one presented in [20].

References

1. Carmel, Y., Paz, S., Jahashan, F., Shoshany, M.: Assessing fire risk using Monte Carlo simulations of fire spread. Forest Ecology and Management 257(1), 370–377 (2009)
2. Ager, A., Finney, M.: Application of wildfire simulation models for risk analysis. In: Geophysical Research Abstracts. EGU2009-5489, EGU General Assembly, vol. 11 (2009)
3. Sullivan, A.: Wildland surface fire spread modelling, 1990-2007. 3: Simulation and mathematical analogue models. International Journal of Wildland Fire 18, 387–403 (2009)
4. Lopes, A.M.G., Cruz, M.G., Viegas, D.X.: Firestation - an integrated software system for the numerical simulation of fire spread on complex topography. Environmental Modelling and Software 17(3), 269–285 (2002)

5. Trunfio, G.A.: Predicting wildfire spreading through a hexagonal cellular automata model. In: Sloot, P.M.A., Chopard, B., Hoekstra, A.G. (eds.) ACRI 2004. LNCS, vol. 3305, pp. 385–394. Springer, Heidelberg (2004)
6. Peterson, S.H., Morais, M.E., Carlson, J.M., Dennison, P.E., Roberts, D.A., Moritz, M.A., Weise, D.R.: Using HFIRE for spatial modeling of fire in shrublands. Technical Report PSW-RP-259, U.S. Department of Agriculture, Forest Service, Pacific Southwest Research Station, Albany, CA (2009)
7. Trunfio, G.A., D'Ambrosio, D., Rongo, R., Spataro, W., Di Gregorio, S.: A new algorithm for simulating wildfire spread through cellular automata. ACM Transactions on Modeling and Computer Simulation 22(1), 1–26 (2011)
8. Avolio, M.V., Di Gregorio, S., Lupiano, V., Trunfio, G.A.: Simulation of wildfire spread using cellular automata with randomized local sources. In: Sirakoulis, G.C., Bandini, S. (eds.) ACRI 2012. LNCS, vol. 7495, pp. 279–288. Springer, Heidelberg (2012)
9. Rothermel, R.C.: A mathematical model for predicting fire spread in wildland fuels. Technical Report INT-115, U.S. Department of Agriculture, Forest Service, Intermountain Forest and Range Experiment Station, Ogden, UT (1972)
10. Alexander, M.: Estimating the length-to-breadth ratio of elliptical forest fire patterns. In: Proc. 8th Conf. Fire and Forest Meteorology, pp. 287–304 (1985)
11. Rongo, R., Spataro, W., D'Ambrosio, D., Avolio, M.V., Trunfio, G.A., Di Gregorio, S.: Lava flow hazard evaluation through cellular automata and genetic algorithms: an application to Mt Etna volcano. Fundamenta Informaticae 87(2), 247–267 (2008)
12. Rongo, R., Lupiano, V., Avolio, M.V., D'Ambrosio, D., Spataro, W., Trunfio, G.A.: Cellular automata simulation of lava flows - applications to civil defense and land use planning with a cellular automata based methodology. In: Proceedings of SIMULTECH 2011 (2011)
13. Filippone, G., Spataro, W., Spingola, G., D'Ambrosio, D., Rongo, R., Perna, G., Di Gregorio, S.: GPGPU programming and cellular automata: Implementation of the SCIARA lava flow simulation code. In: 23rd European Modeling and Simulation Symposium (EMSS), Rome, Italy, September 12-14 (2011)
14. D'Ambrosio, D., Filippone, G., Rongo, R., Spataro, W., Trunfio, G.: Cellular automata and GPGPU: an application to lava flow modeling. International Journal of Grid and High Performance Computing 4(3), 30–47 (2012)
15. CUDA C Programming Guide: v. 3.2 (2010)
16. Anderson, H.: Predicting wind-driven wildland fire size and shape. Technical Report INT-305, U.S Department of Agriculture, Forest Service (1983)
17. Ager, A.A., Vaillant, N.M., Finney, M.A.: A comparison of landscape fuel treatment strategies to mitigate wildland fire risk in the urban interface and preserve old forest structure. Forest Ecology and Management 259(8), 1556–1570 (2010)
18. Xue, H., Gu, F., Hu, X.: Data assimilation using sequential monte carlo methods in wildfire spread simulation. ACM Trans. Model. Comput. Simul. 22(4), 1–25 (2012)
19. Crisci, G.M., Avolio, M.V., Behncke, B., D'Ambrosio, D., Di Gregorio, S., Lupiano, V., Neri, M., Rongo, R., Spataro, W.: Predicting the impact of lava flows at Mount Etna, Italy. Journal of Geophysical Research: Solid Earth 115(B4) (2010)
20. Blecic, I., Cecchini, A., Trunfio, G.A.: A general-purpose geosimulation infrastructure for spatial decision support. Transactions on Computational Science 6, 200–218 (2009)

Controlling Turtles through State Machines: An Application to Pedestrian Simulation

Ilias Sakellariou

Department of Applied Informatics, University of Macedonia,
Egnatia 156, Thessaloniki, Greece
iliass@uom.gr

Abstract. Undoubtedly, agent based modelling and simulation (ABMS) has been recognised as a promising technique for studying complex phenomena. Due to the attention that it has attracted, a significant number of platforms have been proposed, the majority of which target reactive agents, i.e. agents with relatively simple behaviours. Thus, little has been done toward the introduction of richer agent oriented programming constructs that will enhance the platforms' modelling capabilities and could potentially lead to the implementation of more sophisticated models. This paper discusses TSTATES, a domain specific language, together with an execution layer that runs on top of a widely accepted agent simulation environment and presents its application to modelling pedestrian simulation in an underground station scenario.

Keywords: Agent Simulation Platforms, Agent Programming Languages, Crowd Simulation.

1 Introduction

Agent based modelling and simulation has been widely adopted as a new technique for studying complex emergent phenomena in areas such as economics, biology, psychology, traffic and transportation, etc. [1]. This inevitably lead to the introduction of a plethora of agent modelling and simulation platforms (ABMS) [2] [3], that offer modelling environments of different complexity and characteristics, in terms of programming language, scalability, extensibility, ease of use, user support (documentation, tutorials, example models), etc.

The NetLogo multi agent modelling environment [4], has been regarded as one of the most successful and complete ABMS platforms [5,6], offering an IDE with extensive visualization tools, and a simple domain specific agent programming language; this"one-stop" approach allows users to arrive to a simulation experiment with a relatively small effort; a fact that has a definitive advantage toward the adoption of the platform by the community. However, although NetLogo is excellent for "modelling social and emergent phenomena" consisting of a large number of reactive agents, it lacks the modelling facilities to accommodate more complex agent behaviours.

The problem originally has been addressed in [7] that presents an approach toward building higher level intention driven communicating NetLogo agents. That work offers

M.S. Obaidat et al. (eds.), *Simulation and Modeling Methodologies, Technologies and Applications*, Advances in Intelligent Systems and Computing 256,
DOI: 10.1007/978-3-319-03581-9_14,

a framework for message exchange and a simple mechanism for specifying persistent intentions and beliefs, in a PRS like style.

A different approach was adopted in the TSTATES (Turtle-States) domain specific language (DSL) [8]. The latter supports agent behaviour specification through state machines, an approach similar to those that have been mainly used in robotics [9] and RoboCup simulation teams [10]. TSTATES provides a small and simple domain specific language (DSL) on top of NetLogo and an execution layer, and thus allows users to encode and execute more sophisticated agent models. The present paper builds on the work described in [11], by discussing in more detail TSTATES, through a complete specification of the DSL, an evaluation of its performance and its application to metro station simulation scenario.

The rest of the paper is organised as follows: Section 2 acts as a brief introduction to NetLogo and its terminology, necessary for placing the rest of the paper in the right context. Section 3 provides a description of TSTATES by presenting its primitives, its grammar and an evaluation of its performance through a motivating example. In section 4 a complete example of TSTATES to a multi agent model concerning crowd simulation is described. Section 5 presents the work reported in the literature that is closely related to the current approach. Finally, section 6 concludes the paper and discusses future extensions.

2 The NetLogo ABMS Environment

NetLogo is "a cross-platform multi-agent programmable modelling environment" [4] aiming to multi-agent systems' simulation with a large number of agents. There are four entities participating in a NetLogo simulation:

- The *Observer*, that is responsible for simulation initialisation and control.
- *Patches*, i.e. components of a user defined static grid that is a 2D or 3D world, which is inhabited by turtles. Patches are useful in describing environment behaviour, since they are capable of interacting with other agents and executing code.
- *Turtles* that are agents that "live" and interact in the above world. They are organised in *breeds*, i.e. user defined groups sharing some characteristics, such as shape, but most importantly breed specific user defined variables that hold the agents' state.
- *Links* agents that "connect" two turtles representing usually a spatial/logical relation between them.

Patches, turtles and links carry their own *internal state*, stored in a set of system and user-defined variables local to each agent. By the introduction of an adequate set of patch variables, a sufficient description of complex environments can be achieved. The definition of turtle specific variables allows the former to carry their own state and facilitates encoding of complex behaviour.

Agent behaviour can be specified by the domain specific NetLogo programming language, that has a rather functional flavour and supports functions (called *reporters*) and *procedures*. The language includes a large set of primitives for turtles motion, environment inspection, classic program control (ex. branching), etc. NetLogo v5 introduced

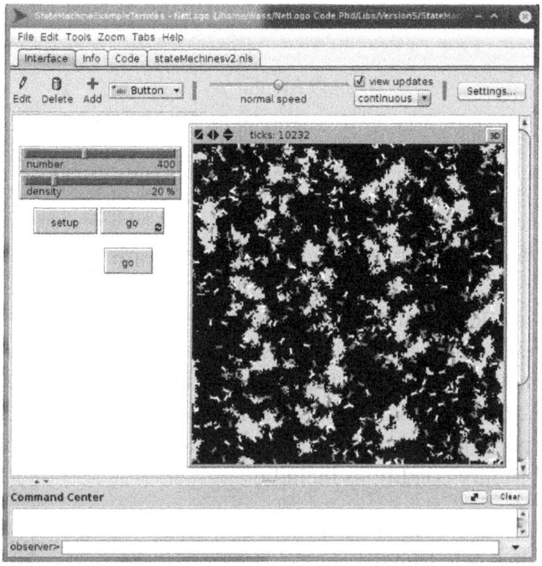

Fig. 1. Termites Model. Yellow patches represent wood-chips, black patches free space, white "bugs" termites not carrying anything, while orange "bugs" termites carrying a wood-chip. Example is taken from the NetLogo models library.

tasks, a significant extension to the language, since through the former users can store code in variables to be executed at a later stage. Reasoning about time is supported through *ticks*, that are controlled by the observer, each tick corresponding to a discrete execution step. Finally, the programming environment offers simple GUI creation facilities that minimizes the time required to develop a simulation. A simple example of a model that can be built in it is shown in figure 1. The model shown is the "termites" model that will serve in the following as the running example in order to present the TSTATES DSL.

3 State Machines for Specifying Behaviour

The requirements for developing TSTATES were a) to provide the modeller with the means to easily encode complex agent behaviours, and b) to seamlessly integrate with the programming environment, maintaining the advantages of the latter. Thus, its was decided that the TSTATES DSL consists of *a set of primitives* to specify turtle behaviour as a state machine, and *an execution layer* for directly executing these specifications in NetLogo. The domain specific language is tightly coupled with the platform's own language, thus allowing the developer to use all the language primitives of the latter in an transparent way.

TSTATES adopts a rather common form of state machines, in which transitions from a state, are labelled with a condition/action pair, i.e

$(State, Condition_1) \Rightarrow (Action_1, Next\_State_1)$.

The library allows directly encoding such transitions in NetLogo in the following form:

```
state <StateName>
  # when <NetLogo Condition 1> do <NetLogo Action 1>  goto <StateName 1>
  ...
  # when <NetLogo Condition i> do <NetLogo Action i>  goto <StateName i>
end-state
```

In the above, a state definition is included in the keywords state and end-state and <StateName> is a string acting as the unique name of the state. Each transition in a state begins with the symbol #. The keywords when, do and goto specify a transition condition, an action and the target state respectively.

A string representation of any valid logical expression of NetLogo reporters preceded by the keyword when can act as a *condition*. Thus, model specific agent "sensors" or platform defined reporters (NetLogo has a large set of the latter) can be used to trigger transitions. Special library conditions include:

- otherwise, in the form of otherwise do <Action> goto <State>, that always evaluates to true.
- for-n-ticks <n>, which evaluates to true for *n ticks* after the state was last entered. This allows agents to perform an action for a certain amount of time upon entering a state.
- after-n-ticks <n>, which constantly evaluates to true *n ticks* after the last entry (activation) of the state. It is useful to encode timeouts related to a state activation.
- Finally, conditions invoked-from <state>, previous-active-state <state>, on-failure <Machine> and on-success <Machine> are special conditions related to machine invocation and will be discussed in section 3.1.

Similarly to conditions, *actions* are string representations of any valid NetLogo sequence of procedures preceded by the keyword do. The special library action ''nothing'' defines transitions that are not labelled with an action.

The keyword goto specifies the transition's target state, one that belongs to the same state machine. There is also another kind of target state transition, that of invoking a different state machine that is discussed in more detail in section 3.1. Two target pseudostates exist success and failure that both represent final states of the machine and have no transitions attached.

The *execution layer* evaluates transition conditions in a state in the order that they appear, firing the first transition in that list whose condition is satisfied (triggered), i.e. imposing a *transition ordering*. Prioritizing transitions based on their order allows behaviour encoding using less complex conditions, at the cost of demanding special care from the user and allows conditions like otherwise to be semantically clear.

A *state machine* is a (NetLogo) list of state definitions, with the first state in this list being the initial state. TSTATES grammar is depicted in figure 2.

Having NetLogo primitives acting as conditions and actions leads to a tight integration with the underlying platform, allows a NetLogo user to easily define all the necessary components of the agents in the model under study and specify the behaviour of the agents using state machines. It also permits adaptation of existing NetLogo models easily. Finally, it should be noted that states (and machines as described later) can communicate information using the turtle's own variables, as for example is reported in [9] as well as through parameter passing of reporters and procedures used in transition definition.

3.1 Callable State Machines

TSTATES supports the concept of *callable* state machines, i.e. state machines that can be invoked by a transition from any state and terminate returning a boolean result. The concept is similar to nested functions, in the sense that when such a machine terminates, control returns to the state that invoked the machine. This feature aims at reducing the number of states required for encoding complex agents, through "code" re-usability. In effect, callable machines allow encoding of a form of agent behaviour templates, i.e. actions to be taken in order to cope with a specific situation that are applied in multiple cases.

A callable state machine, returns whether such a behaviour has succeeded through a boolean value that is signalled by a special transition to a pseudo-state. Thus, each such machine has to include at least a `success` or a `failure` pseudo-state to terminate its execution. Upon termination, the calling state can optionally activate transitions on the result returned by the invoked machine, by employing the special `on-success` <MachineName> and `on-failure` <MachineName> transition conditions. Before the invocation of the corresponding callable machine both these conditions evaluate to false. Machines are invoked through appropriate transition using the `activate-machine` <MachineName> and just as ordinary programming functions, nested invocations for machines can reach any level. The number of different machines that can be invoked from transitions belonging to a single state is unlimited.

To allow encoding of more flexible state machines, i.e. machines the behaviour of which might differentiate based on the "calling" state, two new conditions were introduced:

- `invoked-from <state>`, which evaluates to true if the state that invoked the current state is that stated in the parameter.
- `previous-active-state <state>`, which evaluates to true if the state `<state>` is active (in stack).

The complete TSTATES grammar is depicted in figure 2.

3.2 Coding a Simple Behaviour: The Termites Model

To illustrate the use of TSTATES, a version of the "State Machines" NetLogo library model [4] is employed. The model is an alternative version of the "Termites" model, originally introduced to the platform to illustrate the use of the new concept of tasks, and concerns an example drawn from biology, i.e. simulation of termites gathering wood chips into piles. Termite behaviour is governed by simple rules: each termite wanders

Machine = **'state-def-of-'**MachineName **'report (list'** State+ **')'**
State = StateDef+
StateDef = **'state'** StateName Transition+ **'end-state'**
Transition = **'#'** Condition **'do'** Action StateChange
Condition = **'when'** ReporterExp | **'otherwise'**
| **'for-n-ticks'** N | **'after-n-ticks'** N
| **'invoked-from'** StateName
| **'previous-active-state'** StateName
| **'on-failure'** MachineName
| **'on-success'** MachineName
Action = Procedures | **'nothing'**
StateChange = **"goto'** StateName
| **'activate-machine'** MachineName
| **'success'** | **'failure'**
N = ⟨INTEGER⟩
StateName = ⟨STRING⟩
MachineName = ⟨STRING⟩
Procedures = ⟨STRING⟩
ReporterExp = ⟨STRING⟩

Fig. 2. TSTATES DSL grammar. Please note that Procedures and ReporterExp are string representations of a set of NetLogo procedure and an expression of Netlogo reporters respectively.

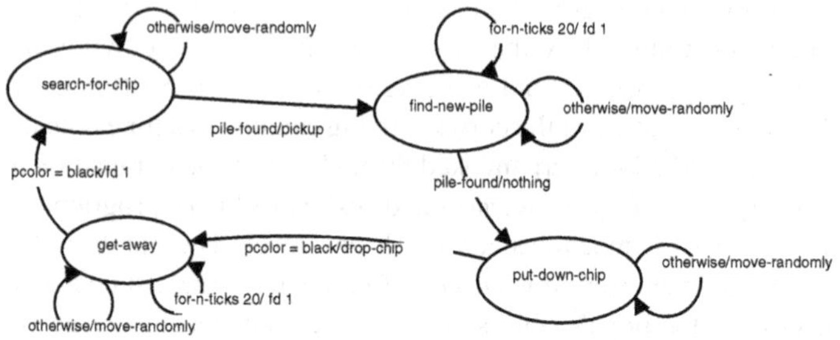

Fig. 3. The Termites State Machine Model. The transitions are labelled by a *condition / action* pair.

randomly until it finds a wood chip, then picks up a chip and carries it until it locates a clear space near another wood pile, where it "drops" the chip its carrying. Eventually, all chips initially scattered in the world are collected in large piles. The state machine model of termites is depicted in figure 3.

The corresponding TSTATES NetLogo code is shown below.

```
to-report state-def-of-turtles
 report (list
  state  "search-for-chip"
   # when "pile-found" do "pick-up" goto "find-new-pile"
   # otherwise do "move-randomly" goto "search-for-chip"
  end-state
  state "find-new-pile"
   # for-n-ticks 20 do "fd 1" goto "find-new-pile"
```

```
    # when "pile-found" do "nothing" goto "put-down-chip"
    # otherwise do "move-randomly" goto "find-new-pile"
  end-state
  state "put-down-chip"
    # when "pcolor = black" do "drop-chip" goto "get-away"
    # otherwise do "move-randomly" goto "put-down-chip"
  end-state
  state "get-away"
    # for-n-ticks 20 do "fd 1" goto "get-away"
    # when "pcolor = black" do "fd 1" goto "search-for-chip"
    # otherwise do "move-randomly" goto "get-away"
  end-state )
end
```

The reader should notice the name of the NetLogo reporter that "stores" the state machine indicates the *breed* of agents whose behaviour is specified (i.e. `state-def-of-turtles` specifies the behaviour of the "turtles" breed). In the model, chips are represented as yellow patches, where free space as black. The conditions "pile-found", corresponds to a simple Netlogo reporter that returns true is the patch the turtle is located on is coloured yellow. Obviously, this could be easily achieved by simply including the `pcolor = yellow` as a condition, as in the case of finding a free space (`pcolor = black`). It was chosen to be included as a reporter in order to demonstrate some aspects of the library and make the model more readable. As seen from the above simple example, encoding state machines in the proposed library is a straightforward task. A comparison and a listing of the two code examples (TSTATES and the original NetLogo) can be found in [11].

3.3 Implementation

A major design choice was to implement TSTATES using the NetLogo programming language. This decision stems from the fact that such a choice allows easy inclusion in any Netlogo model, transparent integration with the underlying platform's language and also easy modification of the primitives offered. The implementation heavily depends on the notion of *tasks* that were introduced in NetLogo version 5. Each such machine specification encoded by the user is compiled at run time to an executable form, that employs directly executable tasks by appropriate function invocations and stored in the corresponding data structures.

For each turtle that uses state machines, four stacks must be defined as turtle's own variables:

- The *active-states* stack that holds the set of states that have not yet terminated along with necessary information concerning the state, i.e. when the state was last entered, results of invoking other machines, etc.
- the *active-states-code* stack that holds the code for each state in the active-states stack,
- the *active-machines* stack that stores the state machine to which each state in the active states stack corresponds to, and finally,

- the *active-machine-names* stack that hold the names of the machines currently active. Obviously the top of each of the stacks is the active state, code and machine respectively that determine the behaviour of the agent.

The DSL includes two procedures. The procedure `initialise-state-machine` performs machine initialisation, i.e. it loads the initial state of the state machine that matches the breed of the turtle. After that, the procedure `execute-state-machine` is the only thing that needs to be "asked" by the turtle in order to execute its specified behaviour. The latter invokes an execution cycle that includes determining all transitions whose condition holds and selecting the first one in order to execute its action part. Although evaluation of all transition conditions seems unnecessary in this step, it was chosen so that transiton prioritisation can be easily implemented in future extension of TSTATES.

Loading a state involves popping the previous state from the active-states stack and pushing the new state. Similar operations occur for the state code in the active-states-code stack. Each normal transition, i.e. to a state that belongs to the current active state machine is loaded in a similar manner. Machine invocation is slightly different in the sense that the execution layer locates the initial state of the invoked machine and structures are simply updated by pushing the new state information.

It should be noted that the `execute-state-machine` procedure selects and executes only one transition at each cycle. This choice was made so that the NetLogo models could be developed more easily since (the *ask turtles*) NetLogo primitive imposes a sequential order on the execution of agents, waiting for one to finish before initiating the next. Additionally, such an approach allows the use of ticks in the simulation.

Obviously, managing such structures impose an overhead to the execution. In order to investigate the former, a set of experiments were performed involving the termites model presented above. The experiments involve running a full simulation, i.e one in which the world is updated on every tick, for a varying number of turtles and allow

Table 1. Experiments with world updates. Columns TSTATES and NetLogo depict the execution time in ms, an AMD Athlon X2, Linux Machine.

Termites (turtles)	Ticks (Run)	TSTATES Time (ms)	NetLogo Time (ms)
300	100	3706.06	3522.7
300	500	17529.38	16968.68
300	1000	35069.38	33930.1
500	100	5274.99	3406.17
500	500	26449.79	16974.99
500	1000	52884.82	33923.64
1000	100	10314.49	5803.46
1000	500	52154.88	29004.4
1000	1000	101169.16	58325.72

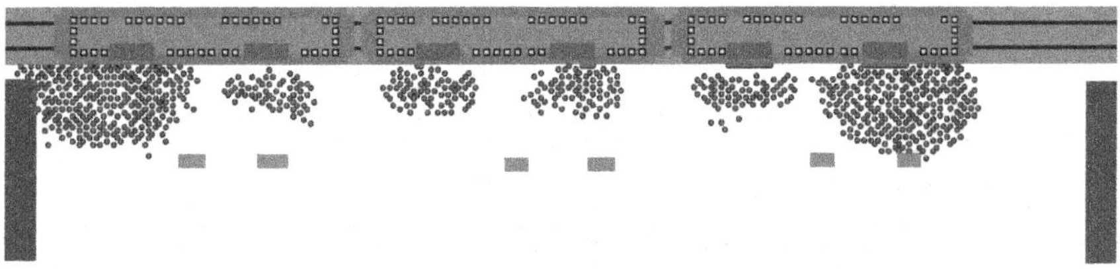

(a) Passengers waiting for doors to open

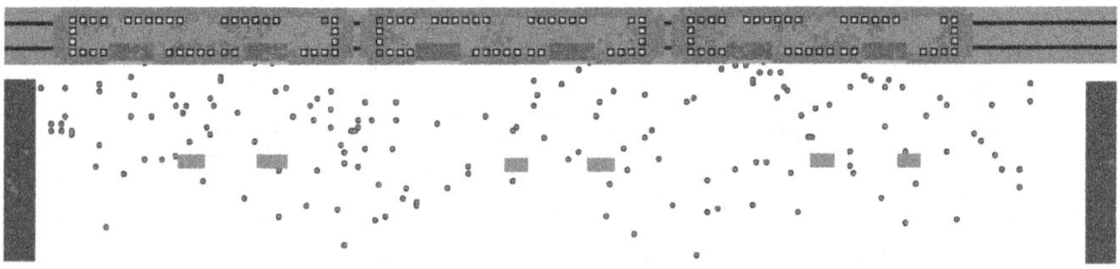

(b) Heading for the exit

Fig. 4. The Underground Station Environment. Different areas are coloured coded in the simulation: the entrance is marked with colour magenta, the wagon area green, and the red coloured patches represent the door area.

each experiment to run for a number of ticks. Time concerns the execution of the top-level procedure that invokes an experiment cycle and was measured using the profiler of NetLogo. The results are shown in table 1, and as indicated when a large number of agents exist in the simulation the execution time is increases, to even a factor of 2 when the number of agents reaches 1000.

It is expected that when NetLogo runs "headless", i.e. without world updates, the overhead introduced would be more significant. Thus, one of the future directions in this research is to provide a more efficient compilation scheme, possibly through directly compiling state machine specifications to NetLogo code. Nevertheless, the aim of this work was to provide modelling facilities for complex behaviours, and in that respect we argue that the modelling complex agents in NetLogo can be greatly facilitated by TSTATES. The case study that follows further supports this claim.

4 Underground Station Simulation

A model of pedestrian simulation in an underground station was developed in order to demonstrate the implementation of more complex behaviours. The example was drawn from [12], where authors use the Situated Cellular Agent model to simulate crowd behaviour while boarding and descending a metro wagon in an underground station. In this paper, as depicted in figure 4, we have tested the passenger model in an environment that has multiple wagons, instead of one as in [11]. Space is discrete, that is agents move on a grid formed by the underlying patches, although continuous space could also be supported.

The simulation concerns a complete passenger cycle, i.e. passenger boarding and descending. This was done, since we wanted to investigate how boarding passengers

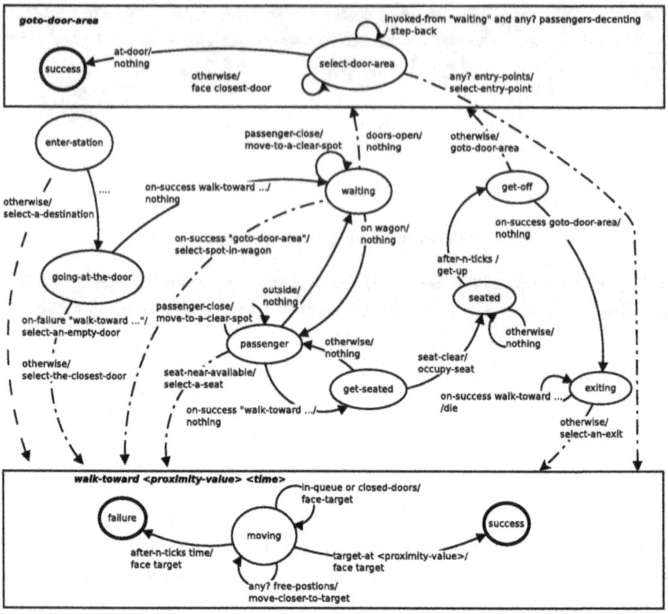

Fig. 5. State Diagram of the Passenger. Please not that dotted arrows indicate machine invocations, where normal arrows simple state transition.

affect the behaviour of passengers descending the wagon, and in order to have a richer state machine to encode.

Passenger behaviour is specified by a state machine, as the latter is depicted in figure 5. This is almost the same machine that we used in [11], that with a few minor modifications copes well with the current environment. Informally and rather briefly, each passenger:

- Upon entering the station walks towards a random platform point and then selects a wagon door to walk to.
- When close to the door and doors open, boards the wagon by selecting a door area to walk toward to. If there are any passengers descending the passenger steps back to facilitate their exit.
- When in the door area, selects a clear spot in the wagon to move to. Upon arriving at the spot, the passenger has completed boarding.
- If the passenger “sees” an empty seat, he/she tries to get seated.
- After a while starts to descend from the wagon. This involves selecting the nearest door area for un-boarding and walks towards that door.
- When at the door area, the passenger selects an exit, walks towards this new target and “leaves” the simulation.

There is a number of interesting points in the state diagram of figure 5. The first point to notice is that the passenger’s walking behaviour is invoked by each state that requires movement: thus it is encoded as a separate state machine (“walk-toward”), reducing the total number of states. There are two things worth mentioning here. Firstly, the state machine is called with two parameters, proximity and time. The first concerns how close to the target should the agent be in order to consider the task successful and the second concerns how long the agent would try to achieve its goal of moving towards the target, before dropping its goal. Thus, the TSTATES allows encoding of parametrised agent plans and a form of intention persistence. Secondly, the target location is communicated between states through an agent (turtle) variable. The sole purpose of this choice was to show that the tight integration of TSTATES with the underlying platform; The same effect could have been easily done by having one more parameter in the state machine. Both the above show how TSATES allows for easy encoding of complex agent behaviour.

A second point to notice concerns the "goto-door-area" state machines. The latter encodes passenger behaviour when moving to the door area, an intermediate target during boarding and descending the wagon. The behaviour is differentiated in the two cases mentioned: if the passenger is boarding, then he must step back to allow other passengers to descend (a polite passenger); if not this behaviour does not occur. This differentiation is achieved by having a transition guarded by a condition that check which state invoked the "goto-door-area" machine, as shown in the code below (numbered (1)):

```
state "select-door-area"
 # when invoked-from "waiting" and any? passengers-descenting"
       do "step-back" goto "select-door-area" (1)
 # when "at-door" do "nothing" success
 # when "any? entry-points" do "select-entry-point"
       activate-machine "walk-toward near 15"
 # otherwise do "face closest-door" goto "select-door-area"
end-state
```

Finally, it should be noted that the "goto-door-area" invokes the "walk-towards" in order the passenger reaches its selected target. Thus, as shown from the example above, TSTATES can indeed meet most of the needs such complex agent simulations demand. Results of the simulation can be viewed in figures 4(a) and 4(b), corresponding to passengers boarding and descending from the wagon.

5 Related Work

The work described in this paper relates both to state machine specification of intelligent agents and programming languages for agent simulation platforms. Thus, in the following we report on the relevant literature on both these areas.

Many approaches reported in the literature adopt finite state machines to control agent behaviour. For example in [10] [13] authors describe a specification language, *XABSL* for defining hierarchies of state machines for the definition of complex agent behaviours in dynamic environments. According to the approach, *options*, i.e. state machines, are organised through successive invocations (one option state can invoke another option) in a hierarchy, an acyclic graph consisting of options, with the leaf nodes being basic behaviours (actions). Traversal of the tree based on external events, state transition conditions and past option activations, leads to a leaf node that is an action. It should be noted that *XABSL* was employed by the German RoboCup robot soccer team with significant success.

COLBERT [9] is an elegant C like language defining hierarchical concurrent state machines. *COLBERT* supports execution of activities (i.e. finite state automata) that run concurrently possibly invoking other activities and communicate through a global store or signals. Agent (robot) actions include robot actions and state changes, and all agent state information is recorded in the Saphira perceptual space.

eXAT [14], models *tasks* of the agent using state machines, that can be "activated" by the rule engine of the agent. eXAT tasks can be combined sequentially or concurrently, allowing re-usability of the defined state machines. Fork and join operators on concurrent state machine execution exist that allow composition of complex tasks.

TSTATES provides some of the above mentioned features and lacks others. State machine invocation is possible through the `activate-machine` primitive, but concurrent execution of state machines, as that is defined in *COLBERT* and *XABSL* is missing. Concurrent actions, although is clearly a desired property in a robotic system that operates in the real world, might not be that suitable for agent simulation platforms and especially for NetLogo. In the latter, fairness among agents in the simulation is provided by ensuring that at each cycle one action is selected and executed in the environment. However, having multiple concurrent active states is a future direction of the TSTATES library, possibly incorporating some sort of priority annotation on the actions that would allow in the end to have a single action as the outcome of the state machine.

There is a large number of agent simulation platforms that have been developed in the past decade [2] [3]. Out of these, state machine like behaviour encoding is offered in two of them, Sesam [15] and RePast [16]. In Sesam a visual approach to modelling agents is adopted, where users develop activities that are organised in using UML-like *activity diagrams*. RePast offers agent behavioural modelling through flowcharts (along with JAVA, Groovy and ReLogo) that allow the user to visually organise tasks. While both approaches are similar to the TSTATES, the latter offers callable states and machine invocation history that, to our opinion, facilitate the development of sophisticated models, as presented above. Furthermore its tight integration with the NetLogo platform and given the latter's simplicity in building simulations, allows users to build models more easily. However, since among some user categories, visual development of state machines is a rather attractive feature, we consider the inclusion of such a facility in the future.

6 Conclusions

This work reports on extensions regarding the TSTATES DSL and on the use of the latter in a more complex example. The approach presents a number of benefits: complex behaviour definition using state transitions and transparent integration with NetLogo platform's language primitives is transparent, thus loosing not expressivity w.r.t. the agent models that can be encoded. We intend to extend the current approach in a number of ways:

- Support the execution of concurrent active states as discussed in section 5 and possibly fork and join composition operators on machine invocation. However, this is a issue that requires further research and outside the scope of this paper.

- Provide facilities for debugging/authoring state machines in NetLogo, as for example visual tools to encode state machines, like in [15] and [16]. The latter we expect to increase the adoption of TSTATES and the platform itself.
- Investigate alternative compilation techniques for more efficient integration with NetLogo.

We are also considering other agent programming language paradigms as well, such as AgentSpeak(L) [17] and their integration to NetLogo. This direction will allow to extend the environment to even more complex simulation scenarios.

Finally, it should be noted that both the library TSTATES and the examples presented in this paper, can be found at http://users.uom.gr/∼ iliass/.

References

1. Davidsson, P., Holmgren, J., Kyhlbäck, H., Mengistu, D., Persson, M.: Applications of agent based simulation. In: Antunes, L., Takadama, K. (eds.) MABS 2006. LNCS (LNAI), vol. 4442, pp. 15–27. Springer, Heidelberg (2007)
2. Nikolai, C., Madey, G.: Tools of the trade: A survey of various agent based modeling platforms. Journal of Artificial Societies and Social Simulation 12, 2 (2009)
3. Allan, R.J.: Survey of agent based modelling and simulation tools. Technical Report DL-TR-2010-007. DL Technical Reports (2010)
4. Wilensky, U.: Netlogo. Center for Connected Learning and Computer-based Modelling. Northwestern University, Evanston, IL (1999), `http://ccl.northwestern.edu/netlogo`
5. Railsback, S.F., Lytinen, S.L., Jackson, S.K.: Agent-based simulation platforms: Review and development recommendations. Simulation 82, 609–623 (2006)
6. Lytinen, S.L., Railsback, S.F.: The evolution of agent-based simulation platforms: A review of netlogo 5.0 and relogo. In: Proceedings of the Fourth International Symposium on Agent-Based Modeling and Simulation, Vienna, Austria (2012)
7. Sakellariou, I., Kefalas, P., Stamatopoulou, I.: Enhancing Netlogo to Simulate BDI Communicating Agents. In: Darzentas, J., Vouros, G.A., Vosinakis, S., Arnellos, A. (eds.) SETN 2008. LNCS (LNAI), vol. 5138, pp. 263–275. Springer, Heidelberg (2008)
8. Sakellariou, I.: Turtles as state machines - agent programming in netlogo using state machines. In: Filipe, J., Fred, A.L.N. (eds.) Proceedings of the 4th International Conference on Agents and Artificial Intelligence, ICAART 2012. Agents, vol. 2, pp. 375–378. SciTePress (2012)
9. Konolige, K.: COLBERT: A language for reactive control in sapphira. In: Brewka, G., Habel, C., Nebel, B. (eds.) KI 1997. LNCS, vol. 1303, pp. 31–52. Springer, Heidelberg (1997)
10. Loetzsch, M., Risler, M., Jungel, M.: Xabsl - a pragmatic approach to behavior engineering. In: 2006 IEEE/RSJ International Conference on Intelligent Robots and Systems, pp. 5124–5129 (2006)
11. Sakellariou, I.: Agent based modelling and simulation using state machines. In: Pina, N., Kacprzyk, J., Obaidat, M.S. (eds.) SIMULTECH, pp. 270–279. SciTePress (2012)
12. Bandini, S., Federici, M.L., Vizzari, G.: Situated cellular agents approach to crowd modeling and simulation. Cybernetics and Systems 38, 729–753 (2007)
13. Risler, M., von Stryk, O.: Formal behavior specification of multi-robot systems using hierarchical state machines in XABSL. In: Workshop on Formal Models and Methods for Multi-Robot Systems, AAMAS 2008, Estoril, Portugal (2008)

14. Stefano, A., Santoro, C.: Supporting agent development in erlang through the exat platform. In: Unland, R., Calisti, M., Klusch, M., Walliser, M., Brantschen, S., Calisti, M., Hempfling, T. (eds.) Software Agent-Based Applications, Platforms and Development Kits. Whitestein Series in Software Agent Technologies and Autonomic Computing, pp. 47–71. Birkhuser, Basel (2005)
15. Klügl, F., Herrler, R., Fehler, M.: Sesam: implementation of agent-based simulation using visual programming. In: Proceedings of the Fifth International Joint Conference on Autonomous Agents and Multiagent Systems, AAMAS 2006, pp. 1439–1440. ACM, New York (2006)
16. North, M.J., Howe, T.R., Collier, N.T., Vos, J.R.: A declarative model assembly infrastructure for verification and validation. In: Advancing Social Simulation: The First World Congress. Springer, Heidelberg (2007)
17. Rao, A.S., Georgeff, M.P.: Modeling rational agents within a BDI-architecture. In: Allen, J., Fikes, R., Sandewall, E. (eds.) Proceedings of the 2nd International Conference on Principles of Knowledge Representation and Reasoning (KR 1991), pp. 473–484. Morgan Kaufmann Publishers Inc. (1991)

Stability Analysis of Climate System Using Fuzzy Cognitive Maps

Carlos Gay García and Iván Paz Ortiz

Programa de Investigacin en Cambio Climático,
Universidad Nacional Autónoma de México, Mexico
cgay@servidor.unam.mx, ivanpaz@ciencias.unam.mx

Abstract. In the present work we developed a soft computing model for the qualitative analysis of the Earth's climate system dynamics throughout the implementation of fuzzy cognitive maps. For this purpose, we identified the subsystems in terms of which the dynamics of the whole system can be described. Then, with these concepts we built a cognitive map via the study of the documented relations among these concepts. Once the map was built, we used the technique of state vector and the adjacent matrix to found the hidden pattens, i.e the feedback processes among system's nodes. Later on, we explored the sensibility of the model to changes in the weights of the edges, and also to changes in the input data values. Finally, we used fuzzy edges to analyze the causality flux among concepts and to explore possible solutions applied in specific edges.

Keywords: Cognitive Maps, Fuzzy Systems, Climate System, Stability Analysis.

1 Introduction

Fuzzy cognitive maps are fuzzy-graph structures for representing causal reasoning between variable concepts [3]. The concepts are represented as nodes $(C_1, C_2, ..., C_n)$ in an interconnected network. Each node C_i represents a variable concept, and the edges e_{ij} which connect C_i and C_j (denoted as $C_i \rightarrow C_j$) are causal connections and express how much C_i causes C_j. These edges can be negatives or positives. A positive relation $C_i \rightarrow {}^+C_j$ states that if C_i grows, also does C_j, and a negative relation $C_i \overset{-}{\rightarrow} C_j$ indicate that as C_i grows C_j decreases. The fuzzy cognitive maps have been used to model different kind of systems, such as control, social and economics systems, and also in robotics, computed assisted learning, expert systems, and many others. Even though they have been applied to many areas, little has been done in atmospheric sciences. Soft computing models are specially useful in cases (such as is the case of the climate system) where the system contains many associated uncertainty and it is impossible to use "hard computing models", or when little is known about the relations among its subsystems components. This is the importance of the present work, it seeks to apply techniques of soft computing (specially fuzzy cognitive maps) for the qualitative analysis of climate systems.

As an example of cognitive maps, consider the following network of 4 nodes. Each node represents a concept and the relations among them are described using positive (+) and negative (-) causality.

M.S. Obaidat et al. (eds.), *Simulation and Modeling Methodologies, Technologies and Applications*, Advances in Intelligent Systems and Computing 256,
DOI: 10.1007/978-3-319-03581-9_15,

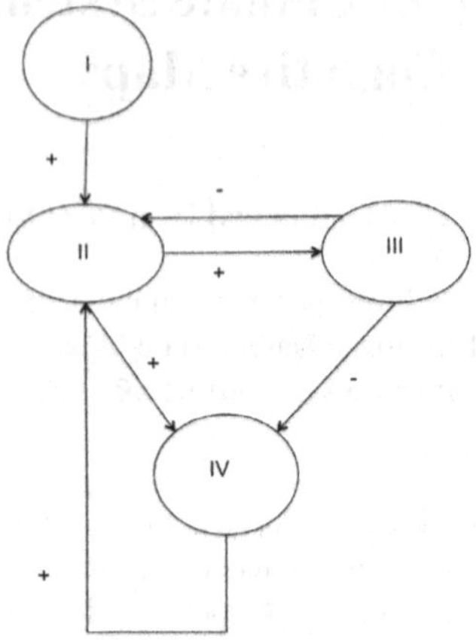

Fig. 1. Cognitive Map of 4-Nodes

The state of the system can be represented as a state row vector, which contains the values for each node at a given time t. For example, the row vector v_0= [1,0,0,0] at t_0 indicates that only the first node in the system is ON at this time. The edges e_{ij} can be represented in an adjacent matrix. This contains the causal relationships among nodes. And in some sense, it contains the system information. In cognitive maps the adjacent matrix is always square. In our example the matrix is:

$$M = \begin{bmatrix} 0 & 1 & 0 & 0 \\ 0 & 0 & 1 & 1 \\ 0 & -1 & 0 & -1 \\ 0 & 1 & 0 & 0 \end{bmatrix}$$

Then, the system is iterated to get v for $t_1, t_2, ..., t_n$. In each iteration, the row vector is multiplied by the matrix $v * M$, so we have:

$$v_t = v_{t-1} * M \quad (1)$$

If we consider v_0= [1,0,0,0], then, when iterating the system we obtained:
v_1= [0,1,0,0] The node I causes II.
v_2= [0,0,1,1] The node II lights on the nodes III and IV.
v_3= [0,0,0,-1] The node III acts negatively over II y IV, in the node II the negative effect vanishes.
v_4= [0,-1,0,0] The node IV acts negatively on II.
v_5= [0,0,-1,-1] Node II acts negatively over III y IV.
v_6= [0,0,0,1].

In the next iteration we obtain v_7= [0,1,0,0], which is equal to v_1= [0,1,0,0]. In this case, the system reaches a limit cycle. In general it can be shown that for every matrix the system converges to a point, to a limit cycle, or diverges in the space of states for the system [4]. If we express the state vector as $X(t) = (x_1(t), ..., x_n(t))$, and the signal value as S_i, we can define a threshold binary function, which limits the signal value by a pre-defined threshold (in this case 0 and 1) for each time t as follows:

$$S_i(x_i^t) = \begin{cases} 1 \ \ if \quad x_i^t > 1 \\ x_i^t \ \ if \ 0 < x_i^t < 1 \\ 0 \ \ if \quad x_i^t < 0 \end{cases} \tag{2}$$

In this way, the node values will be restricted between 0 and 1. Similarly, it can be defined ternary functions with values of {-1,0,1}.

2 Subsystems of the Climate System

The proposed subsystems for describing the dynamic of the climatic system are based on the paper "A safe operating space for humanity" written by Johan Rockström [6], and also on the Intergovernmental Panel on Climate Change Report [2]. These are:

Industrialization
This process comprises: social and economic development, technology tendencies and its applications, industry grown, as well as demographic changes associated with these processes [2].

Climate Change
This concept refers to the increase in the mean temperature of the earth, i.e. changes in climate variability in terms of the extreme and mean values [2]. Specifically, we refer to antropogenic climate change, which is a consequence of the human activity.

Changes in CO_2 Concentration
Defined as the increase in the parts per million of CO_2 molecules in the atmosphere [2].

Biodiversity Loss
Refers to the extinction rate, the number of species loss per million per year. Mace and collaborators [5], define biodiversity as the variability of living organisms, included terrestrial and marine ecosystems, other aquatic ecosystems and the ecological systems in which they reside. It comprises the diversity within species, among species and within ecosystems. Mace emphasizes three levels of biodiversity: genes, species, and ecosystems. Biodiversity loss during the industrial period has grown notably. The species extinction rate is estimated against the fossil record. The extinction rates per million per year varies for marine life between 0.1 and 1 and for mammals between 0.2 and 0.5. Today, the rate of extinction of species is estimated to be 100 to 1,000 times more than what could be considered natural [6].

Phosphorus and Nitrogen Cycles
The changes in P and N cycles are estimated with the quantity of P going to the oceans, measured in million tones per year, and with the amount of N_2 removed from the atmosphere for human use, also in million tones per year.

Ocean Acidification
Defined as the ocean pH increase, mainly in the surface layer. The acidification process is closely related with the CO_2 emission level. When the atmosphere CO_2 concentration increases, the amount of carbon dioxide dissolved in water as carbonic acid increases, which in turn, modifies the surface pH. Normally, the ocean surface is basic with a pH of approximately 8.2. Nevertheless, the observations show a decline in pH to around a value of 8. These estimations are made using the levels of aragonite (a form of calcium carbonate) that is created in the surface layer. This concept has an important relation with biodiversity loss as many organisms (like corals and phytoplankton), basic for the food chain, use aragonite to produce their skeletons or shells. As the aragonite value decreases, the ocean ecosystems weaken. [1].

Land Use (Urban Growth and Agriculture Use)
The IPCC defines the change in land use as the percentage of global land converted into cropland. A general definition of land use change includes any type of human use. This transformation, either to cropland or urban, increases the biodiversity loss, which is associated with the destruction of ecosystems. In order to establish the difference in land use between urban growth and agriculture use, and their different consequences, we include both concepts as nodes in the map.

Increase in Fresh Water Demand
Defined as the increase in its current use. Today, the annual use of freshwater from rivers, lakes and groundwater aquifers is of 2,600 km^3. From that, 70% is destined for irrigation, 20% for industry, and 10% for domestic use. This extraction causes the drying and reduction of body waters [2].

Stratospheric Ozone Depletion
O_3 depletion is estimated according to the ozone concentration in the atmosphere in Dobson Units.[1]

Chemic Pollution
It refers to the emitted quantity, persistence, or concentration of organics pollutants, plastics, heavy metals, chemical and nuclear residues, etc., which affect the dynamic of ecosystems.

Aerosol Loading
Referred as the concentration of particles in the atmosphere. These can be lead, copper, magnesium, iron, traces of fire, ashes, etc.

[1] Dobson unit is a measure of the ozone layer thickness, equal to 0,01 mm of thickness in normal conditions of pressure and temperature (1 atm and 0 C respectively), expressed as the molecule number. DU represents the existence of 2.69 x 10^{16} molecules per square centimeter.

With the use of these definitions we created a cognitive map establishing and weighting the relationships among the concepts (different subsystems). Through the analysis of the map we will create a qualitative global vision of the climatic system dynamics.

3 The Model

Using the subsystems described, we build the cognitive map shown below. The map contains the concepts representing them as nodes, and the edges showing the relations among them. These relations have been established with the consulted references. As many relations between concepts can not be quantifiable, i.e. can not be expressed numerically, we have used linguistic variables taken from experts opinion. These variables will be later used with the *Min-Max* criteria. To analyze the system we start with the adjacent matrix and the state vector, considering values of 0 and 1 (positive causality and no causality) and a binary threshold function. It is important to note that in the map all the relations (edges) are positives, so the associated matrix is:

$$M = \begin{bmatrix} 0&1&0&0&0&0&1&1&1&0&1&1 \\ 0&0&1&0&1&0&0&0&0&0&0&0 \\ 0&0&0&1&0&0&0&0&0&0&1&0 \\ 0&1&1&0&0&0&0&0&0&0&1&0 \\ 0&0&0&1&0&0&0&0&0&0&0&0 \\ 0&0&1&1&0&0&0&1&0&0&0&0 \\ 0&1&0&1&0&0&0&0&0&0&0&0 \\ 0&0&0&1&0&0&0&0&1&1&0&0 \\ 0&0&0&1&0&0&0&0&0&0&0&0 \\ 0&0&0&0&0&0&0&0&0&0&0&0 \\ 0&0&0&1&0&0&0&0&0&0&0&0 \\ 0&0&0&0&0&1&1&0&0&0&1&0 \end{bmatrix}$$

3.1 Hidden Patterns

First, we turn on the first node in the state vector $a_0 = [1, 0, 0, 0, 0, 0, 0, 0, 0, 0, 0, 0]$, then we iterate $a * M$, and reset to 1 the first entrance of the vector a after each iteration i.e. $(a[1] = 1)$ this forcing will show the hidden patterns of the system. In the network the *shortest way* between the first node (C_1) and any other consists of two steps, so when we turn on the node C_1 the system converges to the equilibrium state "a_2" in two steps, with:

$$a_2 = [1, 1, 1, 1, 1, 1, 1, 1, 1, 1, 1, 1]$$

This vector state represents the increase of each node due to the causal action of C_1 but it does not say anything more, i.e. the action of C_1 causally increases the other nodes of the network when is forced with $(a[1] = 1)$. When the first node is not forced $(a[1] = 1$ only in $t_0)$, the system reaches an equilibrium among the nodes that have feedback. In this case we get:

$$a_0 = [1, 0, 0, 0, 0, 0, 0, 0, 0, 0, 0]$$

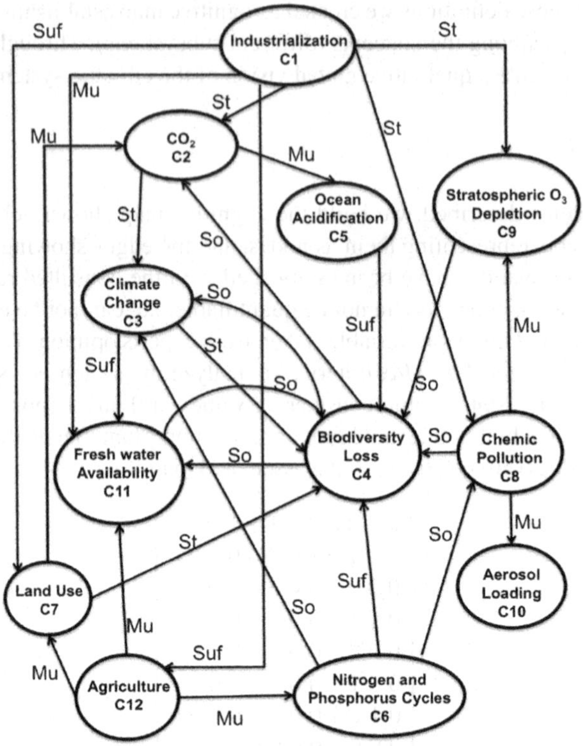

Fig. 2. Cognitive map with fuzzy edges. We have used four causal linguistic quantities: Strong (St), Much (Mu), Sufficient (Suf), and, Some (So).

$$a_1 = [0, 1, 0, 0, 0, 0, 1, 1, 1, 0, 1, 1]$$

The nodes that are connected with C_1 turn on.

$$a_2 = [0, 1, 1, 1, 1, 1, 1, 0, 1, 1, 1, 0]$$

The nodes 3, 4, 5, 6, 10 and 11 are turned on. These are the nodes connected in two steps with C_1. While node 8 is turned off (no node activates it in this iteration). In the fifth iteration the system converges to vector a_5 shown below. This vector exhibits the hidden pattern among nodes with feedback. $a_5 = [0, 1, 1, 1, 1, 0, 0, 0, 0, 0, 1, 0]$

3.2 Changes in the Initial Conditions

Another approximation is to turn on the first node with an intermediate value in $[0, 1]$, which represents the case where the industrialization node decreases from the current level but is not zero. In this case we choose the value 0.5 i.e. assuming an intermediate industrialization activity. We first iterate the system turning on the first node after each iteration. In the following experiment we start the system without turning on the node. In the first case, the system converges in six iterations to $a_6 =$

$[0, 1, 1, 1, 1, 0, 0, 0, 0, 0, 1, 0]$, and in the second case the system converges in four iterations to $a_4 = [0.5, 1, 1, 1, 1, 0.5, 1, 1, 1, 1, 1, 0.5]$. In the first case, we found that the system reaches an equilibrium among nodes 2, 3, 4, 5, and 11, which we will discuss in the next section. In the second case, we found a new state of equilibrium corresponding to the value applied to C_1. These approximations help us to figure out how policies can affect the equilibrium states of the system when they are applied to a specific node.

3.3 Analysis of the Subsystem with Feedback

The equilibrium is reached among nodes 2, 3, 4, 5, and 11 where feedback exists. They remain on, even though the other nodes are turned off. This fact says that once these nodes are turn on, they remain on, and this does not depend on the rest of the system. If we analyse the subsystem of nodes 2, 3, 4, 5 and 11 (that represent CO_2 growth, Climate Change, Biodiversity Loss, Ocean Acidification, and Fresh Water Availability, respectively) renamed as 1, 2, 3, 4, and 5 we have:

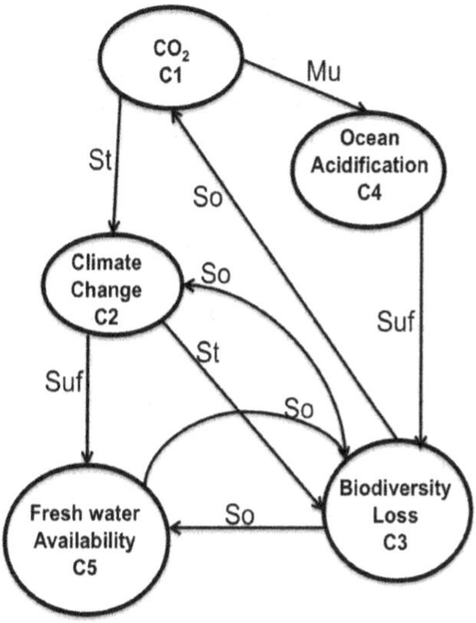

Fig. 3. System of nodes 2, 3, 4, 5, and 11 renamed as 1. CO_2 Increase, 2. Climate Change, 3. Biodiversity Loss, 4. Ocean Acidification, and 5. Fresh Water Availability respectively.

$$M_{subsystem} = \begin{bmatrix} 0\,1\,0\,1\,0 \\ 0\,0\,1\,0\,1 \\ 1\,1\,0\,0\,1 \\ 0\,0\,1\,0\,0 \\ 0\,0\,1\,0\,0 \end{bmatrix}$$

If we force the first node at each step, in two iterations we get $a = [1, 1, 1, 1, 1]$. And when we activate C_1 only in (a_0) the system also converges to $a = [1, 1, 1, 1, 1]$ in five steps. This means that whenever the node C_1 is activated, at any time, the system converges into a point, and the feedback is maintained. To analyze the system behavior we can consider weighted edges. This weights are obtained from the fuzzy linguistic

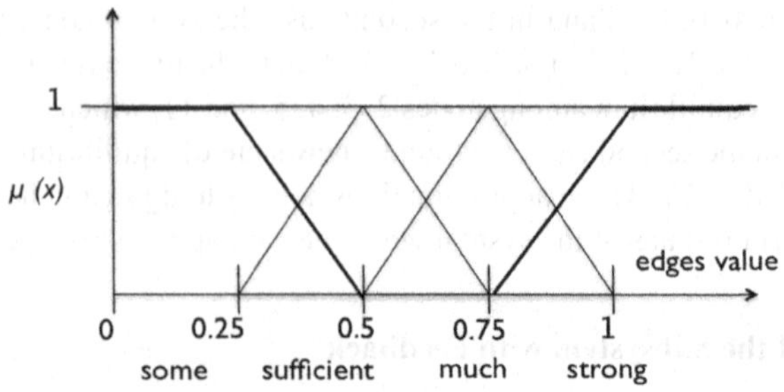

Fig. 4. The x-axis represents the values of the causality between nodes divided in four sets. The y-axis represents the membership degree of each value in the x-axis for each fuzzy set. In this manner, 0.25 belongs to the set of *some* causality with a membership degree $\mu(0.25) = 1$ represented in the y-axis. The value 0.75 belongs to the set *sufficient* with a membership degree of 1 and so on.

variables stablished. We considered (in this case) four possible quantifiers for each edge value *some, sufficient, much, strong*, and associated the values in $[0, 1]$ as follows *some = 0.25, sufficient = 0.5, much = 0.75 and strong = 1*. To do this association we assumed that these numerical values correspond to the maximum membership values for each linguistic fuzzy subset in the universe of causality. This is shown in Figure 4. With these values the system matrix is expressed as:

$$M_{subsystem} = \begin{bmatrix} 0 & 1 & 0 & 0.75 & 0 \\ 0 & 0 & 1 & 0 & 0.5 \\ 0.25 & 0.25 & 0 & 0 & 0.25 \\ 0 & 0 & 0.5 & 0 & 0 \\ 0 & 0 & 0.25 & 0 & 0 \end{bmatrix}$$

Turning ON C_1 and keeping $a[1] = 1$ after each iteration the system converges to the vector:

$$a_3 = [1, 1, 1, 1, 0.5]$$

When we turn on the first node only at the first iteration the state vector shows the following behavior: $a_0 = [1, 0, 0, 0, 0] \rightarrow a_1 = [0, 1, 0, 0.75, 0] \rightarrow a_2 = [0, 0, 1, 0, 0.5]$

$\rightarrow$

$a_3 = [0.25, 0.25, 0.25, 0, 0.25] \rightarrow a_4 = [0.03, 0.28, 0.31, 0.18\ 0.15\ ] \rightarrow$
$a_5 = [0.07, 0.10, 0.41, 0.02\ 0.21] \rightarrow a_6 = [0.10, 0.18, 0.17, 0.05\ 0.15] \rightarrow$
$a_7 = [0.04, 0.14, 0.25, 0.07\ 0.13] \ldots \rightarrow a_{20} = [0.01, 0.02, 0.04, 0.01\ 0.02]$ Which

means that if we force the system only at t_0 the system tends to the equilibrium vector $a_{eq} = [0, 0, 0, 0, 0]$. The system's evolution is shown in Figure 5 left, in the graph the numbers indicate the number of the concept for each line. In this case, the matrix coefficients represent the system damping to the initial perturbation. This fact is important because it shows that once a node in the subsystem is activated, the feedback remains but the final state will depend on the weights of the edges. This specific behavior can be interpreted in terms of the scenarios and mitigation actions, e.g. can the matrix weights

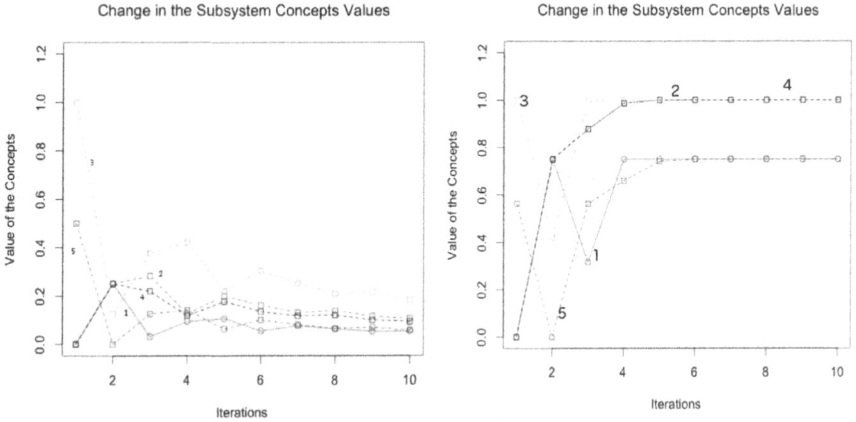

Fig. 5. Left: Evolution of the weighted subsystem of nodes 1. CO_2 Increase, 2. Climate Change, 3. Biodiversity Loss, 4. Ocean Acidification, and 5. Fresh Water Availability. In this case, node 1 has been initiated with a value of 1, but when it is not kept with this value the weighted edges lead the system to zero. Right: Evolution of the subsystem of nodes 1. CO_2 Increase, 2. Climate Change, 3. Biodiversity Loss, 4. Ocean Acidification, and 5. Fresh Water Availability. In this case, node 1 has been initiated with a value of 1 just in the initial vector and we have considered a weight of 0.7 for each edge in the map. We can see that nodes 2, 3, and 4 diverge, while nodes 1 and 5 reach a stable value. This shows the strongly dependence of the system to the weights of the edges.

be modified by specific actions or strategies? Even though with the weights considered, the subsystem converges in the first case, in Figure 5 right, we have turned on the node 1 of the system only in the first state vector, an then, we have iterated the system considering a weight of 0.7 for all the edges. We can see that the system behavior is qualitative different. Nodes 2, 3, and 4 diverge while nodes 1 and 5 reach a stable value. The difference between these two cases shows the strongly dependence of the system to the chosen weights for the edges.

3.4 Fuzzy Weighted System

The matrix associated with the fuzzy system is shown in Figure 6. Turning on the node C_1 we have (a_0 = [1 0 0 0 0 0 0 0 0 0 0 0]), and iterating, the system converges in two steps into vector: a_2^* = [1.00, 1.00, 1.00, 1.00, 0.75, 0.37, 0.87, 1.00, 1.00, 0.75, 1.00, 0.5]

Here we have forced C_1, i.e. at each iteration $a[1] = 1$. We found that the nodes 3, 5, 6, 7, 8, 10, and 12, reach stable values, while the nodes 2, 4, 9, and 11, have been limited to 1 by the threshold function. This offers important information about the network. As nodes 2, 4, 9, and 11 diverge, they could be interpreted as sensitive nodes, i.e. these are the nodes that can not reach an equilibrium with the present conditions. Then, for example, is possible to think about short-term actions for them in order to avoid unwanted scenarios. Furthermore we can say that even thought the forcing node is industrialization, mitigation actions may be focused in more than one node. When we restart C_1 only at t_0 the system is damping again by the matrix coefficients. a_1 = [0,

$$M_{system} = \begin{bmatrix} 0 & 1 & 0 & 0 & 0 & 0 & 0.5 & 1 & 1 & 0 & 0.75 & 0.5 \\ 0 & 0 & 1 & 0 & 0.75 & 0 & 0 & 0 & 0 & 0 & 0 & 0 \\ 0 & 0 & 0 & 1 & 0 & 0 & 0 & 0 & 0 & 0 & 0.5 & 0 \\ 0 & 0.25 & 0.25 & 0 & 0 & 0 & 0 & 0 & 0 & 0 & 0.25 & 0 \\ 0 & 0 & 0 & 0.5 & 0 & 0 & 0 & 0 & 0 & 0 & 0 & 0 \\ 0 & 0 & 0.25 & 0.5 & 0 & 0 & 0 & 0.25 & 0 & 0 & 0 & 0 \\ 0 & 0.75 & 0 & 1 & 0 & 0 & 0 & 0 & 0 & 0 & 0 & 0 \\ 0 & 0 & 0 & 0.25 & 0 & 0 & 0 & 0 & 0.75 & 0.75 & 0 & 0 \\ 0 & 0 & 0 & 0.25 & 0 & 0 & 0 & 0 & 0 & 0 & 0 & 0 \\ 0 & 0 & 0 & 0 & 0 & 0 & 0 & 0 & 0 & 0 & 0 & 0 \\ 0 & 0 & 0 & 0.25 & 0 & 0 & 0 & 0 & 0 & 0 & 0 & 0 \\ 0 & 0 & 0 & 0 & 0 & 0.75 & 0.75 & 0 & 0 & 0 & 0.75 & 0 \end{bmatrix}$$

Fig. 6. Weighted Matrix of the System

1.00, 0, 0, 0, 0, 0.5, 1.00, 1.00, 0, 0.75, 0.5]
$a_2 = [0, 0.37, 1.00, 1.00, 0.75, 0.37, 0.37, 0, 0.75, 0.75, 0.37, 0]$
In a_2 the node 4 has been thresholded form 1.18 to 1, these is because this node has the greatest causality over it in the network.
$a_3 = [0, 0.53, 0.71, 1.00, 0.28, 0, 0, 0.09, 0, 0, 0.75, 0]$
In a_3 the node 4 has been thresholded again.
$a_4 = [0, 0.25, 0.78, 1.00, 0.39, 0, 0, 0, 0.07, 0.07, 0.60, 0]$
In a_4 we also threshold node 4 from 1.07 to 1
$a_5 = [0, 0.25, 0.5, 1, 0.18, 0, 0, 0, 0, 0, 0.64, 0]$
In a_5 node 4 is threshold from 1.15 to 1
$a_6 = [0, 0.25, 0.5, 0.75, 0.18, 0, 0, 0, 0, 0, 0.5, 0]$
In step 6 node 4 need no threshold. From these point the system converges to zero.
$a_7 = [0, 0.18, 0.43, 0.71, 0.18, 0, 0, 0, 0, 0, 0.43, 0] \rightarrow \ldots a_{14} = [0, 0.08, 0.17, 0.29, 0.07, 0, 0, 0, 0, 0, 0.18, 0]$

We can see how the subsystem with feedback remains, but converges to zero as iterations progress. The fact that the node 4 diverges give us important information about the sensibility of this node to the network conditions, i.e. node 4 has strong sensibility.

Another important analysis is what happens when the network is at equilibrium and then the conditions change? In order to analyze this we consider the vector a_2^* that we obtained while maintaining forcing. Then turn off the first node and calculate $a * M_{sist}$.

$a_0 = [1.00, 1.00, 1.00, 1.00, 0.75, 0.37, 0.87, 1.00, 1.00, 0.75, 1.00, 0.5]$
$a_1 = [0, 1.90, 1.34, 3.18, 0.75, 0.37, 0.87, 1.09, 1.75, 0.75, 1.87, 0.5]$
using threshold to limit the values of nodes 2, 3, 4, 8, 9, and 11
$a_1 = [0, 1.00, 1.00, 1.00, 0.75, 0.37, 0.87, 1.00, 1.00, 0.75, 1.00, 0.5]$
in the second iteration we have:
$a_2 = [0, 0.90, 1.00, 1.00, 0.75, 0.37, 0.37, 0.09, 0.75, 0.75, 1.00, 0]$
where we also used the *treshold* function for nodes 3, 4 and 11. At iteration 10 we have:

$a_{10} = [0, 0.13, 0.29, 0.49, 0.12, 0, 0, 0, 0, 0, 0.34, 0]$

Which shows that once forcing is removed, the system reaches the equilibrium at $a_n = [0, 0, 0, 0, 0, 0, 0, 0, 0, 0, 0, 0]$, where the matrix coefficients remain as damping (*Liapunov*) coefficients. Nevertheless, it is important to characterize the feedback processes in order to know if the threshold function is changing (or not) the system's behavior.

3.5 Indirect and Total Effects Using the *(Min-Max)* Criteria

When we assign *fuzzy weights* we can use the Indirect and total effect operators (I and T) to analyze the causal action among nodes. The Indirect effect is the minimum of all the linguistic edges in a trajectory, and the Total effect will be the maximum of all the indirect effects. Let us analyze, for example, the causality between industrialization and biodiversity loss. To go from C_1 to C_4 we have the trajectories: (1,2,3,4), (1,7,4), (1,12,11,4), (1,11,4), (1,8,4), (1,9,4), and (1,12,6,4),(1,12,6,8,4), and (1,12,6,3,4), which are denoted here as $I_1, I_2, I_3, I_4, I_5, I_6, I_7, I_8, I_9, I_{10}$ respectively. If we analyze the linguistic quantifiers among them: $I_1(C_1, C_4) = min\{e_{12}, e_{23}, e_{34}\} = min\{strong, strong, strong\} = strong$. and so: $I_2(C_1, C_4) = min\{e_{17}, e_{74}\} = min\{sufficient, strong\} = sufficient$.
$I_3(C_1, C_4) = min\{sufficient, much, much, sufficient\} = sufficient.$
$I_4(C_1, C_4) = min\{sufficient, much, some\} = some.$
$I_5(C_1, C_4) = min\{much, some\} = some.$
$I_6(C_1, C_4) = min\{strong, some\} = some.$
$I_7(C_1, C_4) = min\{strong, some\} = some.$
$I_8(C_1, C_4) = min\{sufficient, much, sufficient\} = sufficient.$
$I_9(C_1, C_4) = min\{sufficient, much, some, some\} = some.$
$I_{10}(C_1, C_4) = min\{sufficient, much, some, strong\} = some.$
$T(C_1, C_4) = max\{I_1, I_2, I_3, I_4, I_5, I_6 I_7, I_8, I_9, I_{10}\} = max\{strong, sufficient, sufficient, some, some, some, some, sufficient, some, some\} = strong$. Which means that C_1 "strongly causes" C_4. The *(Min-Max)* analysis give us important information about how causality is operating in the system, and which relation or relations among subsystems exert more influence over one specific node. In the example, we found that C_1 strongly causes C_4, however, the only path that posses strong causality as indirect effect is $I_1(C_1, C_4)$. We can use this information for strategic planning, i.e. to modify the action that C_1 exerts over C_4 (the total effect T), we must focus on the relations e_{12}, e_{23}, and e_{34} rather than in the other edges. Changing only one of these relations, the total effect of C_1 over C_4 will change to *sufficient*.

4 Conclusions

Throughout this work we build a model for the earth climate system by relating earth subsystems recognized by experts as those which determine the climate system stability. In the analysis we found that with the established network conditions, and considering only positive causality, the system diverges, which means that all nodes increase without limit leading the system into an undesired state. Then, we turn off the system's forcer in order to identify the feedback processes. Nodes with feedback are specially important for us as the causality persists among them, even though the forcing has been turned off. So, these nodes need special attention because the system behavior depends on the edge's weights, once a node is activated. When we focus on the subsystem with feedback, we discover that without the forcer the system returns to the equilibrium state, even without threshold. Then, we use the threshold function when working with the complete system. It is necessary to explore in greater depth the connection edges and their weights to know if the system is really modeling these interactions.

The system's behavior starting in a particular state exhibits all the behaviors mentioned above, however, the threshold use is stronger in this case, which also suggests the necessity of refining the weights among subsystems.

With respect to the *Min/Max* criteria, we found that the trajectory (I_1) with more weight between the forcer (C_1) and the node with more entries in the net (C_4) is the one that goes from 1 to 2, from 2 to 3 and from 3 to 4. The analysis of all other possible trajectories shows that the indirect effects of each one are less than those of I1. To modify the causality of C_1 over C_4, we have to focus only on the indicated trajectory, i.e. if we change the other trajectories we would not change the weight over the C_4 node. This is one of the examples where the cognitive map allow us to make strategic decisions about a net.

Finally, the use of cognitive maps in climate systems allows the creation of knowledge networks that integrate systems that under any other circumstance would be impossible to integrate to obtain qualitative information of the system that can be easily adapted to decision making schemes.

Acknowledgements. The present work was developed with the support of the Programa de Investigación en Cambio Climático (PINCC) of the Universidad Nacional Autónoma de México (UNAM).

References

1. Foley, J.A.: Boundaries for a Healthy Planet. Scientific American, 54–57 (2010)
2. IPCC Intergovernamental Panel on Climate Change, Cambio climtico 2007: Informe de sntesis. Contribucin de los Grupos de trabajo I, II y III al Cuarto Informe de evaluacin del Grupo Intergubernamental de Expertos sobre el Cambio Climtico. IPCC, Suiza (2007)
3. Kosko, B.: Fuzzy Cognitive Maps. Man-Machine Studies 24, 65–75 (1986)
4. Kosko, B.: Neural Networks and Fuzzy Systems: a dynamical systems approach to machine intelligence. Prentice Hall (1991)
5. Mace, G., Masundire, H., Baillie, J., Ricketts, T., Brooks, T.: Chapter 4: Biodiversity in "Ecosystems and human well-being: Current state and trends". Island Press, USA (2005)
6. Rockström, J., Steffen, W., Noone, K., Persson, C.F.S., Lambin, E.F., Lenton, T.M., Scheffer, M., Folke, C., Schellnhuber, H.J., Nykvist, B., Wit, C.A., Hughes, T., Van der Leeuw, S., Rodhe, H., Srlin, H., Snyder, P.K., Costanza, R., Svedin, U., Falkenmark, M., Karlberg, L., Corell, R.W., Fabry, V.J., Hansen, J., Walker, B., Liverman, D., Richardson, K., Crutzen, P., Foley, J.A.: A Safe Operating Space for Humanity. Nature 461, 472–475 (2009)
7. Stylios, C.D., Georgopoulos, V.C., Groumpos, P.P.: The use of fuzzy cognitive maps in modeling systems. In: 5th IEEE Mediterranean Conference on Control and Systems MED5 paper, vol. 67, p. 7 (1997)

Fuzzy Models: Easier to Understand and an Easier Way to Handle Uncertainties in Climate Change Research

Carlos Gay García[1], Oscar Sánchez Meneses[1], Benjamín Martínez-López[1], Àngela Nebot[2], and Francisco Estrada[1]

[1] Centro de Ciencias de la Atmósfera, Universidad Nacional Autónoma de México, Ciudad Universitaria, Mexico. D.F., Mexico
[2] Grupo de Investigación Soft Computing, Universitat Politécnica de Catalunya, Barcelona, Spain
cgay@unam.mx, {casimiro,benmar,feporrua}@atmosfera.unam.mx, angela@lsi.upc.edu

Abstract. Greenhouse gas emission scenarios (through 2100) developed by the Intergovernmental Panel on Climate Change when converted to concentrations and atmospheric temperatures through the use of climate models result in a wide range of concentrations and temperatures with a rather simple interpretation: the higher the emissions the higher the concentrations and temperatures. Therefore the uncertainty in the projected temperature due to the uncertainty in the emissions is large. Linguistic rules are obtained through the use of linear emission scenarios and the Magicc model. These rules describe the relations between the concentrations (input) and the temperature increase for the year 2100 (output) and are used to build a fuzzy model. Another model is presented that includes, as a second source of uncertainty in input, the climate sensitivity to explore its effects on the temperature. Models are attractive because their simplicity and capability to integrate the uncertainties to the input and the output.

Keywords: Fuzzy Inference Models, Greenhouse Gases Future Scenarios, Global Climate Change.

1 Introduction

There is a growing scientific consensus that increasing emissions of greenhouse gases (GHG) are changing the Earth's climate. The Intergovernmental Panel on Climate Change (IPCC) Fourth Assessment Report [3] states that warming of the climate system is unequivocal and notes that eleven of the last twelve years (1995-2006) rank among the twelve warmest years of recorded temperatures (since 1850). The projections of the IPCCs Third Assessment Report (TAR) [4] regarding future global temperature change ranged from 1.4 to 5.8 °C. More recently, the projections indicate that temperatures would be in a range spanning from 1.1 to 4 °C, but that temperatures increases of more than 6 °C could not be ruled out [3]. This wide range of values reflects the uncertainty in the production of accurate projections of future

M.S. Obaidat et al. (eds.), *Simulation and Modeling Methodologies, Technologies and Applications*, Advances in Intelligent Systems and Computing 256,
DOI: 10.1007/978-3-319-03581-9_16,

climate change due to different potential pathways of GHG emissions. There are other sources of the uncertainty preventing us from obtaining better precision. One of them is related to the computer models used to project future climate change. The global climate is a highly complex system due to many physical, chemical, and biological processes that take place among its subsystems within a wide range of space and time scales.

Global circulation models (GCM) based on the fundamental laws of physics try to incorporate those known processes considered to constitute the climate system and are used for predicting its response to increases in GHG [4]. However, they are not perfect representations of reality because they do not include some important physical processes (e.g. ocean eddies, gravity waves, atmospheric convection, clouds and small-scale turbulence) which are too small or fast to be explicitly modeled. The net impact of these small scale physical processes is included in the model by means of parameterizations [10]. In addition, more complex models imply a large number of parameterized processes and different models use different parameterizations. Thus, different models, using the same forcing produce different results.

One of the main sources of uncertainty is, however, the different potential pathways for anthropogenic GHG emissions, which are used to drive the climate models. Future emissions depend on numerous driving forces, including population growth, economic development, energy supply and use, land-use patterns, and a variety of other human activities (Special Report on Emissions Scenarios, SRES). Future temperature scenarios have been obtained with the emission profiles corresponding to the four principal SRES families (A1, A2, B1, and B2) [7]. From the point of view of a policy-maker, the results of the 3rd and 4th IPCC's assessments regarding the projection of global or regional temperature increases are difficult to interpret due to the wide range of the estimated warming. Nevertheless this is an aspect of uncertainty that scientists and ultimately policy-makers have to deal with. Furthermore, most of the available methodologies that have been proposed for supporting decision-making under uncertainty do not take into account the nature of climate change's uncertainty and are based on classic statistical theory that might not be adequate. Climate change's uncertainty is predominately epistemic and, therefore, it is critical to produce or adapt methodologies that are suitable to deal with it and that can produce policy-useful information. The lack of such methodologies is noticeable in the IPCC's AR4 Contribution of the Working Group I, where the proposed best estimates, likely ranges and probabilistic scenarios are produced using statistically questionable devices [2].

Two main strategies have been proposed for dealing with uncertainty: trying to reduce it by improving the science of climate change a feat tried in the AR4 of the IPCC, and integrating it into the decision-making processes [9]. There are clear limitations regarding how much of the uncertainty can be reduced by improving the state of knowledge of the climate system, since there remains the uncertainty about the emissions which is more a result of political and economic decisions that do not necessarily obey natural laws.

Therefore, we propose that the modern view of climate modelling and decision-making should become more tolerant to uncertainty because it is a feature of the real world [6]. Choosing a modelling approach that includes uncertainty from the start tends to reduce its complexity and promotes a better understanding of the model itself and of its results. Science and decision-making have always had to deal with uncertainty and various methods and even branches of science, such as Probability, have been developed for this matter [5]. Important efforts have been made for developing approaches that can integrate subjective and partial information, being the most successful ones Bayesian and maximum entropy methods and more recently, fuzzy set theory where the concept of objects that have not precise boundaries was introduced [12]. Fuzzy logic provides a meaningful and powerful representation of uncertainties and is able to express efficiently the vague concepts of natural language [12]. These characteristics could make it a very powerful and efficient tool for policy makers due to the fact that the models are based on linguistic rules that could be easily understood.

In this paper two fuzzy logic models are proposed for the global temperature changes (in the year 2100) that are expected to occur in this century. The first model incorporates the uncertainties related to the wide range of emission scenarios and illustrates in a simple manner the importance of the emissions in determining future temperatures. The second incorporates the uncertainty due to climate sensitivity that pretends to emulate the diversity in modelling approaches. Both models are built using the Magicc [11] model and Zadeh´s extension principle for functions where the independent variable belongs to a fuzzy set. Magicc is capable of emulating the behaviour of complex GCMs using a relative simple one dimensional model that incorporates different processes e.g. carbon cycle, earth-ocean diffusivity, multiple gases and climate sensitivity. In our second case we intend to illustrate the combined effects of two sources of uncertainty: emissions and model sensitivity. It is clear that we are leaving out of this paper other important sources of uncertainty whose contribution would be interesting to explore. The GCMs are, from our point of view, useful and very valuable tools when it is intended to study specific aspects or details of the global temperature change. Nevertheless, when the goal is to study and to test global warming policies, simpler models much easier to understand become very attractive. Fuzzy models can perform this task very efficiently.

2 Fuzzy Logic Model Obtained from IPCC Data

The Fourth Assessment Report of the IPCC shows estimates of emissions, concentrations, forcing and temperatures through 2100 [3]. Although there are relationships among these variables, as those reflected in the figure 1 (left panel), it would be useful to find a way to relate emissions directly with increases of temperature. A more physical relation is established between concentrations and temperature because the latter depends almost directly upon the former through the forcing terms. Concentrations are obtained integrating over time the emissions minus the sinks of the GHGs. One way of relating directly emissions and temperature, could be achieved if the emission trajectories were linear and non-intersecting as illustrated in figure 1 (left panel). Here, we perform this task by means of a fuzzy model, which is based on the Magicc model [11] and Zadeh´s extension principle (see Appendix).

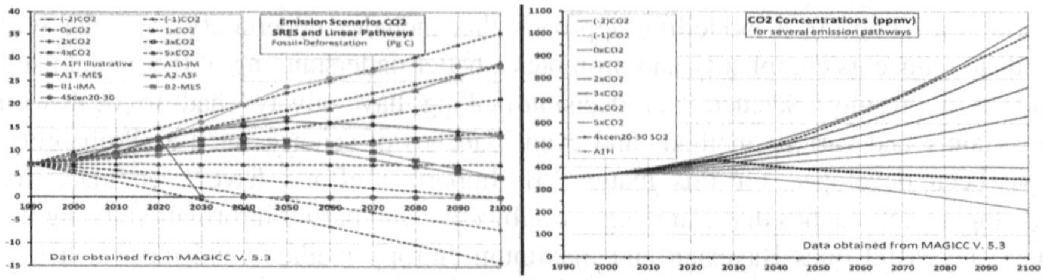

Fig. 1. Left panel: Emissions scenarios CO2, Illustrative SRES and Linear Pathways. (-2) CO2 means -2 times the emission (fossil + deforestation) of CO2 of 1990 by 2100 and so for -1, 0, 1, to 5 CO2. All the linear pathways contain the emission of non CO2 GHG as those of the A1FI. 4scen20-30 scenario follows the pathway of 4xCO2 but at 2030 all gases drop to 0 emissions or minimum value in CH4, N2O and SO2 cases. Right panel: CO2 Concentrations for linear emission pathways, 4scen20-30 SO2 and A1FI are shown for reference. Data calculated using Magicc V. 5.3.

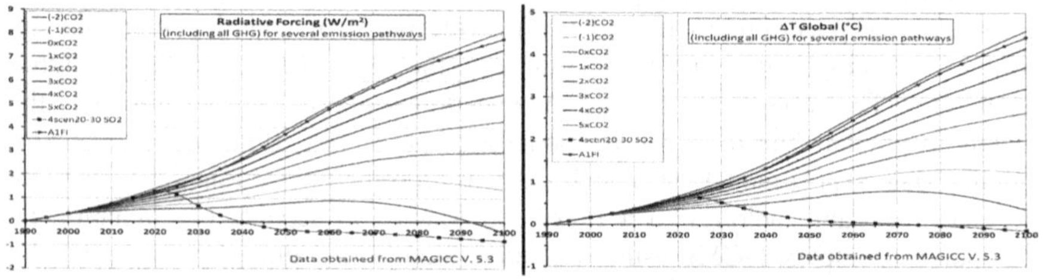

Fig. 2. Left Panel: Radiative Forcings (All GHG Included) for Linear Emission Pathways and A1FI SRES Illustrative, the 4scen20-30 SO2 Only Include SO2. Right Panel: Global Temperature Increments for Linear Emission Pathways, 4scen20-30 SO2 and A1FI as Calculated using Magicc V. 5.3

Using as input for the Magicc model the emissions shown in the previous figure we calculate the resulting concentrations (figure 1 right panel); forcings (figure 2 left panel) and global mean temperature increments (figure 2 right panel).

The set of emissions shown in figure 1 (left panel) has been simplified to linear functions of time that reach by the year 2100 values from minus two times to 5 times the emissions of 1990. The trajectories labelled 5CO2 and (-2) CO2 contain the trajectories of the SRES. We observe that the concentrations corresponding to the 5CO2 and the A1FI trajectories, by year 2100 are very close. The choice of linear pathways allows us to associate emissions to concentrations to forcings and temperatures in a very simple manner. We can say than any trajectory of emissions contained within two of the linear ones will correspond, at any time with a temperature that falls within the interval delimited by the temperatures corresponding to the linear trajectories. This is illustrated for the A1FI trajectory, in figure 2 (right panel) that falls within the temperatures of the 5CO2 and 4CO2 trajectories. We decided to find emission paths that would lead to temperatures of two degrees or less by the year 2100, this led us to the -2CO2, -1CO2 and 0CO2. The latter is a trajectory of constant emissions equal to the emissions in 1990 that gives us a temperature of two degrees by year 2100.

From the linear representation, it is easily deduced that *very high* emissions correspond to *very large* concentrations, forcings and *large* increases of temperature. It is also possible to say that large concentrations correspond to large temperature increases etc. This last statement is very important because in determining the temperature the climate models directly use the concentrations which are the time integral of sources and sinks of the green house gases (GHG). Therefore the detailed history of the emissions is lost. Nevertheless the statement, to large concentrations correspond large temperature increases still holds. These simple observations allows us to formulate a fuzzy model, based on linguistic rules of the **IF-THEN** form, which can be used to estimate increases of temperature within particular uncertainty intervals. Fuzzy logic provides a meaningful and powerful representation of measurement of uncertainties, and it is able to represent efficiently the vague concepts of natural language, of which the climate science is plagued. Therefore, it could be a very useful tool for decision makers. The basic concepts of fuzzy logic are presented in Appendix.

The first fuzzy model one input one output defined for the global temperature change is (quantities between parenthesis were used with Zadeh's principle to generate the fuzzy model, the number 1 means the membership value (μ) of the input variables used in formulating the fuzzy model):

```
1. If (concentration is very low (about -2CO2)) then (deltaT is
very low (1)
2. If (concentration is low (about -1CO2)) then (deltaT is low
(1)
3. If (concentration is medium-low (about 0CO2) then (deltaT is
medium-low (1)
4. If (concentration is medium (about 1CO2)) then (deltaT is
medium (1)
5. If (concentration is medium-high (about 2CO2)) then (deltaT
is medium-high (1)
6. If (concentration is high (about 3CO2)) then (deltaT is high
(1)
7. If (concentration is very high (about 4CO2)) then (deltaT is
very high (1)
8. If (concentration is extremely high (about 5CO2)) then
(deltaT is extremely high (1)
```

The 8 rules for concentration are based on 8 adjacent triangular membership functions (the simplest form) corresponding to linear emission trajectories (-2CO2 to 5CO2). The concentrations were obtained from Magicc model and cover the entire range (210 to 1045 ppmv). The apex of each membership function (μ=1) corresponds with the base (μ =0) of the adjacent one, as we show below (table 1):

Table 1. Concentration (ppmv) corresponding to each linear emission trajectory

Linear emission trajectory	$\mu=0$	$\mu=1$	$\mu=0$
1. -2CO2	210	213	300
2. -1CO2	213	300	401
3. 0CO2	300	401	513
4. 1CO2	401	513	633
5. 2CO2	513	633	762
6. 3CO2	633	762	899
7. 4CO2	762	899	1038
8. 5CO2	899	1038	1045

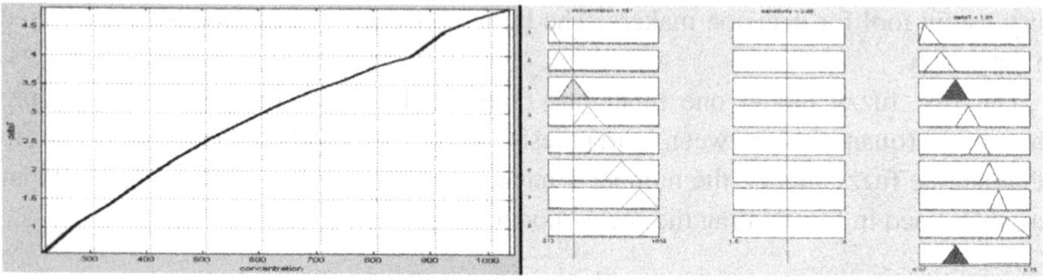

Fig. 3. Fuzzy model based on linguistic rules and Zadeh´s principle. Left panel: increases of temperature at year 2100 for each possible concentration (emission in the case of our linear model) value (solid line). Right panel: Fuzzy rules associated with the different classes of concentrations. (Calculated with MATLAB).

The global temperature changes were obtained through Zadeh's extension principle applied to data from Magicc model. From the point of view of a policy maker, a fuzzy model as the one represented by the previous rules is a very useful tool to study the effect of different policies on the increases of temperature.

The fuzzy rules model can be evaluated by means of the fuzzy inference process in such a way that each possible concentration value is mapped into an increase of temperature value by means of the Mamdani's defuzzification process (see Appendix). The resulting increases of temperature at year 2100 for each possible concentration (emission in the case of our linear model) value (solid line) are shown in the left panel of figure 3.

The right panel illustrates the formulation of the rules by showing the fuzzy set associated with the different classes of concentrations, the antecedent of the fuzzy rule, the IF part and the consequent fuzzy set temperature, the THEN part. The figure 3 right panel also illustrates the uncertainties of one estimation: If the concentration is of 401 ppmv (it fires rule number 3) within an uncertainty interval of (300 to 513 ppmv) 4 then the temperature increment is 1.95 degrees within an uncertainty interval of (1.23 to 2.63 deg C) in this case the temperatures will have uncertainties of one or two times the intervals defined by the expert or the researcher.

The fuzzy model is simpler and obviously less computationally expensive than the set of GCM's reported by the IPCC. The most important benefit, however, is its usefulness for policy-makers. For example, if the required increase of temperature should be *very low* or *low* (-2CO2, -1CO2), then the policy-maker knows, on the basis of this model, that concentrations should not exceed the class *small*.

3 A Simple Climate Model and Its Corresponding Fuzzy Model

Here, we again use the Magicc model but this time we introduce a second source of uncertainty, the climate sensitivity. The purpose is to illustrate the effects of the combination of two sources of uncertainty on the resulting temperatures. The climate model is driven by our linear emission paths. The relationship between concentrations and sensitivity and increases of temperature at year 2100 is then used to construct a fuzzy model following the extension principle of the fuzzy logic approach (see Appendix).

The set of fuzzy rules obtained in this case is the following:

```
1. If (concentration is very very low) and (sensitivity is low)
then (deltaT is low) (1)
2. If (concentration is very low) and (sensitivity is low) then
(deltaT is low) (1)
3. If (concentration is very low) and (sensitivity is high) then
(deltaT is med) (1)
4. If (concentration is medium-low) and (sensitivity is low)
then (deltaT is low) (1)
5. If (concentration is medium-low) and (sensitivity is high)
then (deltaT is high) (1)
6. If (concentration is medium) and (sensitivity is low) then
(deltaT is med) (1)
7. If (concentration is medium) and (sensitivity is high) then
(deltaT is high) (1)
8. If (concentration is medium-high) and (sensitivity is low)
then (deltaT is med) (1)
9. If (concentration is medium-high) and (sensitivity is high)
then (deltaT is high) (1)
10. If (concentration is high) and (sensitivity is low) then
(deltaT is med) (1)
11. If (concentration is high) and (sensitivity is high) then
(deltaT is high) (1)
12. If (concentration is medium-low) and (sensitivity is med)
then (deltaT is med) (1)
```

Note that we have used the same nomenclature as before and the very high and extremely high concentrations are not considered. And the fuzzy sets for the temperature and sensitivity are shown in figure 4. We used this figure to build the rule above in combination with 6 fuzzy sets for concentration similar to those from our first model described in section 2 (table 1).

For sensitivity we built 3 triangular fuzzy sets corresponding to sensitivity values of 1.5, 3 and 6 deg C/W/m2, showed below (table 2):

Table 2. Sensitivity (deg C/W/m2) input fuzzy sets for simple climate model

Sensitivity parameter	μ =0	μ =1	μ =0
1. 1.5 (low)	1.5	1.5	6.0
2. 3.0 (medium)	1.5	3.0	6.0
3. 6.0 (high)	3.0	6.0	6.0

Table 3. Temperature increase (°C) output fuzzy sets for simple climate model

Sensitivity parameter	μ =0	μ =1	μ =0
1. Low	0.07	1.07	2.13
2. Medium	0.36	1.98	3.70
3. High	0.92	3.27	5.75

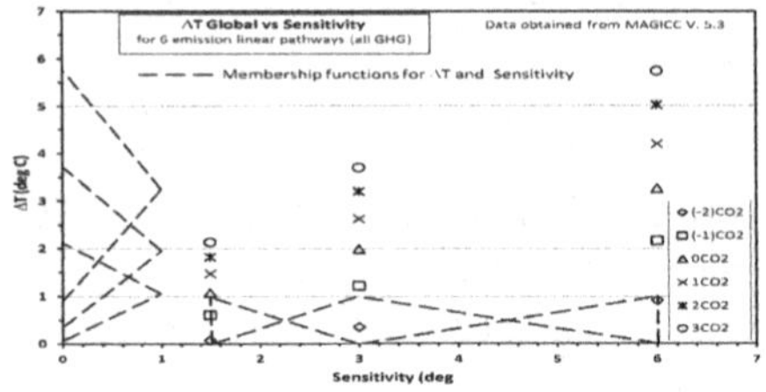

Fig. 4. ΔT Global and sensitivity fuzzy sets for six linear emission pathways at 2100. The dashed lines show the membership functions.

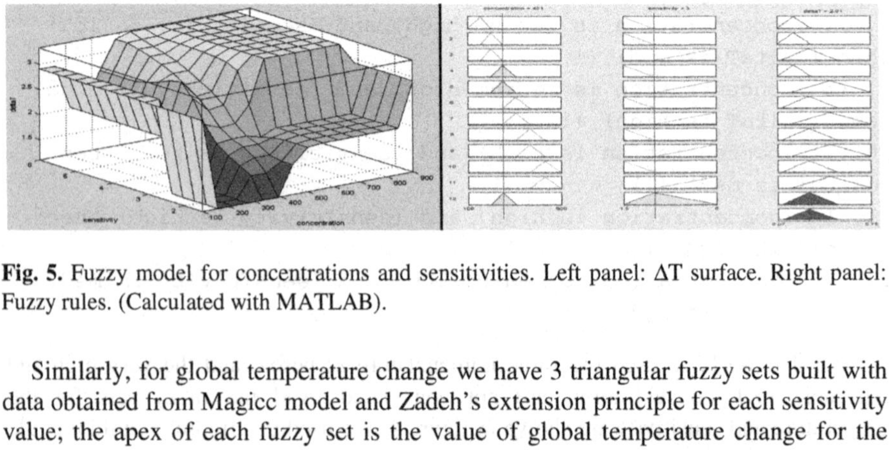

Fig. 5. Fuzzy model for concentrations and sensitivities. Left panel: ΔT surface. Right panel: Fuzzy rules. (Calculated with MATLAB).

Similarly, for global temperature change we have 3 triangular fuzzy sets built with data obtained from Magicc model and Zadeh's extension principle for each sensitivity value; the apex of each fuzzy set is the value of global temperature change for the

0CO2 linear emission path according to the value of sensitivity, the base of the fuzzy sets range from -2CO2 to 3CO2 (assuming global temperature changes below 6 deg C) for each sensitivity value (see table 3 and figure 4):

The Mamdani's fuzzy inference method is used also here as the defuzzification method to compute the increase of temperature values. The results are shown in figure 5. The left panel of figure 5 shows the surface resulting from the defuzzification process. The right panel illustrates again that for the case of a concentration of 401 ppmv and a sensitivity of 3 (medium sensitivity) the temperature is about 2 degrees within an uncertainty interval of (0.36 to 3.70 deg C) where the membership value is different from 0. When we compare our previous result with this one we find that the answers are very close in fact the fall within the uncertainty intervals of both. The uncertainty of concentrations and sensitivity are respectively (300 to 513 ppmv) and (1.5 to 6 deg C/W/m2). The result is to be expected since in our first experiment we used the Magicc model with default value for the sensitivity and this turns to be of 3.

4 An Extended Simple Climate Model

In the previous section the emissions were arranged in 6 fuzzy sets and the sensitiveties in three fuzzy sets. We decided to take just three fuzzy sets for the resulting temperatures; the parameters were calculated according to Zadeh´s extension principle, this means the position of the apex ($\mu=1$) and the feet of the triangles ($\mu=0$) were calculated with Magicc. The choosing of three sets for the temperature was clearly arbitrary since the combination of six emission sets and three sensitivities results in 18 sets for the temperature. Again using Magicc we calculated the parameters of the 18 fuzzy sets. The output fuzzy rules and its corresponding fuzzy sets are shown in tables 4 and 5:

Table 4. Output temperature increase fuzzy rules (IF - THEN form) for combinations of 6 concentrations and 3 sensitivities. Example: **if** concentration is 0CO2 **and** sensitivity is 3.0 **then** temperature increase is T9.

Concentrations of linear	Sensitivity parameter		
emission trajectories	1.5	3.0	6.0
1. -2CO2	T1	T7	T13
2. -1CO2	T2	T8	T14
3. 0CO2	T3	T9	T15
4. 1CO2	T4	T10	T16
5. 2CO2	T5	T11	T17
6. 3CO2	T6	T12	T18

Table 5. Output temperature increase fuzzy sets based on Magicc data

Temperature fuzzy set	μ =0	μ =1	μ =0	Temperature fuzzy set	μ =0	μ =1	μ =0
T1	0.07	0.07	1.23	T10	1.07	2.63	5.02
T2	0.07	0.61	1.98	T11	1.47	3.20	5.75
T3	0.61	1.07	2.63	T12	1.82	3.70	6.41
T4	1.07	1.47	3.20	T13	0.36	0.92	2.17
T5	1.47	1.82	3.70	T14	0.36	2.17	3.27
T6	1.82	2.13	4.16	T15	1.23	3.27	4.20
T7	0.07	0.36	2.17	T16	1.98	4.20	5.02
T8	0.07	1.23	3.27	T17	2.63	5.02	5.75
T9	0.61	1.98	4.20	T18	3.20	5.75	6.41

For this version of the model we use, instead of 100 ppmv of the former version, 210 ppmv as left foot of the triangle (μ=0) of the first concentration fuzzy set (corresponding to the linear emission trajectory -2CO2) because this value can be related to values of temperature increase obtained from Magicc. The results of running this model with 18 rules are shown in figure 6 (left panel, defuzzification surface) and right panel (fuzzy rules). Figures 5 and 6 show the surfaces which are rather similar, the complete set of fuzzy rules leads to the plateaus that are seen in the figure 5 no longer appear.

The fuzzy rules (figure 6, right panel) show again, for comparison, the result of the case of a concentration of 401 ppmv and a medium sensitivity parameter of 3 deg C/W/m2. Now, the defuzzificated value of temperature increase is of 2.26 °C within an uncertainty interval of (0.61 to 4.20 °C).

This illustrates the fact that although the 3 fuzzy sets for the temperature were chosen arbitrarily (or almost) in the first case the result are similar to the most complete model. This is the consequence of considering the uncertainties, and the values of one model lay within the uncertainty of the other. On the other hand the results indicate that a careful selection of rules may reduce the complexity of the model. Other possibility is that within the 18 rule model there is some redundancy.

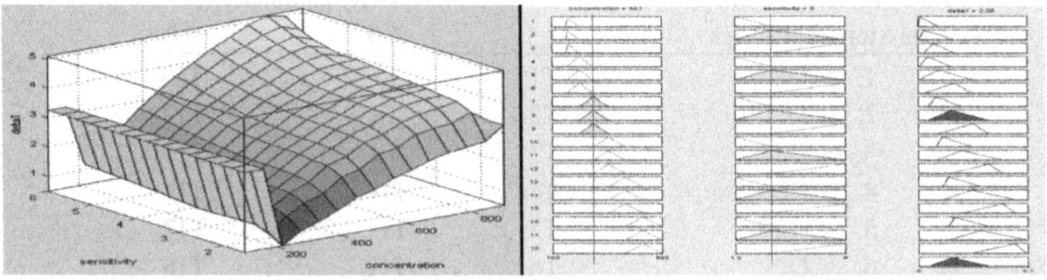

Fig. 6. Extended Fuzzy model for concentrations and sensitivities. Left panel: ΔT surface. Right panel: Fuzzy rules. (Calculated with MATLAB).

5 Discussion and Conclusions

In this work, simple linguistic fuzzy rules relating concentrations and increases of temperatures are extracted from the application of the Magicc model. The fuzzy model uses concentration values of GHG as input variable and gives, as output, the increase of temperature projected at year 2100. A second fuzzy model based on linguistic rules is developed based on the same Magicc climate model introducing a second source of uncertainty coming from the different sensitivities used by the Magicc to emulate more complicated GCMs used in the IPCC reports. An extended version of this simple climate model with 18 fuzzy sets for temperature increase corresponding to each possible combination of concentration and sensitivity input fuzzy sets, has been presented too. These kind of fuzzy models are very useful due to their simplicity and to the fact that include the uncertainties associated to the input and output variables. Simple models that, however, could contain all the information that is necessary for policy makers, these characteristics of the fuzzy models allow not only the understanding of the problem but also the discussion of the possible options available to them. For example going back to the question of stabilizing global temperatures at about 2 degrees or less, we can see the fuzziness of the proposition; we could estate that we should stay well below 400 ppmv by year 2100. The observed emission pictured in figure 6 where the IPCC scenarios are also shown are contained within A1F1 and the A1B therefore we could say that they point to a temperature increase that will surpass the two degrees. In fact to keep temperatures under 2 degrees we have already stated we should remain under 400 ppmv and we are *very very close* (fuzzy concept) to this concentration.

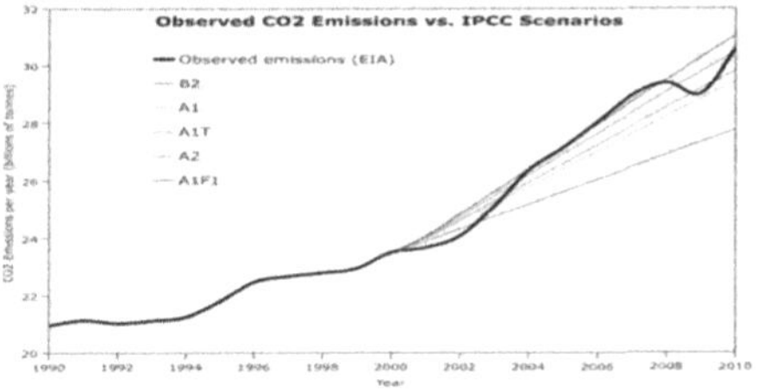

Fig. 7. Observed CO2 emissions against IPCC AR4 scenarios (taken from http://www.Tree hugger.com/ clean-technology/iea-co2-emissions-update-2010-bad-news-very-bad-news.html)

Acknowledgements. This work was supported by the Programa de Investigación en Cambio Climático (PINCC, www.pincc.unam.mx) of the Universidad Nacional Autónoma de México.

References

1. Dubois, D., Prade, H.M.: Fuzzy Sets and Systems: Theory and Applications. Academic Press (1980)

2. Gay, C., Estrada, F.: Objective probabilities about future climate are a matter of opinion. Climatic Change (2009), http://dx.doi.org/10.1007/s10584-009-9681-4
3. Solomon, S., Qin, D., Manning, M., Chen, Z., Marquis, M., Averyt, K.B., Tignor, M., Miller, H.L. (eds.): IPCC-WGI, 2007: Climate Change 2007: The Physical Science Basis. Contribution of Working Group I to the Fourth Assessment Report of the Intergovernmental Panel on Climate Change, p. 996. Cambridge University Press, Cambridge (2007)
4. Houghton, J.T., Ding, Y., Griggs, D.J., Noguer, M., van der Linden, P.J., Dai, X., Maskell, K., Johnson, C.A. (eds.): IPCC-WGI, 2001: Climate Change 2001: The Scientific Basis, Contribution of Working Group I to the Third Assessment Report of the Intergovernmental Panel on Climate Change. Cambridge University Press (2001) ISBN 0-521-80767-0 (pb: 0-521-01495-6)
5. Jaynes, E.T.: Information Theory and Statistical Mechanics: Phys. Rev. 106, 620–630 (1957)
6. Klir, G., Elias, D.: Architecture of Systems Problem Solving, 2nd edn. Plenum Press, NY (2002)
7. Nakicenovic, N., Alcamo, J., Davis, G., de Vries, B., Fenhann, J., Gaffin, S., Gregory, K., Grübler, A., Jung, T.Y., Kram, T., La Rovere, E.L., Michaelis, L., Mori, S., Morita, T., Pepper, W., Pitcher, H., Price, L., Riahi, K., Roehrl, A., Rogner, H.-H., Sankovski, A., Schlesinger, M., Shukla, P., Smith, S., Swart, R., van Rooijen, S., Victor, N., Dadi, Z.: Special Report on Emissions Scenarios: A Special Report of Working Group III of the Intergovernmental Panel on Climate Change, p. 599. Cambridge University Press, Cambridge (2000)
8. Ross, T.J.: Fuzzy Logic with Engineering Applications, 2nd edn. John Wiley & Sons (2004)
9. Schneider, S.H.: Congressional Testimony: U.S. Senate Committee on Commerce, Science and Transportation, Hearing on "The Case for Climate Change Action" (October 1, 2003)
10. Schmidt, G.A.: The physics of climate modeling. Phys. Today 60(1), 72–73 (2007)
11. Wigley, T.M.L.: MAGICC/SCENGEN V. 5.3: User Manual (version 2), p. 80. NCAR, Boulder, CO. (2008), http://www.cgd.ucar.edu/cas/wigley/magicc/
12. Zadeh, L.A.: Fuzzy Sets. Information and Control 8(3), 338–353 (1965)

Appendix

A.1 Fuzzy Logic Basic Concepts

As Klir stated in his book [6], the view of the concept of uncertainty has been changed in science over the years. The traditional view looks to uncertainty as undesirable in science and should be avoided by all possible means. The modern view is tolerant of uncertainty and considers that science should deal with it because it is part of the real world. This is especially relevant when the goal is to construct models. In this case, allowing more uncertainty tends to reduce complexity and increase credibility of the resulting model. The recognition by the researchers of the important role of uncertainty mainly occurs with the first publication of the fuzzy set theory, where the concept of objects that have not precise boundaries (fuzzy sets) is introduced [12].

Fuzzy logic, based on fuzzy sets, is a superset of conventional two-valued logic that has been extended to handle the concept of partial truth, i.e. truth values between completely true and completely false. In classical set theory, when A is a set and x is an object, the proposition "x is a member of A" is necessarily true or false, as stated on equation 1:

$$A(x) = \begin{cases} 1 & for\ x \in A \\ 0 & for\ x \notin A \end{cases} \tag{1}$$

Whereas, in fuzzy set theory, the same proposition is not necessarily either true or false, it may be true only to some degree. In this case, the restriction of classical set theory is relaxed allowing different degrees of membership for the above proposition, represented by real numbers in the closed interval [0,1], i.e.:

$$A : X \rightarrow [0,1]$$

Figure A.1 presents this concept graphically.

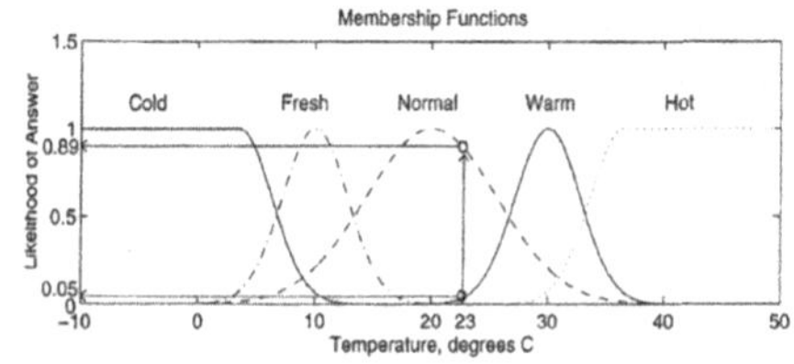

Fig. A.1. Gaussian membership functions of a quantitative variable representing ambient temperature

Figure A.1 illustrates the membership functions of the classes: cold, fresh, normal, warm, and hot, of the ambient temperature variable. A temperature of 23 °C is a member of the class normal with a grade of 0.89 and a member of the class warm with a grade of 0.05. The definition of the membership functions may change with regard to who define them. For example, the class normal for ambient temperature variable in Mexico City can be defined as it is shown in figure A.1. The same class in Anchorage, however, will be defined more likely in the range from -8 °C to -2 °C. It is important to understand that the membership functions are not probability functions but subjective measures. The opportunity that brings fuzzy logic to represent sets as degrees of membership has a broad utility. On the one hand, it provides a meaningful and powerful representation of measurement uncertainties, and, on the other hand, it is able to represent efficiently the vague concepts of natural language. Going back to the example of figure A.1, it is more common and useful for people to know that tomorrow will be hot than to know the exact temperature grade.

At this point, the question is, once we have the variables of the system that we want to study described in terms of fuzzy sets, what can we do with them? The membership functions are the basis of the fuzzy inference concept. The compositional rule of inference is the tool used in fuzzy logic to perform approximate reasoning.

Approximate reasoning is a process by which an imprecise conclusion is deduced from a collection of imprecise premises using fuzzy sets theory as the main tool.

The compositional rule of inference translates the modus ponens of the classical logic to fuzzy logic. The generalized modus ponens is expressed by:

Rule: If X is A then Y is B
Fact: X is A'
Conclusion: Y is B'

Where, X and Y are variables that take values from the sets X and Y, respectively, and A, A' and B, B' are fuzzy sets on X and Y, respectively. Notice that the Rule expresses a fuzzy relation, R, on X × Y.

Then, if the fuzzy relation, R, and the fuzzy set A' are given, it is possible to obtain B' by the compositional rule of inference, given in equation 2,

$$B'(y) = \sup_{x \in X} \min \left[A'(x), R(x,y) \right] \tag{2}$$

Where **sup** stands for supremum (least upper bound) and **min** stands for minimum. When sets X and Y are finite, **sup** is replaced by the maximum operator, **max**. Figure A.2 (left panel) illustrates in a simplified way the compositional rule of inference graphically.

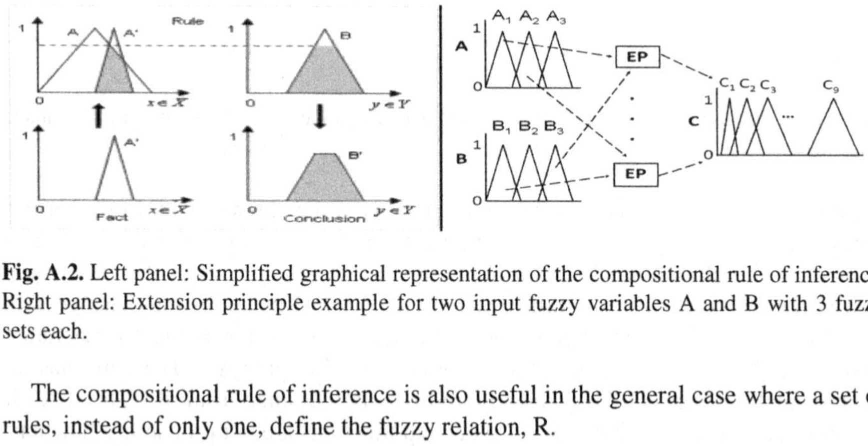

Fig. A.2. Left panel: Simplified graphical representation of the compositional rule of inference. Right panel: Extension principle example for two input fuzzy variables A and B with 3 fuzzy sets each.

The compositional rule of inference is also useful in the general case where a set of rules, instead of only one, define the fuzzy relation, R.

A.2 Extension Principle

Zadeh says that rather than regarding fuzzy theory as a single theory, we should regard the conversion process from binary to membership functions as a methodology to generalize any specific theory from a crisp (discrete) to a continuous (fuzzy) form. The extension principle enables us to extend the domain of a function on fuzzy sets, i.e., it allows us to determine the fuzziness in the output given that the input variables are already fuzzy. Therefore, it is a particular case of the compositional rule of inference. Figure A.2 (right panel) gives a first idea of the extension principle showing an example of two input variables with 3 fuzzy sets each.

The extension principle is applied to transform each fuzzy pair (A_i, B_j), in a fuzzy set of the C output variable. Notice that in the example of figure A.2 (right panel) we have 9 pairs of fuzzy input sets and, therefore, 9 fuzzy sets are obtained representing the conclusion as shown in the right hand side of figure A.2 (right panel). The extension principle when two input variables are available is presented in equation 3. C_k is the kth output fuzzy set extended from the two input fuzzy sets A_i and B_j. In the example at hand, as illustrated in figure A.3, the extension principle is applied 9 times, to obtain each of the output fuzzy sets associated to each fuzzy input pair.

$$C_k = \max_{C_k = f(A_i, B_j)} \min \left[A_i, B_j \right] \tag{3}$$

For instance, the output fuzzy set C9, is obtained when using the extension principle of equation 3 with the input fuzzy sets A1 and B3 [6, 1, 8].

Small-Particle Pollution Modeling Using Fuzzy Approaches

Àngela Nebot and Francisco Mugica

Soft Computing Research Group, Technical University of Catalonia,
Jordi Girona 1-3, Barcelona, Spain
{angela,fmugica}@lsi.upc.edu

Abstract. Air pollution caused by small particles is a major public health problem in many cities of the world. One of the most contaminated cities is Mexico City. The fact that it is located in a volcanic crater surrounded by mountains helps thermal inversion and imply a huge pollution problem by trapping a thick layer of smog that float over the city. Modeling air pollution is a political and administrative important issue due to the fact that the prediction of critical events should guide decision making. The need for countermeasures against such episodes requires predicting with accuracy and in advance relevant indicators of air pollution, such are particles smaller than 2.5 microns ($PM_{2.5}$). In this work two different fuzzy approaches for modeling $PM_{2.5}$ concentrations in Mexico City metropolitan area are compared with respect the simple persistence method.

Keywords: Air Pollution Modeling, $PM_{2.5}$ Pollution, Fuzzy Inductive Reasoning, ANFIS, Persistence, Time Series Analysis.

1 Introduction

The high levels of particulate matter in the air are of high concern since they may produce severe public health effects and are the main cause of the attenuation of visible light. There are very high levels of particles in North Africa, much of the Middle East, Asia and Latin America as well as in the large urban areas. Comparing it with population density maps, the WHO concluded that more than 80% of the world population is exposed to high levels of fine particles ($PM_{2.5}$) [1]. Likewise, identifies $PM_{2.5}$ as an important indicator of risk to health and might also be a better indicator than PM_{10} for anthropogenic suspended particles in many areas [2]). According to the WHO Guidelines, concentrations at this level and higher are associated with an approximately 15% increased risk of mortality, relative to the Air Quality Guideline (AQG) of 10 μg m^{-3} [1].

Regarding the $PM_{2.5}$, it has not yet been identified a threshold below which damage to health does not occur, this has motivated that the limits for the protection of public health are getting lower every year. The geographical characteristics of the Mexico City metropolitan area, i.e. its height, average temperature and terrain, added to the pressure exerted by the growth and intensification of urban activities cause high air pollution episodes that constitute a permanent challenge to the health of its

M.S. Obaidat et al. (eds.), *Simulation and Modeling Methodologies, Technologies and Applications*, Advances in Intelligent Systems and Computing 256,
DOI: 10.1007/978-3-319-03581-9_17,

inhabitants. Although the measures taken over the past 15 years to reduce the impact of air pollution have managed to significantly decrease pollutants such as SO2, CO or the Pb, the concentrations of ozone and fine particles exceed quite often air quality standards.

The monitoring of $PM_{2.5}$ from 2004 to date shows that around 20 million people in Mexico city are exposed to annual average concentrations of this contaminant in between 19 and 25 $\mu g\ m^{-3}$, exceeding by more than double the WHO standard of 10 $\mu g\ m^{-3}$ and substantially exceeding the Mexican norm of 15 $\mu g\ m^{-3}$.

The increase of the concentration of particles in Mexico City is strongly associated with the meteorology of the Valley. During the days of intense wind, resuspension of dust from the ground produces significant increases in the concentrations of total suspended particles (PST) and particles lower than 10 μm (PM_{10}). The presence of surface thermal inversions can contribute to the increase in the concentration of particles smaller than 10 μm and fine particles, due to the lack of dispersion and the accumulation in the atmosphere of the particles emitted by vehicles and industry. Higher concentrations usually occur when the layer trapped under the inversion is not very high and the duration of the thermal inversion is maintained throughout the morning.

The national weather service reported a total of 107 days with surface thermal inversions during 2010, the highest in the past 13 years. The largest part was recorded during the winter months, when the long and cold nights favor its formation. In the dry season months it has been reported a 40% of days with thermal inversion. The months of April and December had the largest number of events with 16 and 17 days, respectively. The influence of high pressure systems during the months of March to May was responsible for the formation of surface thermal inversions [3].

Fuzzy logic-based methods have not been applied extensively in environmental science, however, some interesting research can be found in the area of modeling of pollutants [4-10], where different hybrid methods that make use of fuzzy logic are presented for this task.

In this research we propose prediction models of hourly concentrations of $PM_{2.5}$, based on data registered at downtown Mexico City. In a first study, the concentration of $PM_{2.5}$ is used as input variable, becoming a time series modeling. In a second study, the daily maximum temperature is added to the input data in order to obtain prediction models of $PM_{2.5}$ concentrations.

The fuzzy approaches chosen to perform these tasks are the Fuzzy Inductive Reasoning (FIR) methodology and the Adaptive Neuro Fuzzy Inference System (ANFIS). These two fuzzy approaches for modeling small particles are presented and the prediction results obtained are compared with the results of the persistence simple method.

Sections 2 and 3 introduce the basic concepts of FIR and ANFIS methodologies, respectively. Section 4 presents the methods, i.e. the data, the fuzzy models development and the models evaluation. Section 5 describes the results obtained. Finally the conclusions of this research are given.

2 Fuzzy Inductive Reasoning (FIR)

The conceptualization of the FIR methodology arises of the General System Problem Solving (GSPS) approach proposed by Klir [11]. This methodology of modeling and simulation is able to obtain good qualitative relations between the variables that compose the system and to infer future behavior of that system. It has the ability to describe systems that cannot easily be described by classical mathematics or statistics, i.e. systems for which the underlying physical laws are not well understood.

FIR methodology, offers a model-based approach to predicting either univariate or multi-variate time series [12, 13]. A FIR model is a qualitative, non-parametric, shallow model based on fuzzy logic. Visual-FIR is a tool based on the Fuzzy Inductive Reasoning (FIR) methodology (runs under Matlab environment), that offers a new perspective to the modeling and simulation of complex systems. Visual-FIR designs process blocks that allow the treatment of the model identification and prediction phases of FIR methodology in a compact, efficient and user friendly manner [14].

The FIR model consists of its structure (relevant variables) and a set of input/output relations (history behavior) that are defined as if-then rules. Feature selection in FIR is based on the maximization of the models' forecasting power quantified by a Shannon entropy-based quality measure. The Shannon entropy measure is used to determine the uncertainty associated with forecasting a particular output state given any legal input state. The overall entropy of the FIR model structure studied, H_s, is computed as described in equation 1.

$$H_s = -\sum_{\forall i} p(i) \cdot H_i \tag{1}$$

where $p(i)$ is the probability of that input state to occur and H_i is the Shannon entropy relative to the ith input state. A normalized overall entropy H_n is defined in equation 2.

$$H_n = 1 - \frac{H_s}{H_{max}} \tag{2}$$

H_n is obviously a real-valued number in the range between 0.0 and 1.0, where higher values indicate an improved forecasting power. The model structure with highest H_n value generates forecasts with the smallest amount of uncertainty.

Once the most relevant variables are identified, they are used to derive the set of input/output relations from the training data set, defined as a set of if-then rules. This set of rules contains the behaviour of the system. Using the five-nearest-neighbors (5NN) fuzzy inference algorithm the five rules with the smallest distance measure are selected and a distance-weighted average of their fuzzy membership functions is computed and used to forecast the fuzzy membership function of the current state, as described in equation 3.

$$Memb_{out_{new}} = \sum_{j=1}^{5} w_{rel_j} \cdot Memb_{out_j} \tag{3}$$

The weights w_{rel_j} are based on the distances and are numbers between 0.0 and 1.0.

Their sum is always equal to 1.0. It is therefore possible to interpret the relative weights as percentages. For a more detailed explanation of the FIR methodology refer to [14].

3 Adaptive Neuro-Fuzzy Inference System (ANFIS)

The Adaptive Neuro-Fuzzy Inference System (ANFIS), developed by Jang, is one of the most popular hybrid neuro-fuzzy systems for function approximation [15]. ANFIS represents a Sugeno-type neuro-fuzzy system. A neuro-fuzzy system is a fuzzy system that uses learning methods derived from neural networks to find its own parameters. It is relevant that the learning process is not knowledge-based but data-driven.

The main characteristic of the Sugeno inference system is that the consequent, or output of the fuzzy rules, is not a fuzzy variable but a function, as shown in equation 4.

$$\begin{aligned} &\text{Rule}_1\text{:} \quad \text{If A is } A_1 \text{ and B is } B_1 \text{ then } Z = p_1 {*} a + q_1 {*} b + r_1 \\ &\text{Rule}_2\text{:} \quad \text{If A is } A_2 \text{ and B is } B_2 \text{ then } Z = p_2 {*} a + q_2 {*} b + r_2 \end{aligned} \tag{4}$$

Figure 1 describes graphically how a Sugeno model composed by the two rules described in equation 4 works.

The first step of the Sugeno inference is to combine a given input tuple (in the example of figure 1, a double is used (a=3,b=2)) with the rule's antecedents by determining the degree to which each input belongs to the corresponding fuzzy set (left panel of Fig. 1). The **min** operator is then used to obtain the weight of each rule, w_i, which are used in the final output computation, Z (right panel of Fig. 1). Notice that the Sugeno inference has two differentiated set of parameters. The first set corresponds to the membership functions parameters of the input variables. The second set corresponds to the parameters associated to the output function of each rule, i.e. p_i, q_i and r_i.

ANFIS is the responsible of adjusting in an automatic way these two set of parameters by means of two optimization algorithms, i.e. back-propagation (gradient descendent) and least square estimation. Back-propagation is used to learn about the parameters of the antecedents (membership functions) and the least square estimation is used to determine the coefficients of the linear combinations in the rules' consequents. ANFIS is a function of the Fuzzy toolbox that runs under the Matlab environment. For a more detailed explanation of the ANFIS methodology refer to [15].

4 Methods

4.1 The Data

The data used for this study stems from the Atmospheric Monitoring System of Mexico City (SIMAT in Spanish) that measures contaminants and atmospheric variables from 36 stations distributed through the 5 regions of the Mexico City metropolitan area [16]. The registered variables are the air pollutants, including $PM_{2.5}$, as well as other 10 contaminants, and meteorological variables, 24 hours a day, every day of the year. The web page of SIMAT [16] offers a data base with meteorological and contaminant registers since 1986 up to date, although $PM_{2.5}$ has been registered for the first time in 2004.

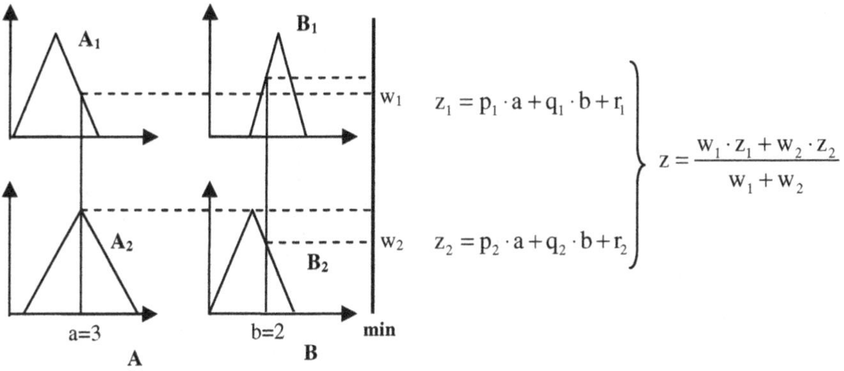

Fig. 1. Example of how a Sugeno model works (evaluation of two fuzzy rules with two input variables or antecedents, i.e. A and B)

A mechanically oscillated mass balance type instrument, TEOM 1400a, is used for the registration of the $PM_{2.5}$. This instrument is very sensitive to changes in concentrations of mass and can provide accurate measurements for samples with less than an hour in length.

This study is centered on the modeling and forecasting of particulate matter with diameter of 2.5 micrometers or less ($PM_{2.5}$) in the Merced station, located in the commercial and administrative district at the downtown of Mexico City Metropolitan Area (MCMA).

$PM_{2.5}$ values are hourly instantaneous observations, not the maximum or the mean of minute registered data. The typical pattern of $PM_{2.5}$ from some city areas, such as for example downtown, suggests that concentrations of this contaminant increase regularly between 8:00 and 16:00 hours, with maximum concentrations around 13:00 hours [17].

It has been decided to use, in this study, data from the half of the year that Mexico City suffers higher $PM_{2.5}$ concentrations, i.e. from December to May. We have used 4 data sets containing 6 month of hourly registers each one, i.e. from the 1st of December until de 31st of May, for years 2007-2008, 2008-2009, 2009-2010 and 2010-2011.

For the first data set, i.e. 1st December 2007 to 31st May 2008, the average concentration is 31.2 μg m^{-3}, the maximum is 147 μg m^{-3} and the standard deviation is 15.6 μg m^{-3}. For the second data set, i.e. 1st December 2008 to 31st May 2009, the average concentration is 26.6 μg m^{-3}, the maximum is 102 μg m^{-3} and the standard deviation is 14.3 μg m^{-3}.

For the third data set, i.e. 1st December 2009 to 31st May 2010, the average concentration is 20.8 μg m^{-3}, the maximum is 101 μg m^{-3} and the standard deviation is

13.4 μg m^{-3}. For the last data set, i.e. 1st December 2010 to 31st May 2011, the average concentration is 32.5 μg m^{-3}, the maximum is 175 μg m^{-3} and the standard deviation is16.5 μg m^{-3}. Figure 2 shows the hourly concentrations of $PM_{2.5}$ during December, 2009.

The data available contains missing values that correspond to data that was not registered due to instrument problems. From the total number of 17496 hourly data registered of $PM_{2.5}$ concentration, 1316 are missing values.

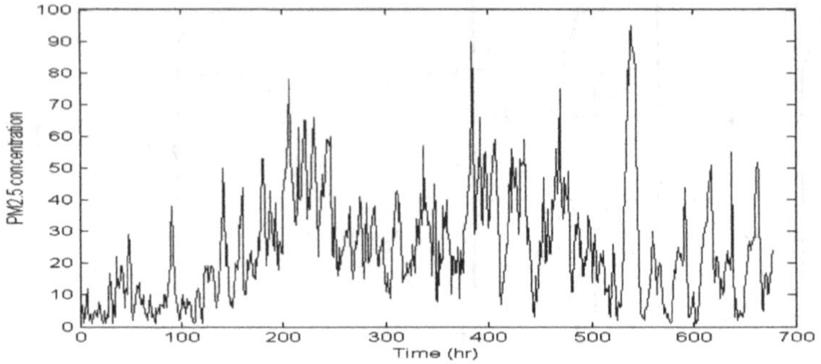

Fig. 2. Hourly concentrations of $PM_{2.5}$ data for December 2009. Units are μg m^{-3}. From the 720 data points, 42 are missing values that are not plotted.

4.2 Models Development

As mentioned before, our goal is to obtain ANFIS and FIR models capable of forecasting the $PM_{2.5}$ concentrations some time in advance, in such a way that efficient actions could be taken in order to protect the citizens of high concentrations episodes.

A study of autocorrelation, both causal and temporal, is first performed. To this end, we used the model structure identification process of the FIR methodology that carries out a feature selection based on the entropy reduction measure, described in section 2.

It has been found that it is possible to relate the concentration of $PM_{2.5}$ at a given time of the day to the sequence of 24 points corresponding to the hourly concentrations of the preceding day. Moreover, the structure of the FIR model has determined that there is a direct causal relation between the level of pollution at present time and the levels at 6 am, 12 pm, 18 pm and 24 pm of the preceding day. That is, there is a positive correlation at 12 pm and 24 pm and a negative correlation at 6 am and 18 pm.

With this information available we think that an interesting and useful approximation to modeling and forecasting $PM_{2.5}$ concentrations is to obtain a specific model for each of the most relevant hours of the day (i.e. 6 am, 12 pm, 18 pm and 24 pm), based on the values of the 6 am, 12 pm, 18 pm and 24 pm hours of the previous day, i.e. hourly models.

Two studies have been performed: the first one uses as models' inputs only the $PM_{2.5}$ concentrations (we refer to them as univariate models) and, the second one, uses also the daily maximum temperature (we refer to them as multi-variate models).

In the first study, we chose to work with the $PM_{2.5}$ scalar time series keeping in mind the idea that if we use a large enough window of data as input, the effect of other pollutants or meteorological data should be implicit in its structure [18].

In the second study, we decide to add a meteorological variable in order to try to enhance the results of the univariate models. Cobourn concludes that the meteorological variables that have a nonlinear relationship with $PM_{2.5}$ statistically significant are daily maximum temperature and wind speed. Moreover, the strongest single relationship between $PM_{2.5}$ and any meteorological variable is the relationship with daily maximum temperature [19]. This is the reason why this variable has been included as input in the multi-variate models.

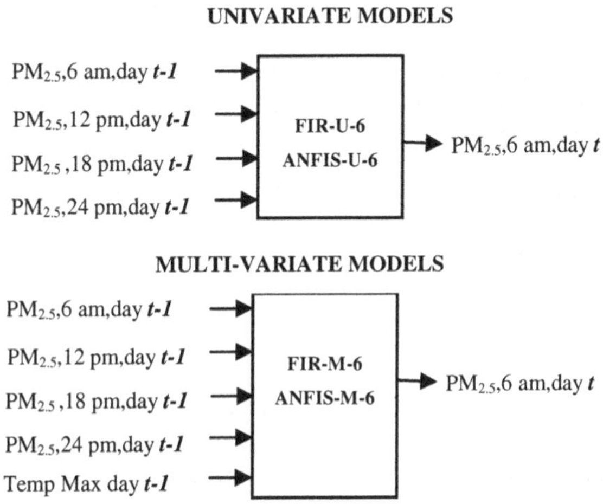

Fig. 3. Input and output variables of the univariate and multi-variate models that predict the $PM_{2.5}$ concentration at 6 am of the subsequent day. The models FIR-U-12, ANFIS-U-12, FIR-M-12 and ANFIS-M-12 have the same input variables described in this figure and as output variable the $PM_{2.5}$ concentration at 12 pm of the subsequent day. Idem for models FIR-U-18, ANFIS-U-18, FIR-M-18, ANFIS-M-18 that have as output variable the $PM_{2.5}$ concentration at 18 pm of the following day, and FIR-U-24, ANFIS-U-24, FIR-M-24, ANFIS-M-24 that have as output variable the $PM_{2.5}$ concentration at 24 pm of the subsequent day.

Both, the univariate and multi-variate models have as input variables the $PM_{2.5}$ concentration at 6 am, 12 pm, 18 pm and 24 pm. The multi-variate models have as additional input variable the daily maximum temperature.

The output variable of each model is the PM2.5 concentration at its corresponding hour. For instance, the output of the 6 am models (i.e. univariate and multi-variate) is the $PM_{2.5}$ concentration at 6 am of the subsequent day. Therefore, for this prediction model, pollutant concentrations are given 6 hours in advance. Figure 3 clarify the inputs and outputs of each of the models developed in this work.

In order to obtain all the models it is necessary to arrange the data in such a way that we have a data stream for each day instead of 24 data streams (one for each hour of that day). The 4 data sets available, and described in section 4.1, have been arranged accordingly, obtaining now a total number of 725 daily data, out of which 220 are missing values.

In this work a 10-fold cross validation is used to assess how the results of the obtained models generalize to an independent data set. The objective is to estimate how accurately the predictive models developed in this study will perform in practice.

FIR Models

The first step in order to obtain the FIR models is to convert quantitative values into fuzzy data. To this end, it is necessary to specify two discretization parameters, i.e. the number of classes for each system variable (granularity) and the membership functions (landmarks) that define its semantics. In this study the granularity and the clustering method used to obtain the landmarks are summarized in table 1. Many

Table 1. Granularity and clustering methods used to discretize the input and output variables in univariate (i.e. FIR-U-6, FIR-U-12, FIR-U-18, FIR-U-24) and multi-variate (i.e. FIR-M-6, FIR-M-12, FIR-M-18, FIR-M-24) FIR models

Number classes	Clustering method	UNIVARIATE MODELS	MULTI-VARIATE MODELS
2	Fuzzy C-means	FOLD 1, 5, 7, 8, 9	FOLD 1, 4, 8, 9
3	Equal Frequency Partition	FOLD 2, 6	FOLD 2, 6, 7
2	Median Linkage	FOLD 3, 10	FOLD 3
2	K-Means	FOLD 4	FOLD 5

folds are discretized into two classes. It is not possible to use more classes in this case because the number of training data (450 points) is not large enough. several clustering methods such are fuzzy c-means, median linkage, k-means and equal frequency partition are used in this study.

In a general way, the univariate FIR models structure can be described using equation 5.

$$y_i(d) = f_q(x_6(d-1), x_{12}(d-1), x_{18}(d-1), x_{24}(d-1)) \tag{5}$$

where $y_i(d)$ is the predicted $PM_{2.5}$ concentration at time i of day d; x_i represent the real concentration at time i of the preceding day $(d\text{-}1)$; and f_q is the qualitative relation of the FIR model. For multi-variate FIR models equation 5 should include the daily maximum temperature as an additional parameter of the function f_q.

ANFIS Models

In order to obtain ANFIS models it is necessary to define the following five parameters: the granularity of each input variable (i.e. number of classes), the shape of the membership functions of the input variables, the type of the output function (i.e. constant or linear), the optimization method to train the fuzzy inference system and the number of training epochs. Several combinations of these parameters have been analyzed in this research and the best results are obtained when two classes with triangular shape membership functions are used for each input variable, a constant output function is defined and a hybrid optimization method is used for training during 200 epochs.

It was expected than a higher granularity (3 for example) and a linear output function would be a better set of parameters to capture system's behavior, however this is not the case. The results obtained when using these parameter values are bad because some prediction points are very big or very low, distorting the whole prediction set. Analyzing the results, we think that this is due to the complex nature of the data. The bad predictions correspond to those real "extreme" situations that do not appear in the actual training data and, therefore, the obtained ANFIS model has not been adapted to this type of data. When the output function is simplified and the number of classes reduced, less partitioned is the space, and these extreme situations are softened.

4.3 Model Evaluation

The normalized root mean square error, described in equation 6, is used to evaluate the performance of each ANFIS, FIR and persistence models.

$$RNMSE = \frac{\sqrt{\frac{1}{N}\sum_{i=1}^{N}(y_i - \hat{y}_i)^2}}{\sum_{i=1}^{N} y_i \Big/ N} \tag{6}$$

where $\hat{y}$ is the predicted output, y is the real output and N is the number of samples.

5 Results and Discussion

The persistence method consists on a very simple principle, i.e. tomorrow at time t the $PM_{2.5}$ mass concentration will be the same as today at time t, as described in equation 7. Therefore, there are no parameters to adjust.

$$y_t(d) = x_t(d-1) \tag{7}$$

The prediction results obtained by ANFIS, FIR and persistence univariate models of the $PM_{2.5}$ contaminant at 6 am, 12 pm, 18 pm and 24 pm of the subsequent day, for each of the 10 folds, are summarized in tables 2 and 3.

From tables 2 and 3 it can be seen that ANFIS and FIR models perform much better than persistence for all the four univariate models, i.e. 6 am, 12 pm, 18 pm and 24 pm. Moreover, both ANFIS and FIR models obtain better results than persistence fold by fold, except for fold 7 of model 12 pm.

The ANFIS models are between 12.5% and 30% better than the persistence models while FIR models between 9.6% and 24%.

If we compare the results of ANFIS vs. FIR models it can be concluded that ANFIS obtains lower average errors than FIR, however the differences are really small. Therefore, both fuzzy methodologies can be considered appropriate approaches to deal with this complex modelling problem.

ANFIS and FIR 6 am models obtain very good results, with average errors of 0.36 and 0.38, respectively. These are low errors if we compare with the errors obtained by the rest of the models, i.e. 12 pm, 18 pm and 24 pm. This makes sense, since models at 6 am are predicting values only 6 hours in advance whereas the rest of the models predict the $PM_{2.5}$ concentration 12, 18 and 24 hours in advance. It is interesting to mention that the higher average error obtained with univariate ANFIS and FIR models in this research are of the order of 0.5. In general, the results obtained are quite good for the problem at hand if we compare them with other results found in the literature that deal with the same problem and use also univariate $PM_{2.5}$ concentration. For instance in [20], the best model obtained that is based on neural networks, has an RNMSE of 0.5, i.e. their better results correspond to the worse results obtained in this research. It should be noted that this comparison is only to point out the complexity of the problem. Both studies use different data and, therefore, it is not possible to perform a rigorous comparison of the different methods used.

The prediction results obtained by ANFIS, FIR and persistence multi-variate models of the $PM_{2.5}$ contaminant at 6 am, 12 pm, 18 pm and 24 pm of the subsequent day, for each of the 10 folds, are summarized in tables 4 and 5. It is important to clarify that the errors of the univariate persistence models are not exactly the same than the errors of the multi-variate persistence models because the inclusion of the daily maximum temperature means an increase in the number of missing values in the test data.

From tables 4 and 5 it can easily be concluded that no enhancement has been pro duced when daily maximum temperature is included as additional input variable to the ANFIS and FIR models. The RNMSE are the same or almost the same for univariate models (tables 2 and 3) and multi-variate models (tables 4 and 5). Therefore, in this case, the use of meteorological information does not help to obtain more accurate and reliable models.

Table 2. Prediction errors (RNMSE) of each fold separately and its average for the $PM_{2.5}$ concentration series. Predictions correspond to 6 am and 12 pm of the subsequent day using ANFIS, FIR and persistence univariate models.

	Univariate Models at 6 am			Univariate Models at 12 pm		
	ANFIS-U-6	FIR-U-6	PERS.-U-6	ANFIS-U-12	FIR-U-12	PERS.-U-12
FOLD 1	0.51	0.54	0.74	0.44	0.46	0.55
FOLD 2	0.37	0.41	0.51	0.54	0.52	0.60
FOLD 3	0.38	0.40	0.54	0.30	0.35	0.42
FOLD 4	0.29	0.34	0.42	0.45	0.44	0.50
FOLD 5	0.26	0.28	0.31	0.56	0.56	0.62
FOLD 6	0.33	0.33	0.52	0.71	0.70	0.92
FOLD 7	0.28	0.33	0.50	0.80	0.82	0.75
FOLD 8	0.48	0.47	0.54	0.53	0.57	0.74
FOLD 9	0.31	0.34	0.39	0.40	0.42	0.51
FOLD 10	0.33	0.40	0.53	0.49	0.50	0.55
AVERAGE	**0.35**	**0.38**	**0.50**	**0.52**	**0.54**	**0.62**

Table 3. Prediction errors (RNMSE) of each fold separately and its average for the $PM_{2.5}$ concentration series. Predictions correspond to 18 pm and 24 pm of the subsequent day using ANFIS, FIR and persistence univariate models.

	Univariate Models at 18 pm			Univariate Models at 24 pm		
	ANFIS-U-18	FIR-U-18	PERS.-U-18	ANFIS-U-24	FIR-U-24	PERS.-U-24
FOLD 1	0.44	0.48	0.55	0.42	0.43	0.52
FOLD 2	0.54	0.52	0.54	0.48	0.46	0.62
FOLD 3	0.41	0.39	0.43	0.35	0.40	0.42
FOLD 4	0.52	0.54	0.68	0.42	0.45	0.46
FOLD 5	0.45	0.48	0.52	0.60	0.65	0.69
FOLD 6	0.65	0.66	0.74	0.55	0.55	0.62
FOLD 7	0.53	0.53	0.63	0.44	0.45	0.45
FOLD 8	0.55	0.57	0.65	0.49	0.54	0.60
FOLD 9	0.38	0.40	0.45	0.36	0.39	0.42
FOLD 10	0.42	0.44	0.44	0.35	0.35	0.42
AVERAGE	**0.49**	**0.50**	**0.56**	**0.45**	**0.47**	**0.52**

On the other hand, ANFIS and FIR multi-variate models obtain similar results and it is not possible to conclude which one has a better performance. Again the prediction errors obtained with 6 am models are much lower than the ones obtained with the rest of the models. ANFIS and FIR models perform much better than persistence models, as already happened in the univariated case. The ANFIS multi-variate models are between 1.9% and 24% better than the persistence models while FIR multi-variate models between 9.4% and 22%.

Table 4. Prediction errors (RNMSE) of each fold separately and its average for the $PM_{2.5}$ concentration series. Predictions correspond to 6 am and 12 pm of the subsequent day using ANFIS, FIR and persistence multivariate models.

	Multi-variate Models at 6 am			Multi-variate Models at 12 pm		
	ANFIS-M-6	FIR-M-6	PERS.-M-6	ANFIS-M-12	FIR-M-12	PERS.-M-12
FOLD 1	0.44	0.46	0.62	0.43	0.48	0.56
FOLD 2	0.34	0.36	0.48	0.46	0.48	0.50
FOLD 3	0.27	0.31	0.36	0.31	0.32	0.43
FOLD 4	0.35	0.41	0.50	0.53	0.47	0.52
FOLD 5	0.35	0.38	0.39	0.58	0.61	0.63
FOLD 6	0.59	0.48	0.73	0.71	0.70	0.94
FOLD 7	0.39	0.44	0.70	0.79	0.66	0.75
FOLD 8	0.42	0.40	0.41	0.47	0.48	0.65
FOLD 9	0.36	0.39	0.44	0.40	0.45	0.50
FOLD 10	0.32	0.29	0.36	0.31	0.34	0.42
AVERAGE	**0.38**	**0.39**	**0.50**	**0.50**	**0.50**	**0.59**

Table 5. Prediction errors (RNMSE) of each fold separately and its average for the $PM_{2.5}$ concentration series. Predictions correspond to 18 pm and 24 pm of the subsequent day using ANFIS, FIR and persistence multivariate models.

	Multi-variate Models at 18 pm			Multi-variate Models at 24 pm		
	ANFIS-M-18	FIR-M-18	PERS.-M-18	ANFIS-M-24	FIR-M-24	PERS.-M-24
FOLD 1	0.46	0.48	0.55	0.45	0.49	0.53
FOLD 2	0.55	0.49	0.56	0.55	0.50	0.62
FOLD 3	0.31	0.38	0.42	0.59	0.39	0.44
FOLD 4	0.52	0.54	0.68	0.43	0.47	0.43
FOLD 5	0.51	0.50	0.53	0.68	0.62	0.70
FOLD 6	0.68	0.68	0.73	0.69	0.56	0.62
FOLD 7	0.57	0.50	0.62	0.46	0.48	0.45
FOLD 8	0.57	0.61	0.66	0.60	0.54	0.63
FOLD 9	0.37	0.40	0.44	0.40	0.42	0.46
FOLD 10	0.39	0.41	0.51	0.35	0.34	0.38
AVERAGE	**0.49**	**0.50**	**0.57**	**0.52**	**0.48**	**0.53**

$PM_{2.5}$ is a difficult contaminant to be predicted due to the fact that there are significant variations of the concentrations of this pollutant from one day to the subsequent day, and, from one hour to the subsequent one, even with similar weather conditions.

Previous works have been focused on the modelling and prediction of mean [21] or maximum [19] $PM_{2.5}$ concentrations. Also, there are studies that perform binary predictions, i.e. if a dangerous level has been reached [22]. Contrarily, we have focused on a short-term $PM_{2.5}$ forecast, although uncertainties in hourly registers pose enormous challenges for developing accurate models.

6 Conclusions

This paper studies the performance of two fuzzy modelling approaches in a complex problem, i.e. the prediction of $PM_{2.5}$ concentration in downtown Mexico City metropolitan area. The first is a neuro-fuzzy approach, i.e. ANFIS, and the second is a hybrid fuzzy-pattern recognition approach, i.e. FIR.

Two studies have been performed: the first one uses as models input only the $PM_{2.5}$ concentrations (called univariate models) and, the second one, uses also the daily maximum temperature (called multi-variate models).

Our approach is based on hourly models. The idea is to obtain a specific model for each of the most relevant hours of the day (i.e. 6 am, 12 pm, 18 pm and 24 pm), based on the values of the 6 am, 12 pm, 18 pm and 24 pm of the previous day. Therefore, eight ANFIS and FIR models have been developed (4 univariate and 4 multi-variate) and its performance compared with persistence models.

The conclusions are that no enhancement has been produced when daily maximum temperature is included as additional input variable to the ANFIS and FIR models. The accuracy of both ANFIS and FIR methodologies are almost the same, so both fuzzy methodologies can be considered appropriate approaches to deal with this complex modelling problem. ANFIS and FIR models perform much better than persistence for all the univariate and multi-variate models.

As a future work we propose to:

- Include other meteorological variables into the model.
- Include additional information such are the day of the week or the hour of the day into the models.
- Use additional hybrid modelling techniques such as FIR with genetic algorithm, which will help to find in an efficient way the number of classes and landmarks parameters of FIR discretization process.

References

1. WHO World healt oranization. Air quality guidelines: the global update 2005 (2006)
2. van Donkelaar, A., Martin, R., Verduzco, C., Brauer, M., Kahn, R., Levy, R., Villeneuve, P.: 2010. A Hybrid Approach for Predicting PM2.5 Exposure: van Donkelaar et al. Respond. Environ. Health Perspect. 118(10), 425 (2010)
3. NWM: National Weather Service of Mexico (2012), http://smn.cna.gob.mx/
4. Mintz, R., Young, B.R., Svrcek, W.Y.: Fuzzy logic modeling of surface ozone concentrations. Computers & Chemical Engineering 29, 2049–2059 (2005)
5. Ghiaus, C.: Linear fuzzy-discriminant analysis applied to forecast ozone concentration classes in sea-breeze regime. Atmospheric Environment 39, 4691–4702 (2005)
6. Morabito, F.C., Versaci, M.: Fuzzy neural identification and forecasting techniques to process experimental urban air pollution data. Neural Networks 16, 493–506 (2003)
7. Heo, J.S., Kim, D.S.: A new method of ozone forecasting using fuzzy expert and neural network system. Sicence of the Total Environment 325, 221–237 (2004)
8. Yildirim, Y., Bayramoglu, M.: Adaptive neuro-fuzzy based modelling for prediction of air pollution daily levels in city of Zonguldak. Chemosphere 63, 1575–1582 (2006)
9. Peton, N., Dray, G., Pearson, D., Mesbah, M., Vuillot, B.: Modelling and analysis of ozone episodes. Environmental Modelling & Software 15, 647–652 (2000)
10. Onkal-Engin, G., Demir, I., Hiz, H.: Assessment of urban air quality in Istanbul using fuzzy synthetic evaluation. Atmospheric Environment 38, 3809–3815 (2004)
11. Klir, G., Elias, D.: Architecture of Systems Problem Solving, 2nd edn. Plenum Press, New York (2002)
12. Nebot, A., Mugica, F., Cellier, F., Vallverdú, M.: Modeling and Simulation of the Central Nervous System Control with Generic Fuzzy Models. Simulation 79(11), 648–669 (2003)
13. Carvajal, R., Nebot, A.: Growth Model for White Shrimp in Semi-intensive Farming using Inductive Reasoning Methodology. Computers and Electronics in Agriculture 19, 187–210 (1998)
14. Escobet, A., Nebot, A., Cellier, F.E.: Visual-FIR: A tool for model identification and prediction of dynamical complex systems. Simulation Modelling Practice and Theory 16, 76–92 (2008)
15. Nauck, D., Klawonn, F., Kruse, R.: Neuro-Fuzzy Systems. John Wiley & Sons (1997)
16. SIMAT (2012), http://www.sma.df.gob.mx/simat/
17. Muñoz, R., Carmona, M.R., Pedroza, J.L., Granados, M.G.: Data analysis of PM2.5 registered with TEOM equipment in Azcapotzalco (AZC) and St. Ursula (SUR) stations of the automatic air quality monitoring network (RAMA). In: National Congress of Medicine Engineering and Ambient Sciences, pp. 21–24 (2000) (in Spanish)

18. Pérez, P., Trier, A., Reyes, A.: Prediction of PM2.5 concentrations several hours in advance using neural networks in Santiago, Chile. Atmospheric Environment 34, 1189–1196 (2000)
19. Cobourn, W.G.: An enhanced PM2.5 air quality forecast model based on nonlinear regression and back-trajectory concentrations. Atmospheric Environment 44, 3015–3023 (2010)
20. Salini, G., Perez-Jara, P.: Time series analysis of atmosphere pollution data using artificial neural networks technique. Revista Chilena de Ingeniería 14(3), 284–290 (2006)
21. Dong, M., Yang, D., Kuang, Y., He, D., Erdal, S., Kenski, D.: PM2.5 concentration prediction using hidden semi-Markov model-based times series data mining. Expert Systems with Applications 36, 9046–9055 (2009)
22. Kang, D., Mathur, R., Trivikrama Rao, S.: Assessment of bias-adjusted PM2.5 air quality forecast over the continental United States during 2007. Geoscience Model Dev. 3, 309–320 (2010)

Stochastic Resonance and Anti-cyclonic Rings in the Gulf of Mexico

Benjamín Martínez-López, Jorge Zavala-Hidalgo, and Carlos Gay García

Center for Atmospherc Sciences, National Autonomous University of Mexico,
Ciudad Universitaria, D.F., Mexico
{benmar,jzavala}@atmosfera.unam.mx, cgay@servidor.unam.mx

Abstract. In this work, we used a nonlinear, reduced gravity model of the Gulf of Mexico to study the effect of a seasonal variation of the reduced gravity parameter on ring-shedding behaviour. When small amplitudes of the seasonal variation are used, the distributions of ring-shedding periods are bi-modal. When the amplitude of the seasonal variation is large enough, the ring-shedding events shift to a regime with a constant, yearly period. If the seasonal amplitude of the reduce gravity parameter is small but a noise term is included, then a yearly regime is obtained, suggesting that stochastic resonance could play a role in the ring-shedding process taking place in the Gulf of Mexico.

Keywords: Gulf of Mexico, Ring Shedding, Reduced Gravity Model, Seasonal Forcing, Stochastic Resonance.

1 Introduction

Anti-cyclonic rings generated by meandering of intense boundary current systems are long-lived, intense near-surface features that dominate the oceanic mesoscale in different regions of the World Ocean. They substantially contribute to determine the water mass characteristics as well as the upper-ocean circulation patterns in these regions and, due to their characteristic self-induced, westward propagation, they often also play an important role in the transfer of chemical and biological properties across frontal zones [8].

The circulation in the Gulf of Mexico (GoM) presents two semipermanent traits: the Loop Current (LC) in the eastern region and a cell of anti-cyclonic circulation on the western boundary [6]. The GoM's LC may profoundly influence the local circulation as it moves northward forming meanders, upwelling regions and eventually detached rings (Fig. 1, left panel), which are large, warm, clockwise-spinning vortex of water that migrate westward across the GoM (Fig. 1, right panel) at speeds that have been estimated to range from 2 to 5 km/day [1].

The large surface temperature anomalies as well as the large surface horizontal velocity shears associated with these rings may profoundly influence human activities. For instance, the passage of warm-core rings detached from the LC is able to disturb oil extraction activities, while it is demonstrated that hurricanes may be intensified by their interaction with the warm ring water (see [3] and references therein).

M.S. Obaidat et al. (eds.), *Simulation and Modeling Methodologies, Technologies and Applications*, Advances in Intelligent Systems and Computing 256,
DOI: 10.1007/978-3-319-03581-9_18,

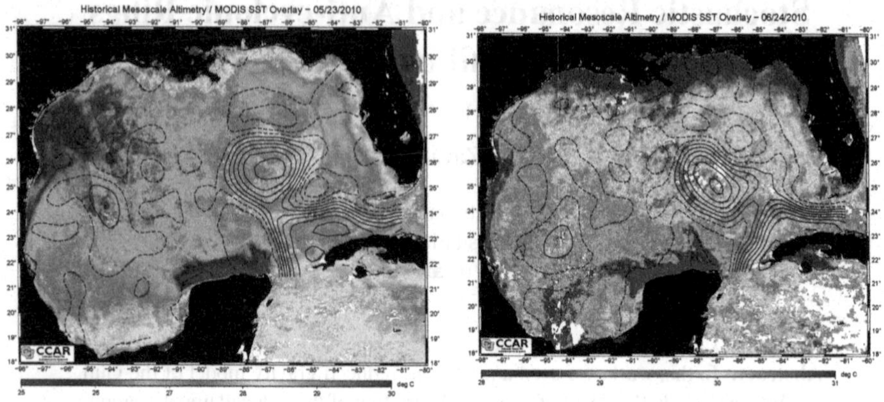

Fig. 1. Satellite measurements of sea surface height overlaid with sea surface temperature MODIS data show a detached ring of the Loop Current which starts to move toward the western Gulf of Mexico. Source: http://eddy.colorado.edu/ccar/ssh/hist_gom_grid_viewer

Predicting the onset and evolution of ring shedding in the GoM may substantially contribute to the understanding of the subtle dynamics involved in the local oceanic phenomena and also to the reduction of the impact on human activities caused, directly or indirectly, by these rings. Observations show a nearly bi-modal distribution (the most evident peaks existing around 6 months and 9 to 11 months). Current full-fledged ocean numerical models can explain some of the observed ring-shedding variability but fail in simulating observed periods [5, 11, 9].

Simple models, on the other hand are only able to reproduce an almost constant period [4, 10, 7], which was called the natural period of the Gulf by Hulbert and Thompson (10-11 months). In part, this deficiency of existing numerical model is undoubtedly the result of inaccuracies induced in the simulated dynamics by the imposed boundary conditions, which unavoidably, tend to introduce in the system an exaggeratedly strong yearly signal. The discrepancy between observations and simulations, however, may be used to gain a deeper understanding of the subtle dynamics governing the process of ring shedding in the GoM. In fact, it results that as the strength of the yearly signal imposed in realistic model simulations decreases, the occurrence of yearly, ring-shedding events also diminishes. This behaviour may indicate that, in numerical models, a synchronization mechanism exists, which is able to shift a natural ring-shedding period toward a yearly one.

Considering that the seasonal cycle of sea surface temperature in the GoM is a natural forcing on the wide spectrum of physical processes taking place there then emerges an attractive possibility: stochastic resonance. If the imposed forcing by the surface temperature is strong enough to drive the system, we can expect that ring-shedding variability will contain spectral energy in the yearly frequency, but this is not the case, or at least not most of the time. Now, if we include noise as a forcing mechanism, then we can expect that a weak signal in the forcing can be amplified and optimized by the assistance of noise [2]. In other words: if in the GoM exists a weak, yearly signal in the forcing, which is not capable of inducing a yearly period in the ring-shedding process,

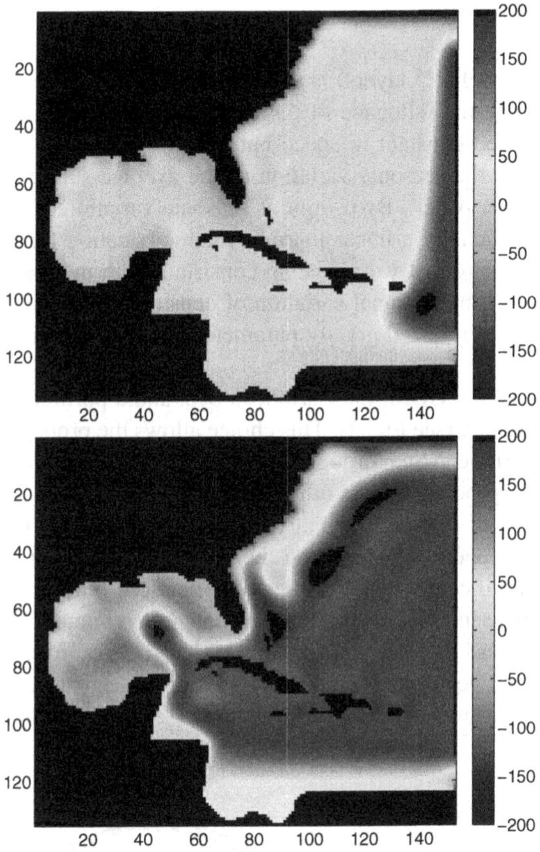

Fig. 2. Domain used in our model simulations. There are two eastern open boundaries, which are forced by a flow induced by a meridional gradient of upper layer height. These input and ouput flows are time invariant. Upper panel: height anomalies at day 50. Lower panel: height anomalies at day 500.

this yearly forcing plus a noisy term could be able to produce ring-shedding events with a yearly period.

Our conjecture may be tested using simple numerical models. To this purpose, we implemented in the GoM a nonlinear, reduced-gravity model simulating an idealized LC through the inflow/outflow of near-surface water through its boundaries (see Fig. 2). This is the simplest model capable of simulating ring shedding [4]. In this study, it is proposed that a reduced gravity model, where the buoyancy term is seasonally forced and the dissipation term is varied, can explain the observed period of shedding behavior. Additionally, we explore the possibility of stochastic resonance in the ring-shedding process by including a noise term in the seasonally forced buoyancy term.

2 Model Simulations

A reduced gravity model (1.5 layers) is used to simulate the ring-shedding process in the GoM and to study the influence of the seasonal cycle upon it. The upper-ocean temperature is assumed constant in space but varies in time. A single sine function is enough to simulate the seasonal variation of the average surface temperature over the GoM (Fig. 3, upper panel). By using this seasonal variation of temperature (while keeping salinity constant) in a linear thermodynamic equation for the density of the upper layer (to make sure that we are being consistent with the planetary geostrophic approximation) we get the seasonal variation of density (Fig. 3, middle panel), which then is used to obtain a reduced gravity parameter that evolves with the time (Fig. 3, lower panel).

The domain covers the GOM, the Caribbean Sea, and a portion of the Gulf Stream area in the North Atlantic (see Fig. 2). This choice allows the propagation of anomalies from the western Caribbean Sea through the Yucatan Channel and their possible effect on the ring-shedding process. On the other hand, the eastern open boundary is located sufficiently far from our region of primary interest, thus assuring that the adjustment taking place near the open boundary, and the unavoidable reflection of some waves, does not interfere significantly with the internal dynamics of the GoM.

In each model experiment, the equations are integrated numerically for a period of 20 years using a forward Euler scheme and an Arakawa C grid for the active layer. All model experiments are performed using a time step of 90 seconds and a horizontal grid with 135×153 points and a spacing of $1/4$ degree (see model domain in Fig. 2). In a

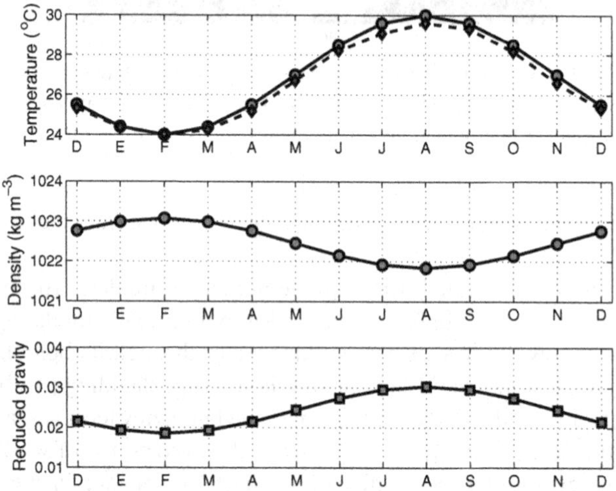

Fig. 3. Upper panel: Average surface temperature over the Gulf of Mexico (95W-84W, 21N-26N) obtained from NOAA Optimum Interpolation Sea Surface Temperature Analysis (dashed line) and seasonal variation of surface temperature simulated by a single sine function (solid line). Middle panel: Density as a linear function of seasonally varying temperature. Lower panel: Seasonal variation of reduced gravity parameter.

first set of experiments, a constant reduced gravity parameter (g') is used with a value of 0.0245, corresponding to the time-averaged value shown in the lower panel of Fig. 3. In the second set, a seasonally evolving $g'(t)$ parameter is considered (lower panel of Fig. 3). Finally, in a third set of experiments, a noisy term is added to the time evolving $g'(t)$ parameter.

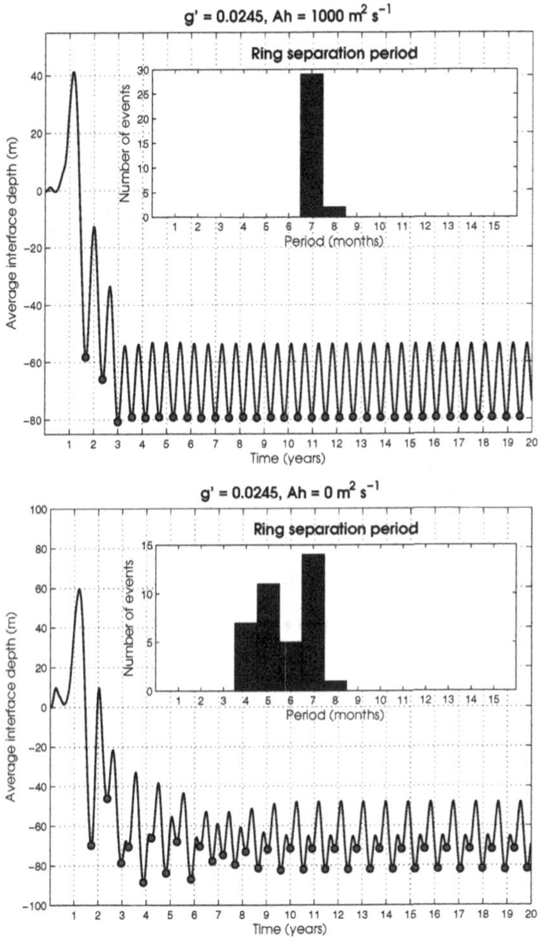

Fig. 4. Histograms and time series of area-averaged height anomalies over the ring-shedding region used to estimate the timing of ring shedding events (red dots). By using a diffusion coefficient of $10^3 m^2 s^{-1}$, a constant period of seven months is obtained (upper panel), while with zero diffusion period doubling is observed (lower panel).

3 Results

Without seasonal forcing, a single peak in the ring shedding period is observed, which depends on model geometry, upper layer thickness, diffusion, and g' value (Fig. 4). A red dot in this figure and the following ones indicate the timing of the detachment of a ring of the LC.

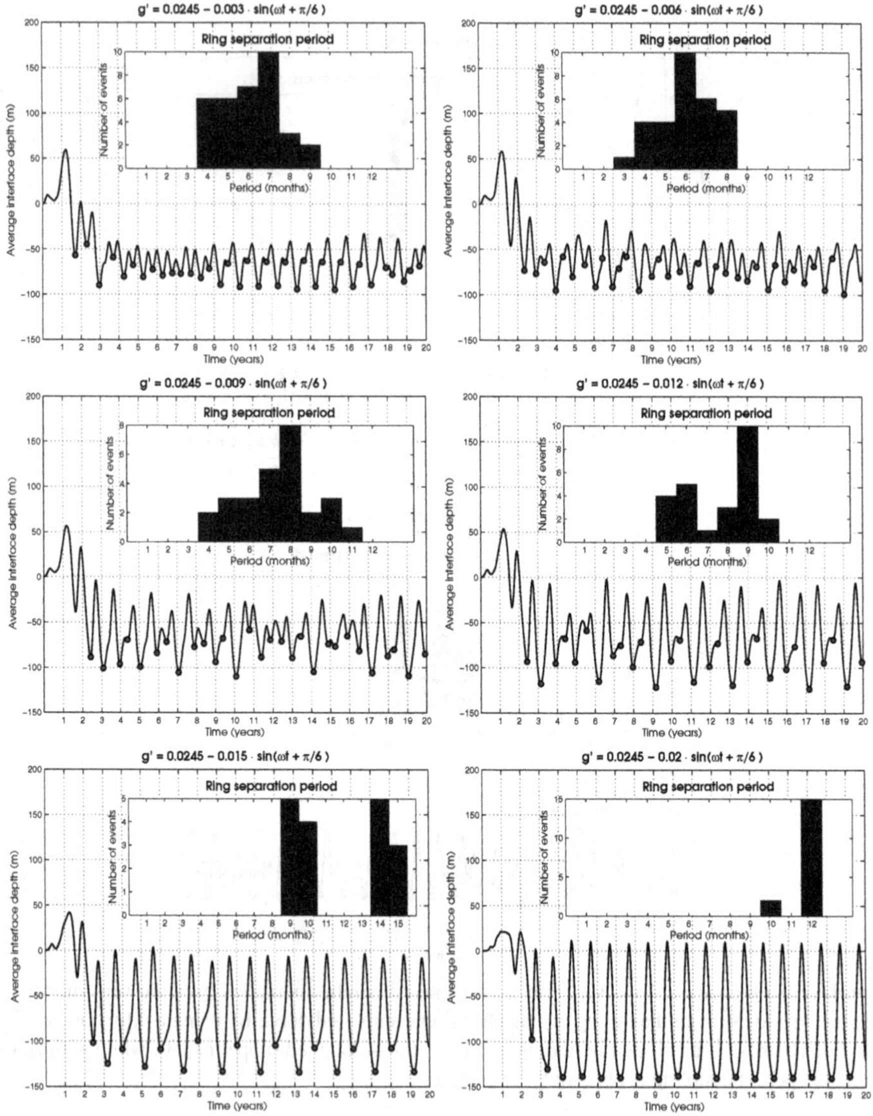

Fig. 5. Ring-shedding distributions obtained when a seasonal variation in $g'(t)$ is included. The phase constant is used to obtain a seasonal variation of density in agreement with the seasonal cycle of surface temperature over the central Gulf of Mexico. The amplitude of $g'(t)$ increases from top to bottom, and from left to right. In all experiments, zero diffusion is used.

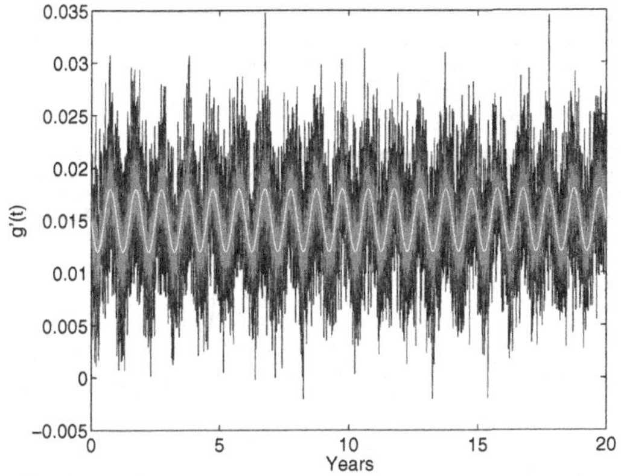

Fig. 6. Stochastic forcing. The white curve represents the seasonal variation of $g'(t)$. In the green curve, a noise term with zero mean and standard deviation of 0.015 is added. This forcing is multiplied by a factor 2 and 3 in the red and blue curves, respectively.

When the amplitude of the variation of g' is increased, a more complex behaviour is observed, characterized by a bi-modal distribution, which crucially depends on the amplitude (Fig. 5). When the amplitude is large enough, the annual signal clearly dominates, but when it is weaker different bi-modal distributions appear, some of which resemble the observed one. In the case of large amplitude of the annual signal, a quite simple ring shedding behaviour emerges: constant, yearly ring-shedding events.

Now, we try to answer what happens when the annual signal is weak but high frequency forcing is present. This kind of forcing could be, for example, similar to that associated with turbulent heat fluxes between atmosphere and ocean. A simple way to simulate them is by using a noise term, which represents heat exchange between the ocean surface and the lower atmosphere. In this case, we used for the stochastic forcing a series of random numbers with zero mean and standard deviation of 0.015. Additionally, we explore the effect of noise magnitude on the ring-shedding process using a factor of two and three in the noise term (Fig. 6).

Fig. 7 shows the ring-shedding period distribution resulting of the seasonal variation of $g'(t)$. The amplitude of the seasonal forcing is not able to produce yearly ring shedding, because it is not large enough (right-upper panel). The inclusion of a small noise term modifies the distribution of ring-shedding periods but it is not able to induce ring-shedding events with a yearly period.

If the noise amplitude is increased three times, then the noise term is able to change this behaviour, inducing a dominance of ring-shedding events with a period close to the yearly one. It is pertinent to emphasize that looking for the right combinations of g' values, diffusion, and noise magnitude, it can be achieve a clear dominance of the annual signal.

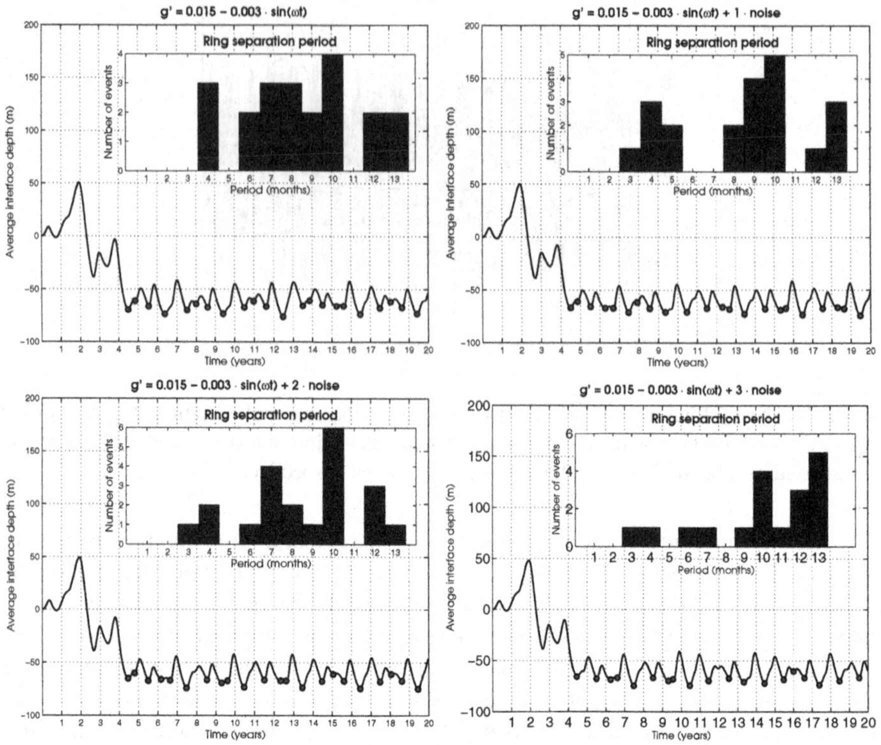

Fig. 7. Ring-shedding period distributions using seasonal forcing (left-upper panel) and seasonal forcing plus noise (right-upper and lower panels). See the text for details.

4 Conclusions

In the case of constant g', it is shown that large values of diffusion lead to ring-shedding events with a constant period of several months, while with zero diffusion the period is doubled.

By considering a seasonal variation of $g'(t)$, a more realistic ring-shedding distribution is obtained, which is characterized by a bi-modal distribution that crucially depends on the amplitude of the seasonal variation.

If the annual signal of $g'(t)$ has large amplitude, a quite simple ring shedding behaviour is observed: a constant, yearly ring-shedding event.

If the amplitude of $g'(t)$ is not large enough to induce yearly variability in the ring-shedding process, such behaviour can be obtained by including stochastic forcing with a proper intensity. Thus, it has been shown that stochastic resonance could play a role in the ring-shedding process taking place in the Gulf of Mexico.

References

1. Elliott, B.A.: Anticyclonic rings in the Gulf of Mexico. Journal of Physical Oceanography 12, 1292–1309 (1982)
2. Gammaitoni, L., Hanggi, P., Jung, P., Marchesoni, F.: Stochastic resonance. Rev. Modern Phys. 70, 223–287 (1998)
3. Halliwell, G.R., Shay, L.K., Brewster, J.K., Teague, W.J.: Evaluation and sensitivity analysis of an ocean model response to hurricane ivan. Mon. Wea. Rev. 139, 921–945 (2011)
4. Hurlburt, H.E., Thompson, J.D.: A numerical study of Loop Current intrusion and eddy shedding. J. Phys. Oceanogr. 10, 1611–1651 (1980)
5. Murphy, S.J., Hurlburt, H.E., OBrien, J.J.: The connectivity of eddy variability in the Caribbean Sea, the Gulf of Mexico, and the Atlantic Ocean. J. Geophys. Res. 104, 1431–1453 (1999)
6. Nowlin Jr., W.D., McLellan, H.J.: A Characterization of the Gulf of Mexico Waters in Winter. Journal of Marine Research 25, 29–59 (1967)
7. Oey, L.-Y., Lee, H.-C., Schmitz Jr., W.J.: Effects of Winds and Caribbean Eddies on the Frequency of Loop Current Eddy Shedding: A Numerical Model Study. J. Geophy. Res. 108(C10) (2003), doi:10.1029/2002JC001698
8. Olson, D.B.: Rings in the ocean. Annu. Rev. Earth Planet. Sci. 19, 283–311 (1991)
9. Romanou, A., Chassignet, E.P., Sturges, W.: The Gulf of Mexico circulation within a high-resolution numerical simulation of the North Atlantic Ocean. J. Geophys. Res. 109(C01003) (2004), doi:10.1029/2003JC001770
10. Sturges, W., Evans, J.C., Holland, W., Welsh, S.: Separation of warm-core rings in the Gulf of Mexico. J. Phys. Oceanogr. 23, 250–268 (1993)
11. Welsh, S.E., Inoue, M.: LC rings and the deep circulation in the Gulf of Mexico. J. Geophys. Res. 105, 16,951–16,959 (2000)

On Low-Fidelity Model Selection for Antenna Design Using Variable-Resolution EM Simulations

Slawomir Koziel, Stanislav Ogurtsov, and Leifur Leifsson

Engineering Optimization & Modeling Center, School of Science and Engineering, Reykjavik University, Menntavegur 1, Reykjavik, IS-101, Iceland
{koziel,stanislav,leifurth}@ru.is

Abstract. One of the most important tools of antenna design is electromagnetic (EM) simulation. High-fidelity simulations offer accurate evaluation of the antenna performance, however, they are computationally expensive. As a result, employing EM solvers in automated antenna design using numerical optimization techniques is quite challenging. A possible workaround are surrogate-based optimization (SBO) methods. In case of antennas, the generic way to construct the surrogate is through coarse-discretization EM simulations that are faster but, at the same time, less accurate. For most SBO algorithms, quality of such low-fidelity models may be critical for performance. In this work, we investigate the trade-offs between the speed and the accuracy of the low-fidelity antenna models as well as the impact of the model selection on the quality of the design produced by the SBO algorithm as well as the computational cost of the optimization process. Our considerations are illustrated using examples.

Keywords: Computer-Aided Design (CAD), Antenna Design, Electromagnetic Simulation, Surrogate-based Optimization.

1 Introduction

Nowadays, antenna design depends heavily on electromagnetic (EM) simulation tools. While in some cases it is possible to resort to simple theoretical models to estimate initial values for antenna dimensions, further tuning of the antenna geometry parameters has to be based on repetitive EM simulations. This is particularly the case for ultra-wideband (UWB) antennas [1] as well as dielectric resonator antennas (DRAs) [2]. In general, whenever antenna environment has to be taken into account (e.g., housing, platform, connectors), EM-based design in the only reliable option.

The use of numerical optimization techniques to automate the antenna design process seems to be very appealing, however, it turns out to be also quite challenging. The major obstacle is a high computational cost of accurate, high-fidelity EM simulation. This often makes utilization of conventional optimization methods (such as the gradient-based algorithms) prohibitive due to the large number of simulations necessary to converge to an optimized design. In some applications, so-called population-based search techniques (or metaheuristics) have gained some popularity [3]-[6]. These methods are helpful in handling certain issues such as multiple local

M.S. Obaidat et al. (eds.), *Simulation and Modeling Methodologies, Technologies and Applications*, Advances in Intelligent Systems and Computing 256,
DOI: 10.1007/978-3-319-03581-9_19, © Springer International Publishing Switzerland 2014

optima [7] and they are mostly applied to problems prone to this kind of difficulties, e.g., synthesis of array patterns [8]. Unfortunately, the computational cost of population-based algorithms is normally very high due to processing a number of potential solutions simultaneously, which makes these methods not suitable for antenna design in general.

Computationally efficient EM-simulation-based design can be realized using surrogate-based optimization (SBO) techniques [9]-[11]. In SBO, the direct optimization of the expensive EM model is replaced by iterative updating and optimization of a surrogate model, which is a fast and (at least locally) accurate representation of the structure of interest. There are two main types of the surrogates, function approximation and physics-based ones. The first type based on an approximation of high-fidelity model data and includes a number of techniques, such as artificial neural networks [12], [13], support vector regression [14], [15], radial-basis function [16], or kriging [17], [18]. Approximation-based surrogates are normally fast, however, their setup costs may be very high because of the large number of training samples (and, consequently, EM simulations) necessary to ensure decent accuracy. Physics-based surrogates exploit the underlying low-fidelity model that is corrected to become a more accurate representation of the high-fidelity model. Various approaches differ in the way the correction is realized. Examples of SBO techniques exploiting physics-based surrogates include space mapping [9], [19], [20], tuning methodologies [21], [22], [23], manifold mapping [24], and shape-preserving response prediction [25]. Methods from this group seem more attractive for ad-hoc optimization because they are capable to yield a satisfactory design after a limited number of high-fidelity EM simulations [9].

An important prerequisite of the efficiency of the physics-based SBO techniques is that the underlying low-fidelity model is computationally cheap. For that reason low-fidelity models comprising circuit equivalents or analytical formulas are particularly suitable [9]. Unfortunately, such models are not available for antennas in many cases so that the most universal way to create the low-fidelity model is through coarse-discretization EM simulations. Low-fidelity EM models are, however, relatively expensive and their evaluation time has to be taken into account while designing and executing algorithms for surrogate-based antenna optimization.

There is a clear trade-off between the computational cost and the accuracy of the simulation based low-fidelity antenna models. Coarser models are obviously faster but, at the same time, less accurate. In the context of surrogate-based optimization, faster models result in reducing the CPU cost of single algorithm iteration, however, because of accuracy limitations, a larger number of iterations may be necessary to obtain a satisfactory design. The overall design may be about the same or even higher than when the finer low-fidelity model is employed. Also, a coarser model increases a risk of failure of the entire process. On the other hand, a finer low-fidelity model is more expensive (per model evaluation); however, it is not obvious that its longer simulation time would turn into a higher overall design cost because a smaller number of iterations may be sufficient to find a satisfactory design.

In general, the problems related to the computational cost of the low-fidelity model can be alleviated to some extent by a careful development of the SBO algorithms. In particular, it is more advantageous to replace surrogate models that require extraction of parameters through nonlinear optimization (cf. parameter extraction in space mapping [10]) by response correction techniques (e.g., [24]-[26]) where the surrogate model parameters can be obtained analytically.

In this work, the importance of the antenna model fidelity selection and its impact on the performance of the SBO design process are investigated. We consider the two main criteria, namely the computational cost of the design process and the quality of the design produced by the SBO algorithm. The benefits of utilizing several models of various fidelities are also discussed. Finally, we formulate recommendations regarding the surrogate model selection for EM-based antenna design.

2 Simulation-Driven Antenna Design: Low-Fidelity Model Selection

In this section, we discuss surrogate-based optimization (SBO) of antennas. We formulate the design problem as well as a generic SBO algorithm. The main part is devoted to low-fidelity antenna models and construction of surrogate models suitable for antenna optimization.

2.1 Antenna Design Problem Formulation

Let $\boldsymbol{R}_f \in R^m$ denote a response vector of a high-fidelity (or fine) model of the antenna structure of interest. We assume that $\boldsymbol{R}_f$ represents, e.g., antenna reflection (and/or other relevant characteristics such as gain) evaluated over a certain frequency band of interest. The antenna is parameterized using a vector of designable variables $\boldsymbol{x} \in R^n$ (e.g., geometry and/or material parameters). The high-fidelity model response is obtained using accurate (and, therefore, expensive) EM simulation.

The design goal is to solve the following nonlinear minimization problem

$$\boldsymbol{x}_f^* \in \arg\min_{\boldsymbol{x}} U\left(\boldsymbol{R}_f(\boldsymbol{x})\right) \tag{1}$$

where U is an objective function (typically a norm or a minimax function with upper and lower specifications [9]). U encodes the design requirements and it is formulated so that a better design corresponds to a smaller value of U. $\boldsymbol{x}_f^*$ is the optimal design to be determined.

2.2 Surrogate-Based Optimization Basics

In this work, we are focused on a specific class of design methods, so-called surrogate-based optimization (SBO) techniques [11]. The main purpose of SBO is to reduce the computational cost of the design process. It is performed by shifting the optimization burden into a cheap yet reasonably accurate representation of the high-fidelity model, a surrogate.

Many surrogate-based algorithms utilize the following iterative procedure [11]

$$\boldsymbol{x}^{(i+1)} = \arg\min_{\boldsymbol{x}} U(\boldsymbol{R}_s^{(i)}(\boldsymbol{x})) \tag{2}$$

which generates a sequence of approximate solutions $\boldsymbol{x}^{(i)}$, $i = 0, 1, \ldots$, to the original problem (1). Here, $\boldsymbol{R}_s^{(i)}$ is the surrogate model at iteration i, whereas $\boldsymbol{x}^{(0)}$ is the initial design. In the above scheme, the surrogate model is updated at each iteration using the high-fidelity model data accumulated during the optimization process. Normally, the high-fidelity model is only evaluated once per iteration, at a newly found design vector $\boldsymbol{x}^{(i+1)}$. It is essential that the process of finding the new design is entirely based on the surrogate. Consequently, the overall design cost can be greatly reduced in comparison with direct solving of the original problem (1).

As mentioned in the introduction, the surrogate model can be constructed using various methods that fall into two main categories: approximation-based and physics-based models. Approximation-based modeling techniques are pretty matured and include a number of methods such as artificial neural networks [12], kriging interpolation [18] or support-vector regression [14]. Once developed, this type of models are fast and generic, and, therefore, easily transferrable from one problem to another. There are strategies available for iterative model improvement by incorporating new training samples (so-call infill criteria) that can aim at design space exploration (improvement of global model accuracy) or local exploitation of a given region [27]. From the point of view of one-time design of a given antenna, all of these methods have an important drawback: it is quite expensive to construct an approximation model with decent accuracy, particularly for a larger number of design variables. In practice, approximation models require hundreds or even thousands of data samples (and, a corresponding number of EM simulations).

For the sake of efficiency, we focus here on physics-based surrogates created from an underlying low-fidelity model $\boldsymbol{R}_c$, which is a fast yet reasonable representation of the high-fidelity model $\boldsymbol{R}_f$. In general, the surrogate $\boldsymbol{R}_s^{(i)}$ is obtained by aligning $\boldsymbol{R}_c$ with $\boldsymbol{R}_f$ at the current design $\boldsymbol{x}^{(i)}$ using $\boldsymbol{R}_f$ data accumulated in previous iterations. The advantage of the physics-based surrogate is that the low-fidelity model embeds some "knowledge" about the structure under consideration, so that only a limited amount of high-fidelity model data is necessary to correct $\boldsymbol{R}_c$. Also, the generalization capability of the physics-based surrogates (i.e., the ability to represent the high-fidelity model outside the training set) is much better than for the approximation models.

2.3 Low-Fidelity EM-Simulation Models for Antenna Structures

A universal way to create physics-based low-fidelity models of realistic antennas involves coarse-discretization simulation, in particular using available EM software packages such as [28] and [29]. This approach is also appealing from the practical point of view and allows obtaining reliable antenna responses with respect to environment and feeds. Examples of antenna structures for which coarse-discretization simulation is the only modeling possibility include but is not limited to ultra-wideband (UWB) antennas [1], as well as dielectric resonator antennas (DRAs) [2]. In discussions and examples of this work the low-fidelity model $\boldsymbol{R}_c$ and the high -fidelity model of the antenna are evaluated with the same EM solver.

A straightforward way to create a low-fidelity model of the antenna is coarser mesh settings compared to those of the high-fidelity antenna model, e.g., as illustrated in Fig. 1. In addition to a coarser mesh, other computational simplifications can be made, including (a) shrinking the computational domain and applying simple absorbing boundaries [28]-[31](with the finite-volume methods implemented in the EM software in use); (b) using low order basis functions [31]- [33] (with the finite-element and moment methods); (c) using more relaxed solution termination criteria such as the S-parameter error [28], [29] (for the frequency domain methods with adaptive meshing) and residue energy [28] (for the time-domain methods). Simplification of the physics of the models can be: (a) ignoring dielectric and metal losses as well as material dispersion if their impact to the simulated response is up to moderate; (b) setting metallization thickness to zero for of traces, strips, and patches; (c) ignoring moderate anisotropy of substrates; (d) energizing the antenna with discrete sources rather than waveguide ports, etc. Rigorously speaking, computational and physical simplifications are closely related and listed in two groups for classification purposes mostly. For example, ignoring dielectric losses and material dispersion in the model simulated with the time-domain finite-volume method turns in to a much simpler formulation of the solution process with a smaller number of unknowns for the same mesh discretization [30].

Because of possible simplifications, the low-fidelity model $\boldsymbol{R}_c$ is faster than $\boldsymbol{R}_f$, typically it can be made 10 to 50 times; however, model $\boldsymbol{R}_c$ is obviously not as accurate as $\boldsymbol{R}_f$. Therefore, the low-fidelity model cannot replace the high-fidelity model in the design optimization process. Figure 2 shows the high- and low-fidelity model responses at a specific design for the antenna of Fig. 1 obtained with different meshes, as well as the relationship between mesh coarseness and simulation time. Selection of the model coarseness strongly affects the simulation time and performance of the design optimization process. Coarser models are faster, and it turns into a lower cost of design per iteration. The coarser models, however, are less accurate, which may results in a larger number of iterations necessary to find a satisfactory design. Also, there is an increased risk of failure for the optimization algorithm to find a good design. Finer models, on the other hand, are more expensive but they are more likely to produce a useful design with a smaller number of iteration.

The major focus of this work is to study the relationship between the performance of the surrogate-based antenna design process and the underlying coarse model fidelity.

It is worth to point out that at the present stage of research, a visual inspection of the model response and the relationship between the high- and low-fidelity models is an important part of the model selection process. It is essential that the low-fidelity model captures all important features of the high-fidelity model response.

One can infer from Fig. 2 that the two "finest" coarse-discretization models (with ~400,000 and ~740,000 mesh cells) represent the high-fidelity model response (shown as a thick solid line) quite properly. The model with ~270,000 cells can be considered as a borderline one. The two remaining models could be considered as poor ones, particularly the model with ~20,000 cells; its response is essentially unreliable.

2.4 Construction of the Surrogate Model. Response Correction and Frequency Scaling

Surrogate model can be constructed from a physics-based low-fidelity model in various ways. However, having in mind that the methods that do not involve multiple evaluation of $\boldsymbol{R}_c$ to find the model parameters are more attractive from the point of view or reducing the computational cost of the design process. Examples of such methods include manifold mapping [24], adaptive response correction [35], or

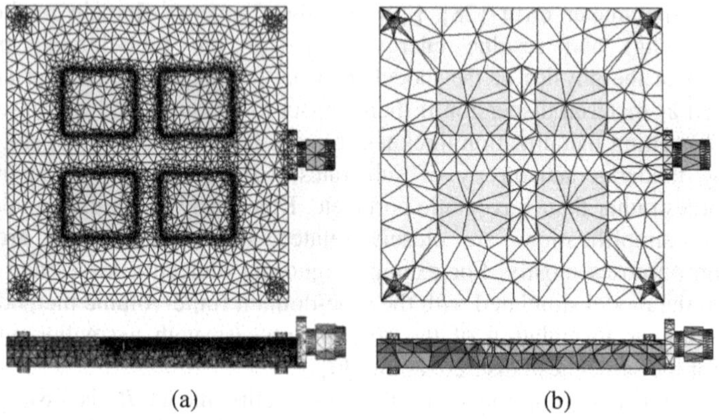

Fig. 1. Microstrip antenna [34]: (a) high-fidelity model shown with a fine tetrahedral mesh; and (b) low-fidelity model shown with a much coarser mesh.

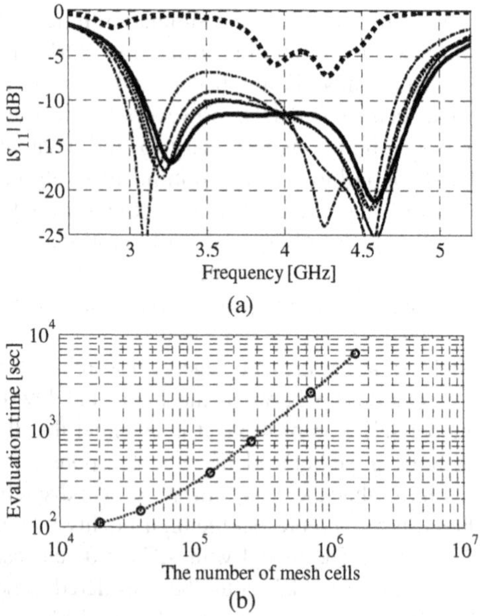

Fig. 2. Microstrip antenna of Fig. 1 at a selected design simulated with the CST MWS transient solver [28]: (a) reflection response of different discretization densities, 19,866 cells (▪▪▪), 40,068 cells (· — ·), 266,396 cells (– –), 413,946 cells (···), 740,740 cells (—), and 1,588,608 cells (▬); and (b) the antenna simulation time versus the number of mesh cells.

shape-preserving response prediction [25]. For the sake of illustration, we focus here on some more basic methods which are sufficient for our considerations, i.e., response correction and frequency scaling.

Response correction can be generally formulated as a modification of the response of the low-fidelity model:

$$\boldsymbol{R}_s(\boldsymbol{x}) = \boldsymbol{C}(\boldsymbol{R}_c(\boldsymbol{x})) \tag{3}$$

Here, $\boldsymbol{C} : R^m \to R^m$ is a response correction function. In practice, the correction may change from one iteration of the SBO process (2) to another so that, in general, the surrogate model is defined as $\boldsymbol{R}_s^{(i)}(\boldsymbol{x}) = \boldsymbol{C}^{(i)}(\boldsymbol{R}_c(\boldsymbol{x}))$, where $\boldsymbol{C}^{(i)}$ is the correction function at iteration i. The minimum requirement for the surrogates constructed using response correction is so-called zero-order consistency between the surrogate and the fine model, i.e., $\boldsymbol{R}_s^{(i)}(\boldsymbol{x}^{(i)}) = \boldsymbol{R}_f(\boldsymbol{x}^{(i)})$. If the derivative information for both the surrogate and the high-fidelity model is available, the first-order consistency condition can also be satisfied, i.e., $\boldsymbol{J}[\boldsymbol{R}_s^{(i)}(\boldsymbol{x}^{(i)})] = \boldsymbol{J}[\boldsymbol{R}_f(\boldsymbol{x}^{(i)})]$ (here, $\boldsymbol{J}[\cdot]$ denotes the Jacobian of the respective model), which may guarantee convergence of $\{\boldsymbol{x}^{(i)}\}$ to a local optimum of $\boldsymbol{R}_f$ [36], assuming that the algorithm (2) is embedded in the trust region framework [37] and the models involved are sufficiently smooth.

Here, we consider the most basic response correction formulated as follows

$$\boldsymbol{C}(\boldsymbol{R}_c(\boldsymbol{x})) = \boldsymbol{R}_c(\boldsymbol{x}) + [\boldsymbol{R}_f(\boldsymbol{x}^{(i)}) - \boldsymbol{R}_c(\boldsymbol{x}^{(i)})] \tag{4}$$

which—by definition—ensures zero-order consistency, i.e., $\boldsymbol{R}_s^{(i)}(\boldsymbol{x}^{(i)}) = \boldsymbol{R}_f(\boldsymbol{x}^{(i)})$.

Frequency scaling (or frequency space mapping [9]) is another basic way of enhancing the low-fidelity model. It applies to the situations when the model response is evaluation of the same quantity, e.g., S-parameters over certain range of some free parameters such as signal frequency. Its usefulness lies in the fact that in many cases, the discrepancy between the high- and low-fidelity model is a frequency shift. This kind of discrepancy can be easily reduced by means of simple scaling functions parameterized by just a few coefficients. The most popular type of frequency scaling is an affine scaling defined as $F(\omega) = f_0 + f_1\omega$ [10], where f_0 and f_1 are unknown parameters to be determined. Speaking more rigorously, let $\boldsymbol{R}_c(\boldsymbol{x}) = [R_c(\boldsymbol{x},\omega_1), \ldots, R_c(\boldsymbol{x},\omega_m)]^T$. The frequency scaled model is then defined as

$$\boldsymbol{R}_{c.F}(\boldsymbol{x}) = [R_c(\boldsymbol{x}, F(\omega_1)), \ldots, R_c(\boldsymbol{x}, F(\omega_m))]^T \tag{5}$$

$$\sum\nolimits_{k=1}^{m} [R_f(\boldsymbol{x}^{(i)}, \omega_k) - R_c(\boldsymbol{x}^{(i)}, f_0 + f_1\omega_k)]^2 \tag{6}$$

As indicated by (6), finding the scaling parameters requires the knowledge of the low-fidelity model response at any frequency (not necessarily the original discrete set ω_1 to ω_m). In practice, this does not require referring to an EM simulation because all the necessary responses $R_c(\boldsymbol{x}^{(i)}, f_0 + f_1\omega_k)$ can be obtained by interpolating/ extrapolating the know values $R_c(\boldsymbol{x}^{(i)},\omega_k)$, $k = 1, \ldots, m$.

3 Case Study 1: Selecting Model Fidelity

We consider two design cases where the optimized antenna designs are found using an SBO algorithm of the type (2). For each antenna we use three low-fidelity models of different mesh density. Further we investigate performance of the SBO algorithm, which works with these models, in terms of the computational cost and the quality of the final design.

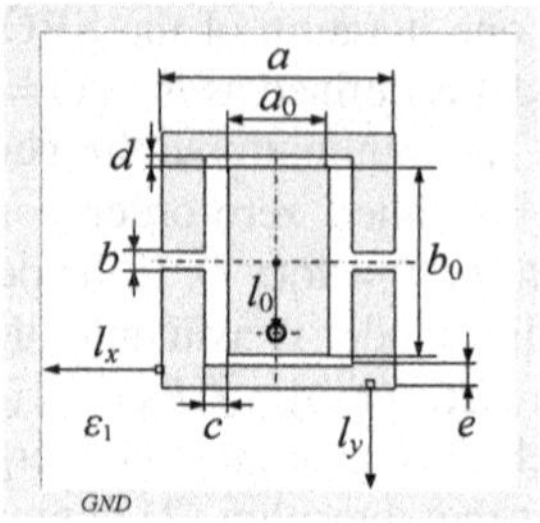

Fig. 3. Microstrip antenna [38]: (a) perspective view; (b) layout, model responses at the initial design, $\boldsymbol{R}_{c1}$ (···), $\boldsymbol{R}_{c2}$ (·-·-), $\boldsymbol{R}_{c3}$ (- - -), and $\boldsymbol{R}_f$ (—), high-fidelity model response at the final design found using the low-fidelity model $\boldsymbol{R}_{c3}$.

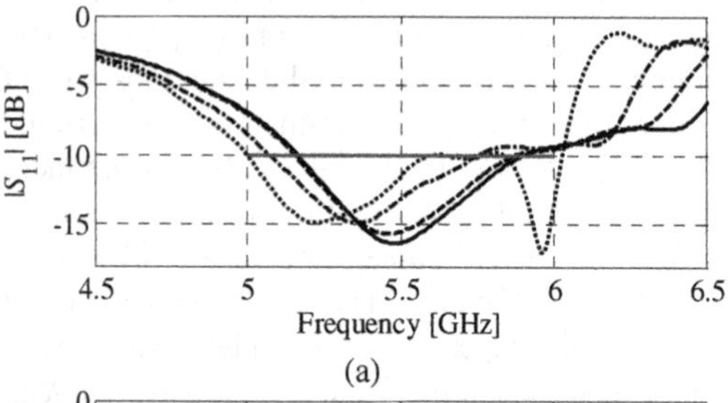

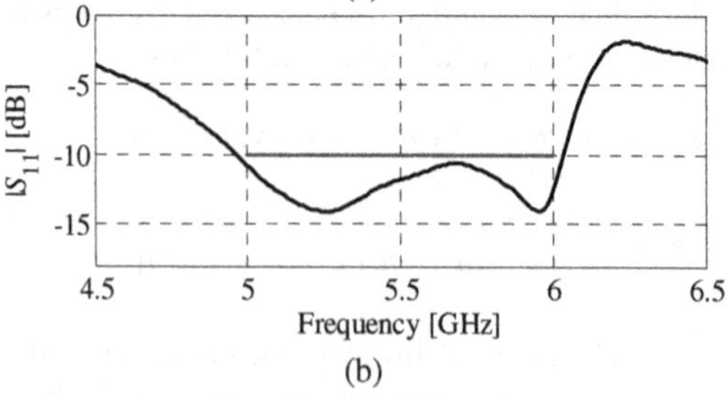

Fig. 4. Responses of the microstrip antenna: (a) with models $\boldsymbol{R}_{c1}$ (···), $\boldsymbol{R}_{c2}$ (·-·-), $\boldsymbol{R}_{c3}$ (- - -), and $\boldsymbol{R}_f$ (—) at the initial design; (b) high-fidelity model at the final design found using the low-fidelity model $\boldsymbol{R}_{c3}$.

Consider a microstrip antenna of Figs. 3(a) and 3(b) [38]. It comprises a driven patch, two parasitic patches, 3.81 mm thick Rogers TMM4 dielectric (ε_1 = 4.5 at 10 GHz) of finite dimensions (l_x= l_y= 6.75 mm), and ground plane of infinite extends. The driven patch is energized with a probe. The probe diameter is 0.8 mm. The connector's inner conductor is 1.27 mm in diameter.

The design variables, dimension parameters of the antenna, are $\boldsymbol{x} = [a\ b\ c\ d\ e\ l_0\ a_0\ b_0]^T$. Design specifications are $|S_{11}| \leq -10$ dB for 5 GHz to 6 GHz. The high-fidelity model $\boldsymbol{R}_f$ is evaluated with the CST MWS transient solver [28] (704,165 mesh cells, evaluation time 60 min). We consider three coarse models: $\boldsymbol{R}_{c1}$ (41,496 cells, 1 min), $\boldsymbol{R}_{c2}$ (96,096 cells, 3 min), and $\boldsymbol{R}_{c3}$ (180,480 cells, 6 min). The initial design is $\boldsymbol{x}^{(0)} = [6\ 12\ 15\ 1\ 1\ 1\ 1\ -4]^T$ mm. Figure 4 shows the responses of all the models at the approximate optimum of $\boldsymbol{R}_{c1}$. The major misalignment between the responses is due to the frequency shift so that the surrogate is created here using frequency scaling (5), (6) [10] as well as output SM (4) [9]. The results, Table 1 and Fig. 5, indicate that the model $\boldsymbol{R}_{c1}$ is too inaccurate and the SBO design process using it fails to find a satisfactory design. The designs found with models $\boldsymbol{R}_{c2}$ and $\boldsymbol{R}_{c3}$ satisfy the specifications and the cost of the SBO process using $\boldsymbol{R}_{c2}$ is slightly lower than while using $\boldsymbol{R}_{c3}$.

Table 1. Microstrip Antenna of Fig. 3 – Design Results

Low-Fidelity Model	Design Cost: Number of Model Evaluations[1]		Relative Design Cost[2]	max$\|S_{11}\|$ for 2 GHz to 8 GHz at Final Design
	$\boldsymbol{R}_c$	$\boldsymbol{R}_f$		
$\boldsymbol{R}_{c1}$	385	6	12.4	–8.0 dB
$\boldsymbol{R}_{c2}$	185	3	12.3	–10.0 dB
$\boldsymbol{R}_{c3}$	121	2	14.1	–10.7 dB

[1] Number of $\boldsymbol{R}_f$ evaluations is equal to the number of SBO iterations in (2).
[2] Equivalent number of $\boldsymbol{R}_f$ evaluations.

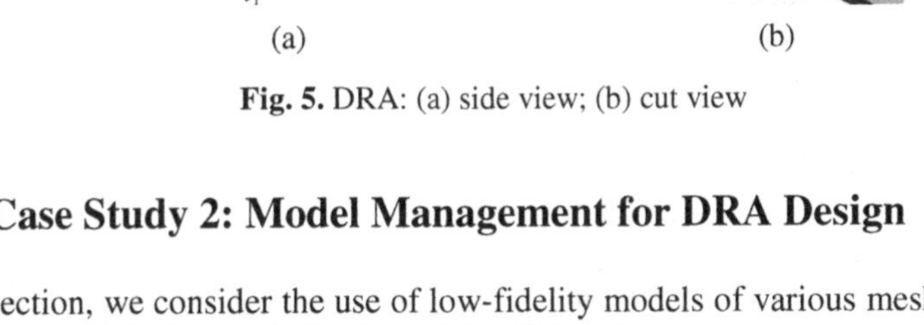

(a) (b)

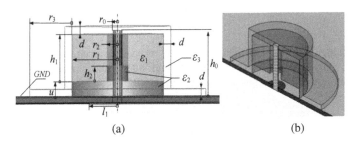

Fig. 5. DRA: (a) side view; (b) cut view

4 Case Study 2: Model Management for DRA Design

In this section, we consider the use of low-fidelity models of various mesh density for surrogate-based design optimization of the dielectric resonator antenna (DRA). Also we investigate potential benefits of using two models of different fidelity within a single optimization process.

Consider a hybrid DRA shown in Fig. 5. The DRA is fed by a 50 ohm microstrip terminated with an open-end section. Microstrip substrate is 0.787 mm thick Rogers RT5880. The design variables are $\boldsymbol{x} = [h_0\ r_1\ h_1\ u\ l_1\ r_2]^T$. Other dimensions are fixed: r_0=0.635, h_2=2, d=1, r_3= 6, all in mm. Permittivity of the DRA core is 36, and the loss tangent is 10^{-4}, both at 10 GHz. The DRA support material is Teflon (ε_2=2.1). The radome is of polycarbonate (ε_3 = 2.7 and tanδ = 0.01). The radius of the ground plane aperture, shown in Fig. 5(b), is 2 mm.

The high-fidelity antenna model $\boldsymbol{R}_f(\boldsymbol{x})$ is evaluated with the transient solver of CST Microwave Studio [28] (~1,400,000 mesh cells, evaluation time of 60 minutes). The optimization goal is to adjust geometry parameters so that the following specifications are met: $|S_{11}| \leq -12$ dB for 5.15 GHz to 5.8 GHz. The initial design is $\boldsymbol{x}^{(0)} = [7.0\ 7.0\ 5.0\ 2.0\ 2.0\ 2.0]^T$ mm.

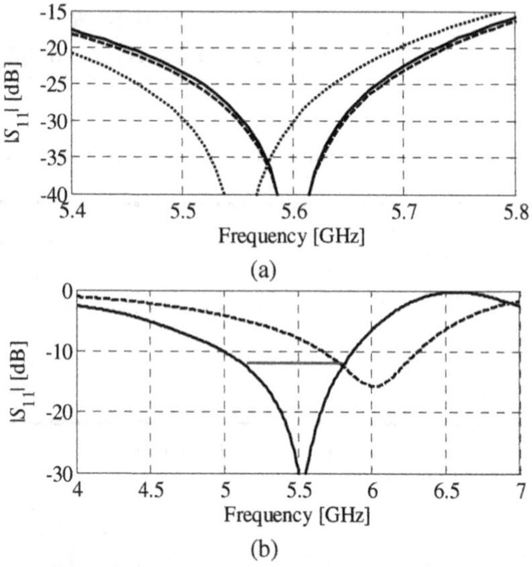

Fig. 6. DRA: (a) high- (—) and low-fidelity model $\boldsymbol{R}_{c2}$ response at a certain design before (····) and after (- - -) applying the frequency scaling, (b) high-fidelity model response at the initial design (- - -) and at the final design obtained using the SBO algorithm with the low-fidelity model $\boldsymbol{R}_{c2}$ (—).

We consider two auxiliary models of different fidelity, $\boldsymbol{R}_{c1}$ (~45,000 mesh cells, evaluation time of 1 min), and $\boldsymbol{R}_{c2}$ (~300,000 mesh cells, evaluation time of 3 min). We investigate algorithm (2) using either one of these models or both ($\boldsymbol{R}_{c1}$ at the initial state and $\boldsymbol{R}_{c2}$ in the later stages). The antenna surrogate is constructed using both output SM (4) and the frequency scaling (5), (6). Figure 6(a) shows the importance of the frequency scaling, which, due to the shape similarity of the high- and low-fidelity model responses, allows substantial reduction of the misalignment between the models.

DRA optimization has been performed three times: (i) with the surrogate constructed using $\boldsymbol{R}_{c1}$ – cheaper but less accurate (Case 1), (ii) with the surrogate constructed using $\boldsymbol{R}_{c2}$ – more expensive but also more accurate (Case 2), and (iii) with the surrogate constructed using $\boldsymbol{R}_{c1}$ at the first iteration and using $\boldsymbol{R}_{c2}$ for subsequent iterations (Case 3). The last option allows us to locate the approximate high-fidelity model optimum faster and then refine it with the more accurate model.

The limit of surrogate model evaluations was set to 100 in the first iteration which involves the largest design change and to 50 in the subsequent iterations where smaller design modifications are required.

Table 2 shows the optimization results for all three cases. Figure 6(b) shows the high-fidelity model response at the final design obtained using the SBO algorithm working with the low-fidelity model $\boldsymbol{R}_{c2}$. The quality of the final designs found in all cases is the same. However, the SBO algorithm exploiting the low-fidelity model $\boldsymbol{R}_{c1}$ (Case 1) requires more iterations than the algorithm using the model $\boldsymbol{R}_{c2}$ (Case 3), which is because $\boldsymbol{R}_{c2}$ is more accurate. In this particular case, the overall computational cost of the design process is still lower for $\boldsymbol{R}_{c1}$ than for $\boldsymbol{R}_{c2}$.

Among all cases the cheapest approach is Case 2. It is when the model $\boldsymbol{R}_{c1}$ is utilized in the first iteration that requires the largest number of EM analyses, and then the algorithm switches to $\boldsymbol{R}_{c2}$ in the second iteration. This strategy allows us to reduce the number of iterations and the number of evaluations of $\boldsymbol{R}_{c2}$ at the same time. The total design cost is the lowest overall.

Table 2. DRA Design Results

Case	Number of Iterations	Number of Model Evaluations[1]			Total Design Cost[2]	max$\|S_{11}\|$ for 5.15 GHz to 5.8 GHz at Final Design
		$\boldsymbol{R}_{c1}$	$\boldsymbol{R}_{c2}$	$\boldsymbol{R}_f$		
1	4	250	0	4	8.2	−12.6 dB
2	2	0	150	2	9.5	−12.6 dB
3	2	100	50	2	6.2	−12.6 dB

[1] Number of $\boldsymbol{R}_f$ evaluations is equal to the number of SBO iterations in (1).
[2] Equivalent number of $\boldsymbol{R}_f$ evaluations.

5 Discussion

The results presented in this work confirm certain common sense predictions on the selection of the low-fidelity models for surrogate-based antenna optimization. In particular, it might be advantageous to use less accurate but faster models, because it may translate into reduced design cost. On the other hand, less accurate models increase the possibility of the algorithm failure. When using higher-fidelity models, the overall design cost may be higher but the SBO design process becomes more robust. Also, the number of iterations necessary to find a satisfactory design is usually reduced. For now, visual inspection, specifically a comparison of the low- and high-fidelity model responses remains the most important way of accessing the model quality. Visual inspection may also help deciding which type of model correction should be applied while creating the surrogate.

Based on our results one can formulate a few rules of thumb that might be helpful in the model selection process:

- It is recommended to conduct an initial parametric study regarding meshing setup in order to find the "coarsest" low-fidelity model that still adequately represents all the important features of the high-fidelity model response. It seems sufficient if this study is performed at the initial design only. The assessment should be done by visual inspection of the model responses having in mind that the critical factor is not the absolute model discrepancy but the similarity of the response shape. For example, even relatively large frequency shift can be easily reduced by a proper frequency scaling.
- It is usually safer to use a slightly finer low-fidelity model rather than a coarser one. This prevents from losing a potential design cost reduction due to a possible algorithm failure (as a consequence of a low-fidelity model inaccuracy) in finding a satisfactory design.
- It is recommended that the type of the low-fidelity model correction used to construct the surrogate is based on the observed type of the misalignment between the low- and high-fidelity model responses. The two methods considered in this work (additive response correction and frequency scaling) can be considered as safe choices for most situations.

It should be emphasized that for certain antenna structures, such as some narrow-band antennas or wideband travelling wave antennas, it is possible to obtain quite good ratio between the simulation times of the high- and low-fidelity models (e.g., up to 50). This is because even for relatively coarse mesh, the low-fidelity model may still be a good representation of the high-fidelity one. For other structures, particularly multi-resonant antennas, only much lower ratios (e.g., 5 to 10) may be possible, which limits the design cost savings that can be achieved by using the surrogate-based optimization techniques.

6 Conclusions

A problem of selecting the simulation models for surrogate-based optimization of antenna structures has been investigated. We have discussed the trade-off between the computational complexity and accuracy of the low-fidelity EM antenna models and their effects on the performance of the surrogate-based optimization process. Our analysis has been mostly qualitative and based on numerical experiments. The presented results illustrate the importance of the low-fidelity model selection. We also demonstrate the potential benefits of using multiple low-fidelity models in the same design optimization run. One of the open issues is automation of the model selection process, which may be the key for wider acceptance of the surrogate-based methods in antenna design community.

References

1. Schantz, H.: The art and science of ultrawideband antennas. Artech House (2007)
2. Petosa, A.: Dielectric Resonator Antenna Handbook. Artech House (2007)
3. Haupt, R.L.: Antenna design with a mixed integer genetic algorithm. IEEE Trans. Antennas Propag. 55(3), 577–582 (2007)
4. Kerkhoff, A.J., Ling, H.: Design of a band-notched planar monopole antenna using genetic algorithm optimization. IEEE Trans. Antennas Propag. 55(3), 604–610 (2007)
5. Pantoja, M.F., Meincke, P., Bretones, A.R.: A hybrid genetic algorithm space-mapping tool for the optimization of antennas. IEEE Trans. Antennas Propag. 55(3), 777–781 (2007)
6. Jin, N., Rahmat-Samii, Y.: Parallel particle swarm optimization and finite- difference time-domain (PSO/FDTD) algorithm for multiband and wide-band patch antenna designs. IEEE Trans. Antennas Propag. 53(11), 3459–3468 (2005)
7. Halehdar, A., Thiel, D.V., Lewis, A., Randall, M.: Multiobjective optimization of small meander wire dipole antennas in a fixed area using ant colony system. Int. J. RF and Microwave CAE 19(5), 592–597 (2009)
8. Jin, N., Rahmat-Samii, Y.: Analysis and particle swarm optimization of correlator antenna arrays for radio astronomy applications. IEEE Trans. Antennas Propag. 56(5), 1269–1279 (2008)
9. Bandler, J.W., Cheng, Q.S., Dakroury, S.A., Mohamed, A.S., Bakr, M.H., Madsen, K., Søndergaard, J.: Space mapping: the state of the art. IEEE Trans. Microwave Theory Tech. 52(1), 337–361 (2004)
10. Koziel, S., Bandler, J.W., Madsen, K.: A space mapping framework for engineering optimization: theory and implementation. IEEE Trans. Microwave Theory Tech. 54(10), 3721–3730 (2006)
11. Koziel, S., Ciaurri, D.E., Leifsson, L.: Surrogate-based methods. In: Koziel, S., Yang, X.-S. (eds.) COMA. SCI, vol. 356, pp. 33–59. Springer, Heidelberg (2011)
12. Rayas-Sánchez, J.E.: EM-based optimization of microwave circuits using artificial neural networks: the state-of-the-art. IEEE Trans. Microwave Theory Tech. 52(1), 420–435 (2004)
13. Kabir, H., Wang, Y., Yu, M., Zhang, Q.J.: Neural network inverse modeling and applications to microwave filter design. IEEE Trans. Microwave Theory Tech. 56(4), 867–879 (2008)
14. Smola, A.J., Schölkopf, B.: A tutorial on support vector regression. Statistics and Computing 14(3), 199–222 (2004)
15. Meng, J., Xia, L.: Support-vector regression model for millimeter wave transition. Int. J. Infrared and Milimeter Waves 28(5), 413–421 (2007)
16. Buhmann, M.D., Ablowitz, M.J.: Radial Basis Functions: Theory and Implementations. Cambridge University (2003)
17. Simpson, T.W., Peplinski, J., Koch, P.N., Allen, J.K.: Metamodels for computer-based engineering design: survey and recommendations. Engineering with Computers 17(2), 129–150 (2001)
18. Forrester, A.I.J., Keane, A.J.: Recent advances in surrogate-based optimization, Prog. Aerospace Sciences 45(1-3), 50–79 (2009)
19. Amari, S., LeDrew, C., Menzel, W.: Space-mapping optimization of planar coupled-resonator microwave filters. IEEE Trans. Microwave Theory Tech. 54(5), 2153–2159 (2006)

20. Koziel, S., Cheng, Q.S., Bandler, J.W.: Space mapping. IEEE Microwave Magazine 9(6), 105–122 (2008)
21. Swanson, D., Macchiarella, G.: Microwave filter design by synthesis and optimization. IEEE Microwave Magazine 8(2), 55–69 (2007)
22. Rautio, J.C.: Perfectly calibrated internal ports in EM analysis of planar circuits. In: IEEE MTT-S Int. Microwave Symp. Dig., Atlanta, GA, pp. 1373–1376 (2008)
23. Cheng, Q.S., Rautio, J.C., Bandler, J.W., Koziel, S.: Progress in simulator-based tuning—the art of tuning space mapping. IEEE Microwave Magazine 11(4), 96–110 (2010)
24. Echeverria, D., Hemker, P.W.: Space mapping and defect correction. CMAM The International Mathematical Journal Computational Methods in Applied Mathematics 5(2), 107–136 (2005)
25. Koziel, S.: Shape-preserving response prediction for microwave design optimization. IEEE Trans. Microwave Theory and Tech. 58(11), 2829–2837 (2010)
26. Koziel, S.: Adaptively adjusted design specifications for efficient optimization of microwave structures. Progress in Electromagnetic Research B (PIER B) 21, 219–234 (2010)
27. Couckuyt, I., Declercq, F., Dhaene, T., Rogier, H., Knockaert, L.: Surrogate-based infill optimization applied to electromagnetic problems. Int. J. RF and Microwave CAE 20(5), 492–501 (2010)
28. CST Microwave Studio, 2012. CST AG, Bad Nauheimer Str. 19, D-64289 Darmstadt, Germany (2012)
29. HFSS. Release 13.0, ANSYS (2010), `http://www.ansoft.com/products/hf/hfss/`
30. Taflove, A., Hagness, S.C.: Computational electrodynamics: the finite-difference time-domain method, 3rd edn. Artech House (2006)
31. Lin, J.-M.: The Finite Element Method in Electromagnetics, 2nd edn. Wiley-IEEE Press (2002)
32. Harrington, R.F.: Field Computation by Moment Methods. Wiley-IEEE Press (1993)
33. Makarov, S.: Antenna and EM modeling with Matlab. Wiley-Interscience (2002)
34. Chen, Z.N.: Wideband microstrip antennas with sandwich substrate. IET Microw. Ant. Prop. 2(6), 538–546 (2008)
35. Koziel, S., Bandler, J.W., Madsen, K.: Space mapping with adaptive response correction for microwave design optimization. IEEE Trans. Microwave Theory Tech. 57(2), 478–486 (2009)
36. Alexandrov, N.M., Dennis, J.E., Lewis, R.M., Torczon, V.: A trust region framework for managing use of approximation models in optimization. Struct. Multidisciplinary Optim. 15(1), 16–23 (1998)
37. Conn, A.R., Gould, N.I.M., Toint, P.L.: Trust Region Methods. MPS-SIAM Series on Optimization (2000)
38. Wi, S.-H., Lee, Y.-S., Yook, J.-G.: Wideband Microstrip Patch Antenna with U-shaped Parasitic Elements. IEEE Trans. Antennas Propagat. 55(4), 1196–1199 (2007)

An X-FEM Based Approach for Topology Optimization of Continuum Structures

Meisam Abdi, Ian Ashcroft, and Ricky Wildman

Faculty of Engineering, University of Nottingham, University Park, Nottingham, U.K.
{eaxma5,Ian.Ashcroft,Ricky.Wildman}@nottingham.ac.uk

Abstract. In this study, extended finite element (X-FEM) is implemented to represent topology optimization of continuum structures in a fixed grid design domain. An evolutionary optimization algorithm is used to gradually remove inefficient material from the design space during the optimization process. In the case of 2D problems, evolution of the design boundary which is supper-imposed on the fixed grid finite element framework is captured using isolines of structural performance. The proposed method does not need any remeshing approach as the X-FEM scheme can approximate the contribution of boundary elements in the finite element frame work of the problem. Therefore the converged solutions come up with clear and smooth boundaries which need no further interpretation. This approach is then extended to 3D by using a 3D X-FEM scheme implemented on isosurface topology design.

Keywords: Topology Optimization, X-FEM, Isoline, Isosurface, Evolutionary.

1 Introduction

The aim of structural optimization is to minimize the amount of material required for a structure in a design space to be used in a specific application defined by a set of loads and constrains. Structural optimization can be classified into several categories including size, shape and topology optimization. In a size optimization problem, the goal could be to find optimal size of members of a structure while its shape is fixed throughout the optimization process. In a shape optimization problem, the shape of boundaries changes during optimization, whilst topology remains unchanged. Topology optimization presents the most challenges as the material distribution within the design domain, including the shape, number and location of holes and connectivity of design domain, may change throughout the optimization process (Fig. 1).

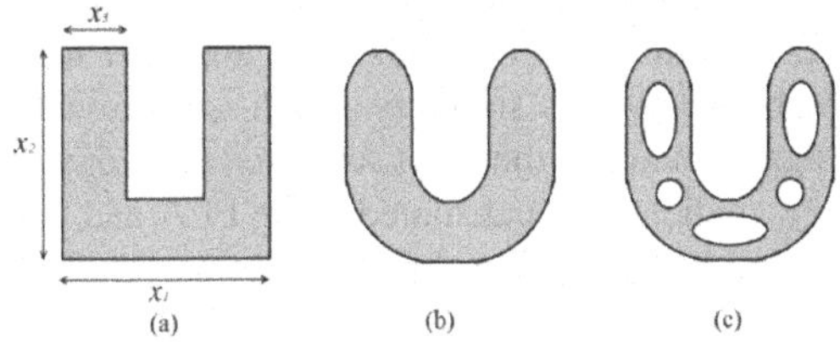

(a) (b) (c)

Fig. 1. (a) Size (b) shape and (c) topology optimization of a typical structure

Advances in Intelligent Systems and Computing 256,
DOI: 10.1007/978-3-319-03581-9_20, © Springer International Publishing Switzerland 2014

Since late 80s, several mathematical based and heuristic approaches have been proposed to address shape and topology optimization of continuum structures. Bendsøe and Kikuchi [6] developed a homogenization method for shape and topology optimization of continuum structures. In this approach, the design domain consists of periodically distributed perforated microstructures. This assumption reduces the complex topology optimization problem into a simple size optimization problem of finding the size parameters of the microstructures. Homogenization has been a widely accepted approach as it provides mathematical bounds to theoretical performance of the structure. However, determination and evaluation of the optimal microstructures is cumbersome. Also, the obtained solutions cannot be built directly since no definite length scale is associated with the microstructures [14]. Solid Isotropic Material with Penalization (SIMP) [5, 21] is an alternative approach in which the material properties are assumed constant within each element of the design domain. The intermediate densities between void and solid are penalized to obtain a near binary solution. Although penalization makes it easier to implement the final solutions into practice, the obtained topology is dependent on the value of penalization. Evolutionary structural optimization (ESO) first proposed by Xie and Steven [19], is a heuristic method, based on the assumption that the structure evolves to an optimum by gradually removing its inefficient material. Bidirectional Evolutionary Structural Optimization (BESO) [13, 20, 3] is an extension of ESO in which the efficient/inefficient material can be added/removed simultaneously. These heuristic methods are easy to program and provide a clear topology (no grey regions of intermediate density as seen following SIMP). As these methods are less dependent on mathematical properties of the objective function, they are suitable for complicated problems. The main drawbacks of these methods are that they are strongly dependent on the mesh size and may result in a non-convergent solution. All the above mentioned topology optimization methods are element based, which means a property of an element or the element itself is considered as a design variable. Element based methods can be easily combined with the finite element model of the structure. However a drawback is that the boundary representation is limited to the elements shape. The smoother solutions can be obtained by using finer elements near the boundary (remeshing or moving mesh approaches) which is cumbersome. Otherwise the final solutions need further interpretation and post-processing.

Fixed grid finite element method (FG-FEM) is a possible tool for the problems where the geometry of a structure or a physical property of it changes with time. In the FG-FEM the geometry of the structure is super-imposed on the fixed position FE framework of the design space. Therefore, the boundary can evolve independent of the elements shape and position. This technique has been used in level set based and isoline based structural optimization methods. The level set method is a numerical technique to track the boundary of a structure which is represented by zero level set function over the design domain. In the level set based topology optimization [17, 2], the shape and topology gradients are calculated using FEA and the evolution of shape and topology is controlled by a Hamilton-Jacobi equation. This method combines the advantages of having a high numerical efficiency and smooth boundary representation using a fixed-grid mesh. A drawback is that nucleation of new holes in the solid region is difficult, especially for 2D design domains. The isoline based topology design [8, 16] is also capable of generating smooth topologies in a fixed grid domain and unlike the level-set method, it doesn't require the solution of a specific partial differential equation. The isoline topology design (ITD) is an evolutionary based optimization method (like ESO/BESO methods) which operates by slowly removing inefficient material from the design domain.

A major issue in the fixed-grid design approaches is how to treat the boundary elements (partially solid elements which have the boundary super-imposed on them). A conventional approach in the FG-FEM is to assume the stiffness of the boundary element proportional to the area fraction of solid material within the element (also known as a density scheme). This approach has been used in many works [2, 16], however studies have shown that it cannot provide accurate results for the boundary elements as it does not consider how the material is distributed within the element [7, 18]. Extended Finite Element Method (X-FEM), initially proposed to represent crack growth in a fixed-grid domain [12], is an alternative approach to model void/solid interfaces without remeshing the boundaries. X-FEM extends the conventional finite element approach by adding special shape functions which can represent the discontinuity inside an element. Therefore it has the potential to approximate the stiffness of boundary elements by considering the material distribution within the element. There are a number of studies that have successfully used X-FEM combined with level-set description of the boundary in topology optimization [18, 11, 9]. However, up to now there have been limited numbers of studies where the focus has been the application of X-FEM in isoline based topology optimization. Abdi et al [1] proposed a combined XFEM-isoline approach for topology optimization of 2D continuum structures. This method has the advantage of high accuracy and the solutions were represented with smooth boundaries. The current paper is a review on the XFEM-isoline optimization approach followed by a study on the extension of this method to 3D and the possible advantages of a 3D topology optimization compared to the 2D case.

2 Methodology

2.1 Problem Statement

A conventional problem in topology optimization design is to find the stiffest possible structure for a given volume of material. In the standard evolutionary optimization methods such as ESO/BESO, the optimal design can be achieved by sequential element addition and rejection. Therefore an element itself is considered as a design variable. In the isoline topology design method the boundary can cross over the finite elements, thus the material distribution inside each element could be considered as a design variable. In this case, the optimization problem where the objective is to minimize the strain energy (maximize the stiffness) can be written as:

$$\text{Minimize: } c = \frac{1}{2} U^T K U \tag{1}$$

$$\text{Subject to: } \frac{\sum_{e=1}^{N} v_S^{(e)}}{V_0} = V^* \tag{2}$$

where c is the total strain energy, and U and K are the global displacement and global stiffness matrices, respectively. N denotes the number of finite elements in the design domain, $v_{e,s}$ the volume of the solid part of the element, V_0 the design domain volume and V^* the prescribed volume fraction. Choosing strain energy density (SED) as a criterion for finding inefficient use of material in the design domain, the material will be removed from the regions having low values of SED and will be added to those with high values of SED during the optimization process. The strain energy density of the elements can be calculated from

$$SED_e = \frac{1}{2} u_e^T k_e u_e / v_e \tag{3}$$

where u_e is the element displacement vector and k_e represents the element stiffness matrix which is calculated using an X-FEM scheme.

2.2 Isoline Design

Isoline design is a method of representing the shape and topology of a structure using contours of a desired structural performance such as Von-Mises stress and SED. Isoline design method has been used in several studies in order to generate designs with smooth boundaries [10, 8]. The isoline design approach used in this paper can be summarized into the following steps [1]:

1. Find the SED distribution over the design domain by performing an extended finite element analysis over the fixed-grid design domain.
2. Determine a minimum SED level (MSL) and calculate the relative strain energy which is defined by $SED_{rel} = SED - MSL$.
3. Assign the solid material property to the regions of the domain having positive values of SED_{rel}, and the weak material property to the regions with negative value of SED_{rel}. The boundary is determined by $SED_{rel} = 0$.
4. Repeat steps 1-3 by gradually increasing MSL, until a desired optimum is obtained.

Figure 2 shows boundary representation of a bridge structure in a fixed grid domain using isoline design method.

2.3 X-FEM

X-FEM proposed by Belytschko et al [4] is an extension of the classical finite element method which allows modeling a discontinuity inside an element. Although it was initially developed to represent crack growth in a structure, later this approach developed to include other sorts of discontinuities such as fluid/solid interaction, void/solid interface, and inclusion problems. For the cracked structures, the main idea of X-FEM is to extend the classical FE approximation by adding an enrichment function to the continuous displacement field to represent the discontinuity:

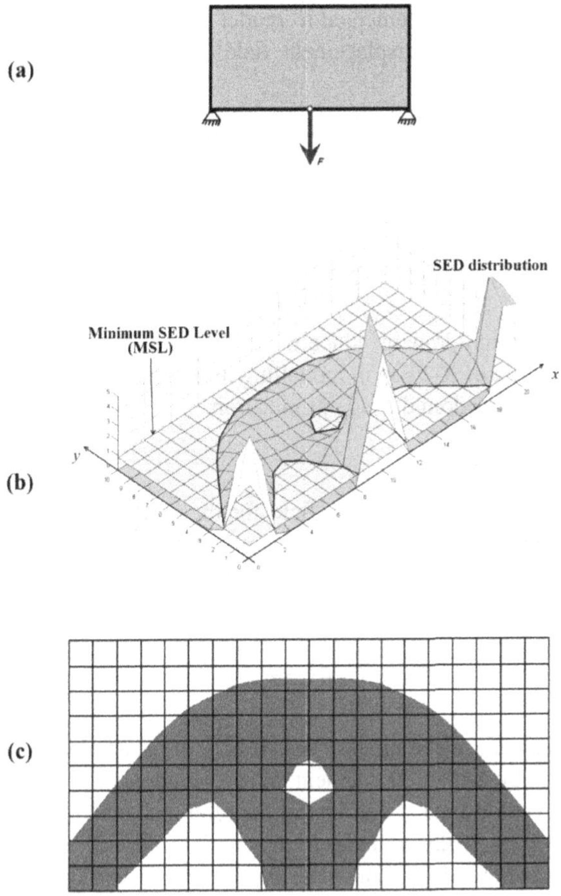

Fig. 2. (a) initial design domain, load and boundary condition for topology optimization of a 2D bridge structure. (b) Boundary of design represented by intersection of SED distribution and MSL. (c) geometry of the design, superimposed on the fixed grid finite elements of the design domain.

$$u(x) = \sum_i u_i N_i(x) + \sum_j a_j N_j(x)\, H(x). \tag{4}$$

In the above equation, the first term on the right hand side shows the conventional finite element approximation of the displacement field in an element where N_i are the classical shape functions associated to the nodal degrees of freedom, u_i. In the X-FEM, the additional term $N_j(x)H(x)$ which is supported by enriched degrees of freedom, a_j, are discontinuous shape functions constructed by multiplying a classical $N_j(x)$ shape function with a Heaviside function $H(x)$ presenting a switch value where the discontinuity lies. It can be noticed that the X-FEM with enrichment introduces additional degrees of freedom to the FE framework of the structure. For the case of shape and topology optimization, the X-FEM scheme for modeling holes and inclusions [15] can be implemented to model solid/void material interface of the design. In this approach, the displacement field is approximated by the following equation

$$u(x) = \sum_i N_i(x)\, H(x) u_i. \tag{5}$$

where the Heaviside function *H(x)* has the following properties

$$(x) = \begin{cases} 1 & if\ x\ \in \Omega_S \\ 0 & if\ x\ \notin \Omega_S \end{cases}. \tag{6}$$

where Ω_S is the solid sub-domain. Since there is no enrichment in the displacement approximation equation of X-FEM in modelling holes and inclusions, there will be no augmented degrees of freedom during optimization.

Integration Scheme

Equation 6 defines a zero displacement field for the void part of the element, which means that only the solid part of the element contributes to the element stiffness matrix. Thus we can use the same displacement function as FEM and simply remove the integral in the void sub-domain of the element.

$$K_e = \int_{\Omega_S} B^T D_S B t d\Omega \tag{7}$$

with B the displacement differentiation matrix, D_S the elasticity matrix for the solid material and t the thickness of the element. When an element is cut by the boundary, the remaining solid sub-domain is no longer the reference rectangular element. So we partition the solid part of the boundary element into several sub-triangles and use Gauss quadrature to numerically calculate the integral given by equation 7. In this paper, 4-node quadrilateral elements are used for the finite element modeling of 2D design domains, and the X-FEM numerical integration scheme is realized using second order 3-point quadrature on triangles (as shown in figure 3).

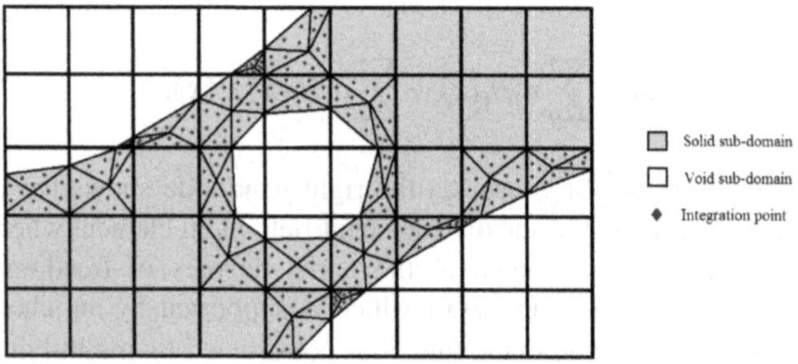

Fig. 3. Representation of the boundary elements in the X-FEM scheme

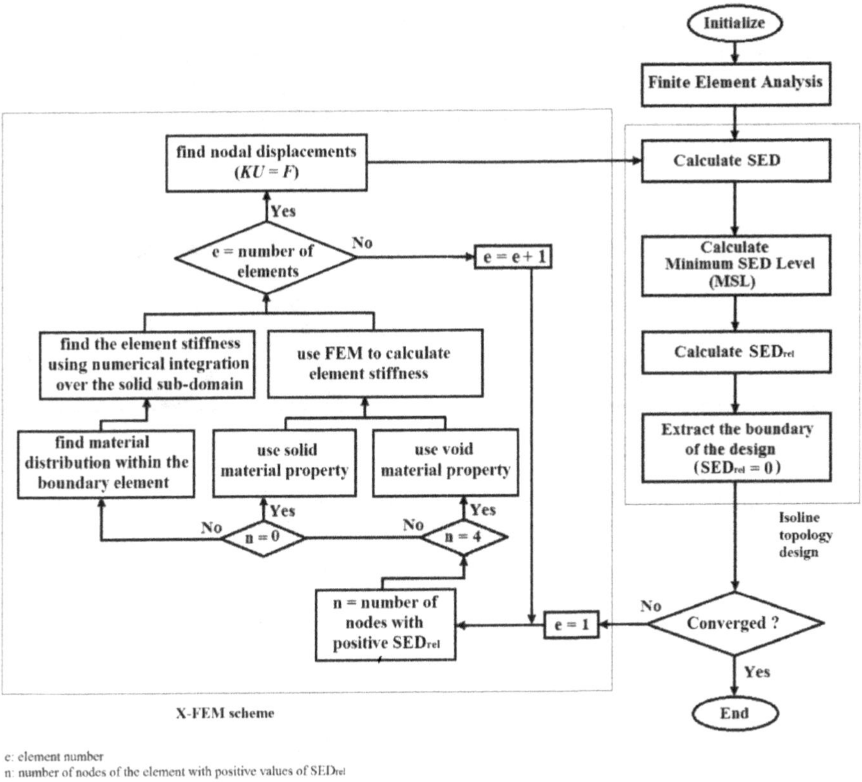

Fig. 4. Flow chart of the proposed optimization method

2.4 Application of X-FEM in Isoline Topology Optimization

Figure 4 illustrates the topology optimization procedure used which in general consists of initialization, X-FEM structural analysis, and isoline update scheme. In initialization, the initial material distribution within the design domain and the discretization of the design domain, as well as the necessary parameters for the isoline topology design are defined.

In the X-FEM structural analysis, by using nodal SED_{rel} numbers, the elements are categorized into three groups: solid elements in which all the 4 nodal SED_{rel} numbers have positive values, void elements in which all the 4 nodal SED_{rel} numbers have negative values and boundary elements which have at least one node with positive SED_{rel} and one with negative SED_{rel}. Solid and void elements are treated using classical finite element approximation. The stiffness matrix of the boundary elements are calculated by partitioning the solid sub-domain into several sub-triangles and applying the Gauss quadrature integration scheme described in the previous section.

The minimum SED level (MSL) is determined by increasing the value from the last iteration and the nodal values of SED_{rel} will be calculated. The new structure is obtained from the intersection of the MSL and current criteria distribution (where we have $SED_{rel} = 0$). The process is continued until the convergence condition is achieved.

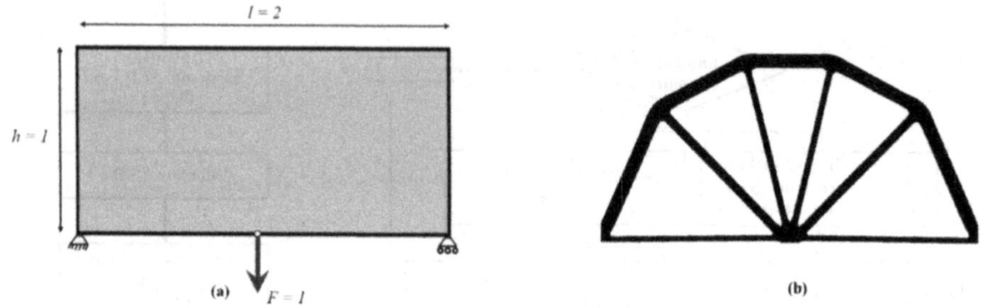

Fig. 5. (a) Initial design domain, load and boundary conditions. (b) Topology optimized solution for final volume of 20% of initial volume.

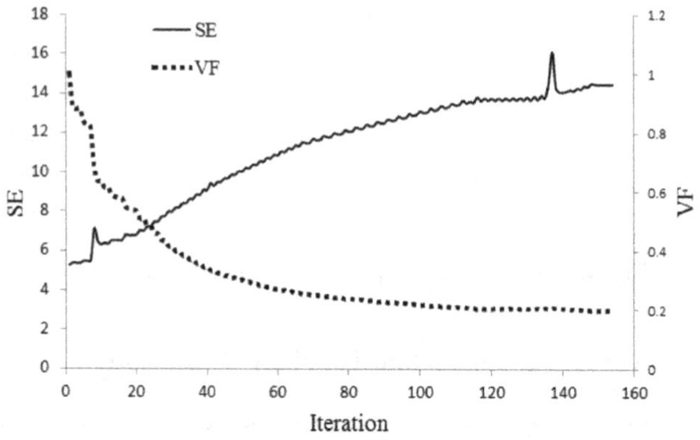

Fig. 6. Evolution history of objective function (SE) and volume fraction (VF) of the 2D test case

2.5 2D Numerical Example

The proposed method of combining X-FEM and evolutionary optimization algorithm was implemented in a MATLAB code to present the topology optimization of 2D rectangular domains. A set of dimensionless parameters are used for all test cases of this study. The 2D test case used in this study is a bridge structure with the design domain an boundary conditions shown in figure 5(a). A 120x60 mesh is used for the finite element model of the design. The optimized final design for a volume fraction of 20% of the initial volume is shown in figure 5(b). It can be seen that the proposed X-FEM based approach has resulted in a final solution having smooth and clearly defined boundaries. The evolution histories of the objective function and volume fraction is shown in figure 6. It can be seen that as material is slowly removed from the design domain, the strain energy increases smoothly, then reaches a constant value at convergence. It can also be seen a few jumps in the graph of objective function, which can be attributed to removing a link from the body of the structure.

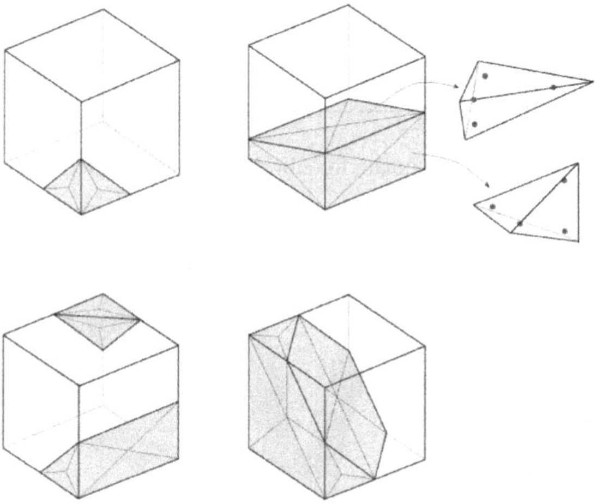

Fig. 7. 3D X-FEM integration scheme: solid sub-domain of a boundary element decomposed into sub-tetrahedra

3 Extension to 3D

The proposed optimization method can be extended to 3D considering the following changes:

1. The boundary of the design is represented with isosurfaces of strain energy density. It will be a surface that represent the points of a constant value ($SED_{rel} = 0$) in a 3D design space.
2. An 8-node hexahedral element is used for finite element model of the structure. The X-FEM scheme is realized by decomposition of solid part of a boundary element into several sub-tetrahedra and performing numerical integration on tetrahedra using 4-point Gauss quadrature, as shown in figure 7. The stiffness matrix of the element is obtained by summing the integrations of all sub-tetrahedra in the elements.

3.1 3D Test Case

The 3D test case is a bridge structure having length l=*40*, height h=*20* and thickness t=*5* where a unique distributed force is applied to the middle of the bottom surface as shown in figure 8. The design domain is discretized with 40x20x5 hexahedral finite elements and the final volume fraction is set to 20% of the initial volume. Figure 9 shows the final optimized design, obtained after 231 evolutionary iterations. Evolution history of the objective function and volume fraction is illustrated in figure 10.

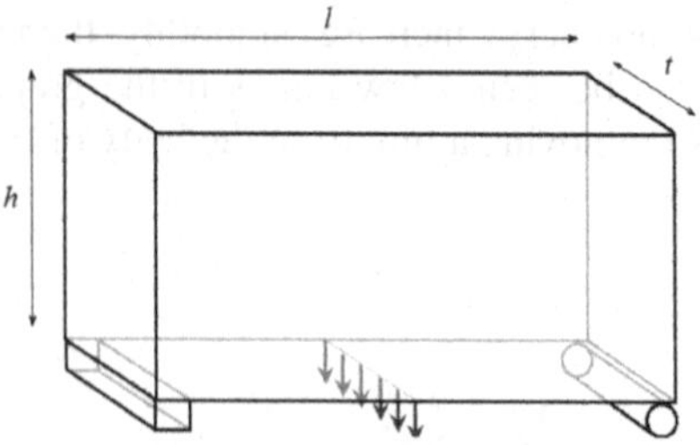

Fig. 8. The initial design domain for a bridge problem used as a 3D test case

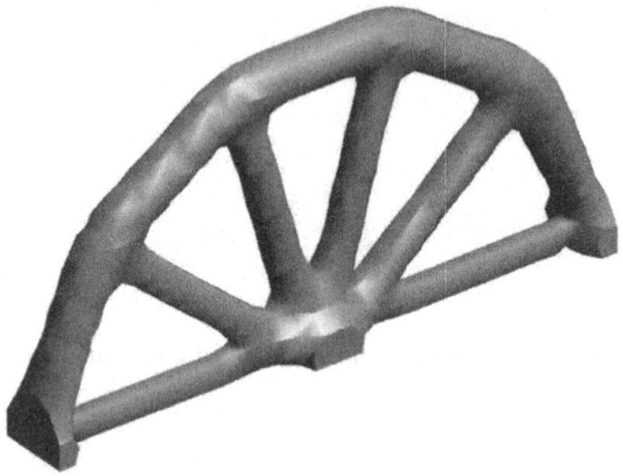

Fig. 9. The optimization result of the 3D bridge structure

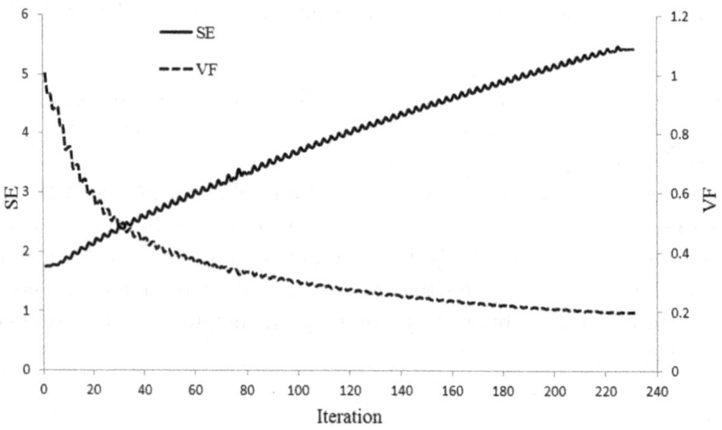

Fig. 10. Evolution history of the objective function (SE) and volume fraction (VF) for the 3D test case

3.2 Comparison of Solutions for Different Thicknesses

A simple experiment was performed to determine the need for 3D analysis. For this purpose, the same structure as section 3.1 was considered and only the thickness, as well as number of elements in the thickness direction was changed. For example the domain with a size 40x20x10 was discretized with 40x20x10 hexahedral elements. The optimized design topologies as a function of thickness were then compared to the design identified using a 2D analysis. Figure 11 shows the final designs for final volume fraction of 20% of initial design domain and thickness of 5, 10, 12 and 15, obtained using the 3D optimization approach, whereas the equivalent results for 2D would be the result shown in figure 5 extruded to the appropriate thickness. It can be seen in figure 11, however, that the final topologies are dependent on the thickness of the structure. When the thickness is small compared to height and length, the final solution are close to that obtained using a 2D optimization approach, but more complex solutions including separated and parallel arch members arise when a 3D approach is employed, suggesting a need for full 3D analysis when the thickness to length ratio becomes appreciable.

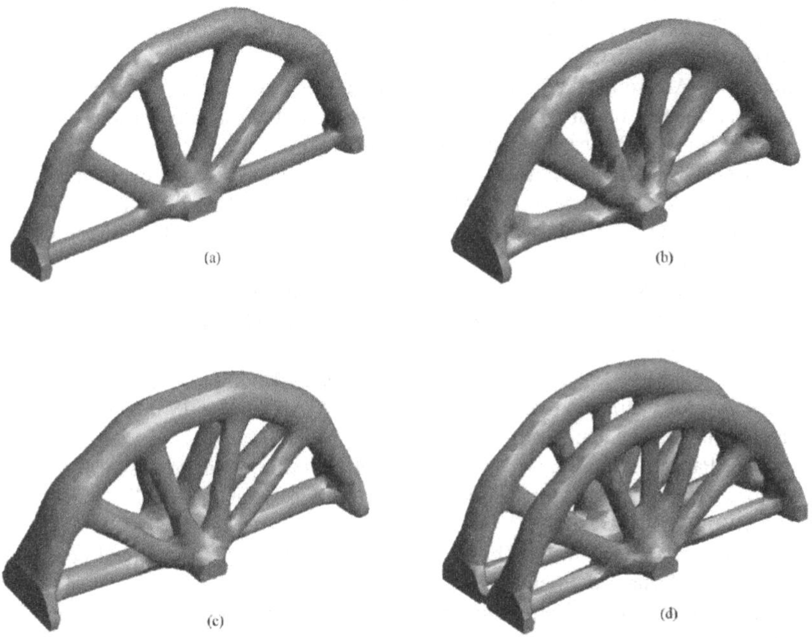

Fig. 11. The optimized topologies for volume fraction of *VF=0.20* and a range of thicknesses: (a) *t=5* (b) *t=10* (c) *t=12* (d) *t=15*

4 Conclusions

This study presents an X-FEM based approach for topology optimization of continuum structures. The X-FEM is implemented in an evolutionary isoline/ isosurface topology optimization approach, in order to increase the accuracy of FE approximations on the boundary of the design and improve the boundary representation. Our results suggest that both 2D and 3D optimization approaches can be used to generate topologies having smooth and clearly defined boundaries. The 2D optimization approach is suggested for when the thickness to length ratio is small, due to its low computational cost, but when the thickness to length ratio is large, a 3D optimization approach becomes necessary to realize optimal structures. Therefore we suggest that further work is required to address the full 3D X-FEM based topology optimization approach.

Acknowledgements. The authors are grateful for the funding provided by the university of Nottingham.

References

1. Abdi, M., Wildman, R., Ashcroft, I.: Evolutionary topology optimization using X-FEM and isolines. Engineering Optimization (in press, 2013)
2. Allaire, G., Jouve, F., Toader, A.M.: Structural optimisation using sensitivity analysis and a level set method. J. Comp. Phys. 194, 363–393 (2004)
3. Aremu, A., Ashcroft, I., Wildman, R., Hague, R., Tuck, C., Brackett, D.: The effects of BESO parameters on an industrial designed component for additive manufacture. Proc. IMechE Part B: J. Engineering Manufacture (2012) (in press)
4. Belytschko, T., Black, T.: Elastic crack growth in finite elements with minimal remeshing. International Journal for Numerical Methods in Engineering 45(5), 601–620 (1999)
5. Bendsøe, M.P.: Optimal shape design as a material distribution problem. Struct. Optim. 1, 193–202 (1989)
6. Bendsøe, M.P., Kikuchi, N.: Generating optimal topologies in structural design using a homogenization method. Computer Methods in Applied Mechanics and Engineering 71, 197–224 (1988)
7. Dunning, P., Kim, H.A., Mullineux, G.: Error analysis of fixed grid formulation for boundary based structural optimisation. In: 7th ASMO UK / ISSMO Conference on Engineering Design Optimisation, Bath, UK, July 7-8 (2008)
8. Lee, D., Park, S., Shin, S.: Node-wise topological shape optimum design for structural reinforced modeling of Michell-type concrete deep beams. J. Solid Mech. Mater. Eng. 1(9), 1085–1096 (2007)
9. Li, L., Wang, M.Y., Wei, P.: XFEM schemes for level set based structural optimization. Frontiers of Mechanical Engineering 7(4), 335–356 (2012)
10. Maute, K., Ramm, E.: Adaptive topology optimisation. Struct. Optim. 10, 100–112 (1995)
11. Miegroet, L.V., Duysinx, P.: Stress concentration minimization of 2D Filets using X-FEM and level set description. Structural and Multidisciplinary Optimisation 33, 425–438 (2007)
12. Moës, N., Dolbow, J., Belytschko, T.: A finite element method for crack growth without remeshing. International Journal for Numerical Methods in Engineering 46, 131–150 (1999)
13. Querin, O.M., Steven, G.P., Xie, Y.M.: Evolutionary structural optimisation (ESO) using a bidirectional algorithm. Engineering Computations 15(8), 1031–1048 (1988)
14. Sigmund, O.: A 99 line topology optimisation code written in Matlab. Struct. Multidiscipl. Optim. 21, 120–127 (2001)

15. Sukumar, N., Chopp, D.L., Moës, N., Belytschko, T.: Modeling Holes and Inclusions by Level Sets in the Extended Finite Element Method. Computer Methods in Applied Mechanics and Engineering 190, 6183–6200 (2001)
16. Victoria, M., Martí, P., Querin, O.M.: Topology design of two-dimensional continuum structures using isolines. Computer and Structures 87, 101–109 (2009)
17. Wang, M.Y., Wang, X., Guo, D.: A level set method for structural topology optimisation. Comput. Meth. Appl. Eng. 192, 227–246 (2003)
18. Wei, P., Wang, M.Y., Xing, X.: A study on X-FEM in continuum structural optimization using level set method. Computer-Aided Design 42, 708–719 (2010)
19. Xie, Y.M., Steven, G.P.: A simple evolutionary procedure for structural optimization. Computers & Structures 49, 885–896 (1993)
20. Yang, X.Y., Xie, Y.M., Steven, G.P., Querin, O.M.: Bidirectional evolutionary method for stiffness optimisation. AIAA J. 37(11), 1483–1488 (1999)
21. Zhou, M., Rozvany, G.I.N.: The COG algorithm, Part II: Topological, geometrical and general shape optimisation. Comp. Meth. Appl. Mech. Eng. 89, 309–336 (1991)

Collaborative Optimization Based Design Process for Process Engineering

Mika Strömman, Ilkka Seilonen, and Kari Koskinen

School of Electrical Engineering, Aalto University, Finland
{mika.stromman,ilkka.seilonen,kari.o.koskinen}@aalto.fi

Abstract. Traditionally, the paper mills have been designed using mostly engineering experiences and rules of thumbs. The main target of the design has been the structure of the plant, whereas finding the optimal operation of the plant has been left for the operators. Bi-level multi-objective optimization (BLMOO) offers a method for optimizing both the structure and the operation during the design phase. In order to use BLMOO in design projects, the business process has to be re-engineered. This research defines a process for applying BLMOO in process design in multi-organizational projects. The process is then evaluated by interviewing experts.

Keywords: Process Design, Continuous Process, Optimization, Modeling, Collaboration, Pulp and Paper Industry.

1 Introduction

Profitability in paper making has decreased and therefore also the competition in paper mill design is getting harder. The mill should be constructed with minimal capital expenses and at the same time the facility should be optimal for the current market situation. Compared to other chemical processes, where the whole design can be simulation driven, modeling, simulation and optimization is currently not very efficiently used in pulp and paper sector.

As the design of a paper mill consist of both structural design and operational design of the mill, it is useful to apply bi-level multi-objective optimization (BLMOO) [1] to the design. In our previous research, a process model for applying BLMOO in pulp and paper facility design has been presented. In this paper, this model is expanded with the multi-organizational aspects. Also the workflow has been developed and the model has been evaluated by expert interviews.

2 Design Processes

2.1 Optimization and Modeling in Pulp and Paper Mill Design

In paper mills the modeling has been usually used for two things: mass balance calculations and logistics problems. These simulations are similar in that sense, that

M.S. Obaidat et al. (eds.), *Simulation and Modeling Methodologies, Technologies and Applications*, Advances in Intelligent Systems and Computing 256,
DOI: 10.1007/978-3-319-03581-9_21,

the basic phenomena are simple and the challenge is to understand the system as a whole. [2] Logistic problems are simulated with event based models, which are outside of the scope of this paper.

The dynamic process models can be divided into first principle models, statistical models and the combination of those. Also terms white model, black model and gray model are used [3].

White, or first principle models are directly based on physical laws. For example the modeling of mass and energy flows is quite straightforward and they are also applicable outside the originally designed area if they are not excessively simplified [3]. A framework for representation of mathematical models in chemical processes has been developed in [4].

When modeling quality issues of paper or probabilities of web breaks, simplified statistical models, or black models are used. The downside of statistical data is that it is often gathered in normal operating situation, where some variables are kept constant. This can lead to omission of important variables in the model. Also the model cannot be extrapolated over the limits of the gathered data.

Hybrid, or gray models combine the physical model with the empirical, statistic model. As the hybrid model can be extrapolated, it is important to assess the validity of the model. In [5] two methods for validating hybrid models are presented.

Certain phenomena are difficult to model in process industry. For example quality properties of paper combined with the control of a mill are hard to model.

When using models in an optimization loop, the computational time can become too large. Computational time requirements have been tackled by using simpler surrogate models A model of the papermaking process can also be constructed of several types of sub models. The sub models have to be chosen in such a way that they meet the requirements of the optimization problem. A decision support system using a process line model consisting different types of sub models is presented in [6].

As both the process structure design and process control design are essential parts of the paper mill design, they should be designed simultaneously [7]. The problem can be formulated as a bi-level multi-objective optimization problem as in [8]. The dynamic model of the papermaking line and the dynamic multi-objective optimization can be coupled [9].

2.2 Conceptual Design of Continuous Processes

The optimizing design can be utilized in two kinds of project; in a product development project or in a conceptual design phase of a delivery project. The difference between these project types is that in product development project, the goal and timetable can be more freely defined. A product development project often has a stage-gate kind of process, which means that the project consists of several phases. After every phase, the feasibility of the project is evaluated and the project is continued only if certain criteria are fulfilled. Therefore, more risks can be taken at the early phases of the project.

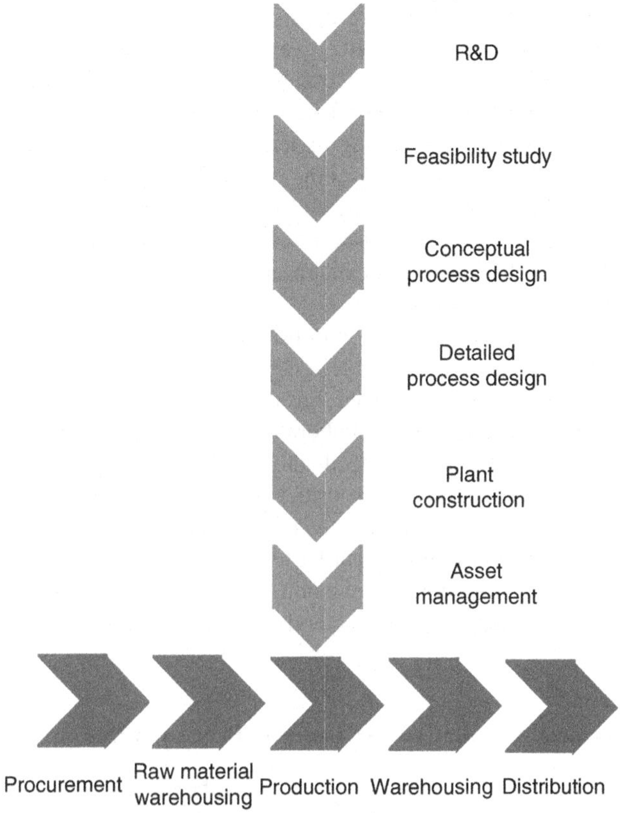

Fig. 1. Two major business processes in pulp and paper industry. Modified from [10].

In delivery project however, the delay or cancellation of the project often leads to substantial expenses. The goal of the project is a feasible concept of a functional plant.

Whereas in the product development project, there can be only one organization involved, the delivery project always have several. The customer asks bids for the project from one or several engineering enterprises. The bids can contain the whole project as a turnkey project, just the conceptual design phase or anything between. Before the bidding phase, there are usually unofficial negotiations between the participants about the higher level concepts. The bidding request should define the project so well that the bids can be made.

At an aggregate level the business process of process industry can be modeled as a combination of two major processes; the one containing the manufacturing of the product and the other one containing the design of the product and production plant (Fig. 1). The design process starts with a feasibility study containing economical impacts and is then followed by conceptual design and front-end engineering and design (FEED).

The early phases of the project are considered important, because the decisions made have a large impact on the life cycle costs of the plant. The nature of multidiscipline collaborative creative work makes it difficult to model the design process and to develop common tools. The process design approaches can be divided into three: 1) heuristic and engineering experience based methods, 2) optimization based methods and 3) case-based reasoning methods [11]. The combination of the methods and the usability of a certain method should also be taken into account.

The conceptual design process has been researched by several research groups. In University of Edinburgh, Bañares-Alcantara et al. have developed a design support system for chemical engineering [12].

In RWTH University Aachen, the workflows of the conceptual phase have been studied as well as the modeling of the design process [10], [13]. A specific modeling language, WPML, has been developed for modeling design processes [14]. Also requirements for tools in distributed collaborative engineering has been specified

Some researchers have emphasized the importance of creativeness in the conceptual design phase and therefore criticized too strict process definitions [15]. Without any documented process, the collaboration, common tools and the improvement of the process is though hardly feasible.

The usage of simulation and optimization methods have been researched before, but the effects of the usage of the methods in business processes is not so well studied.

2.3 Improvement of Collaborative Business Process of Design

Adopting a new method in process plant design can be seen as a business process re-engineering project. Kettinger and Grover present a Business Process Change Model, which divides the required changes in an organization into five areas: Management, IT, Business Processes, People and Organization Structure. [16]

In process redesign, a focus of development has to be chosen so that it is safe and productive enough. Schein [17] uses process consultation to define the focus. However, when introducing a significantly different new method in the process, the participants don't have experience of the new process beforehand. In such a change, the evaluation of the new method, workflow and tools have to be done by evaluating first the current method, process and tools, suggesting new process and evaluating the process in experimental pilot project. A framework for BPR presented in [18] divides the BPR process into six steps, namely Envision, Initiate, Diagnose, Redesign, Reconstruct and Evaluate.

The viewpoint of process improvement in general process improvement methods is often top down; the first step is to develop business vision, then the critical processes are being identified and after that, IT and methods are considered. e.g. in [19]. When applying optimizing design, the starting point is the optimization method, but large parts of the process should be redesigned.

As the evaluation process is likely to be iterative, so is also this research: the initial version of the process described in this paper is published in [20], and is updated and expanded according to the new knowledge gained from a new case study and expert interviews.

Collaborative design process applying bi-level multiobjective optimization with usage of limited models.

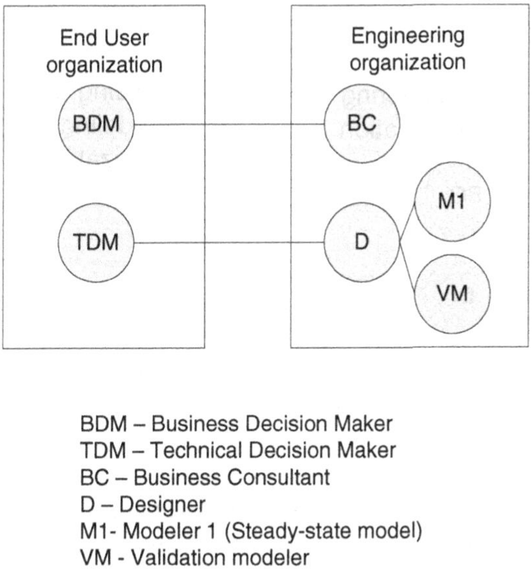

Fig. 2. Organizations and roles in traditional design

2.4 Research Focus

As the use of collaborative optimizing design in industrial projects is not straightforward, the applicability has been researched in case studies. Our previous publication [20] presented a process model for optimizing design. In our recent research, the model has been widened to taking organizational interfaces into account. This expansion was made because the comments from industrial experts showed that the optimizing design is likely to change also the customer interface.

Here the design process is represented according to the classification by [16]. The management and IT parts have been left out here and left for further research. It should be pointed out that before applying collaborative optimizing design in enterprise, it is extremely important to define e.g. process measuring and risk propensity and IT tools.

2.5 Collaborative Optimizing Design

People and Organization Structure

A typical organization in a delivery project is shown in Fig. 2. The decision making in customer organization is divided into business decision making and technical decision making. Depending on the size of the investment, the business decision making can be in corporate level or on local site. Business decision makers are interested mostly in return of the investment, but also in some other issues which have direct or indirect influence on earnings like good image, the green values of the corporation, investment risks taken, future of the markets and prices of the raw material and energy etc Technical decision makers are interested on the feasibility of the design, life-cycle costs, easiness of maintenance and flexibility for the changes.

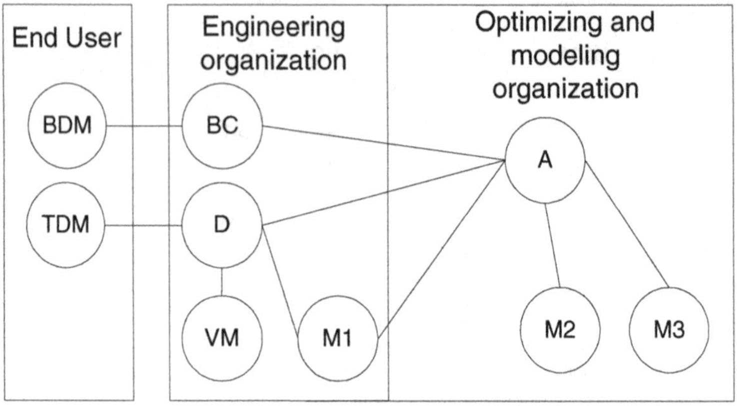

BDM – Business Decision Maker M1- Modeler 1 (Steady-state)
TDM – Technical Decision Maker M2 – Modeler 2 (Nominal model)
BC – Business Consultant M3 – Modeler 3 (Predictive model)
D – Designer VM - Validation modeler
A – Analyst

Fig. 3. Organizations and roles in optimizing design

The end-user organization purchases such services it doesn't have. As the end-user organizations have become trimmed, in investment projects they are more dependent on the engineering organizations than before. The business decision makers need consulting services about market situation, risks and expectation about future development. Large engineering enterprises can offer also the business consulting services or they can be bought elsewhere.

The actual engineering organization consists of designers from different disciplines. In the conceptual design phase, the main responsibility is on process designer. The process is modeled e.g. with flow chart containing the static balances of mass and energy flows in a typical operating point.

In Fig. 3, the optimizing and modeling organization is added. Though these roles can also be in the same organization they are here separated from the engineering organization in order to emphasize the interfaces between these roles. The added organization consists of an analyst and modeler roles. In optimizing design, the designer has the main responsibility for the design. He also is in the key role in identifying possible optimization targets. The analyst is responsible for mathematical representation of the optimizing problem, coordination of the model building and for solving the optimization problem.

Business Processes

The optimization activities take place in a few stages as an extension to conceptual process design phase as illustrated in Fig. 4. The process starts from a feasibility

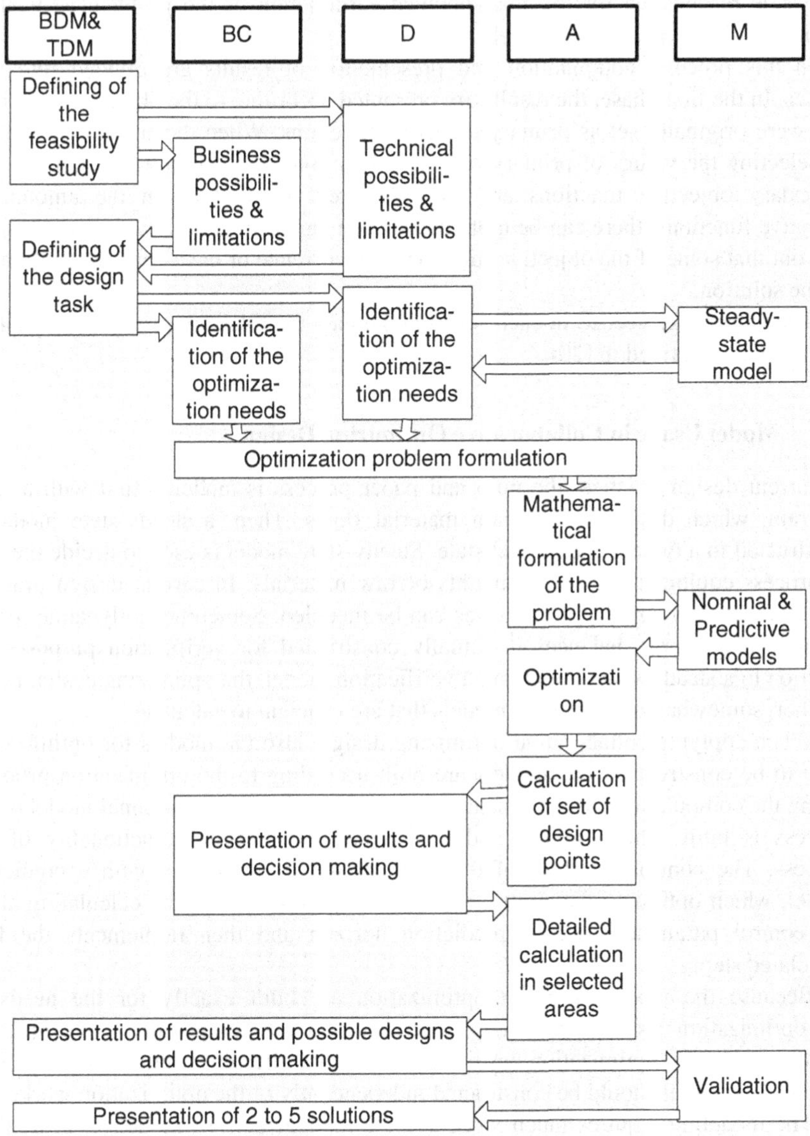

Fig. 4. Workflow in optimizing design

study, where economical and technical possibilities and limitations are evaluated. Then, a conceptual design phase can be started by giving the design task to the engineering organization. When the designer and/or business consult identifies a need for optimization in his conceptual design, he initiates cooperation with the analyst. As a result of this cooperation, a definition of an optimization problem is made. The analyst is responsible for the mathematical formulation of the problem as well as finding a solution to it using models he chooses.

In this process, computation and presentation of results are divided into two phases. In the first phase, the results are presented according to the objective functions that were originally set as primary objective functions. When the solution is limited by selecting the values of primary objective functions, the solution according to the secondary objective functions are the calculated. Depending on the amount of objective functions, there can be more phases that can also be iterative, because it can turn out that some of the objective functions e.g. correlate or have very little influence on the solution.

The knowledge needed in each role and the data the roles are producing, updating and using is described in [20].

2.6 Model Usage in Collaborative Optimizing Design

In current design practice, the pulp and paper process is modeled first with a flow diagram, which describes the main material flows. Then, a steady-state model is constructed in a typical operational state. Steady-state model is used to decide the size of process equipment and the amounts of raw materials. In current design practice dynamics of only some unit processes can be modeled. Sometimes a dynamic model of mass and energy balances is actually constructed for verification purposes. In addition to a steady-state model and a verification model, the optimizing design needs another, somewhat more limited, models that are efficient to calculate.

When applying collaborative optimizing design, also the models for optimization have to be constructed. The models are built according to the optimization problem taking the computational requirements also into account. First, a nominal model of the process is built. The nominal model describes the essential functionality of the process. The control variables of the nominal model are solved with a predictive model, which optimizes the control variables in every time step by calculating all of the control parameters over a prediction horizon and then implements the first calculated step.

Because the models used in optimization are built exactly for the needs of the optimization problem and they are simplified for computational reasons, the verification of the optimization has to be done with a different simulation model. The verification model should be constructed independently of the optimization work.

The modeling requires much work and the management of models is difficult; a change in the design requires changes in every model and in the cases in this project, the changes are made manually in each model. As the computational time also is a problem, the detail level of the models should be appropriate and consistent.

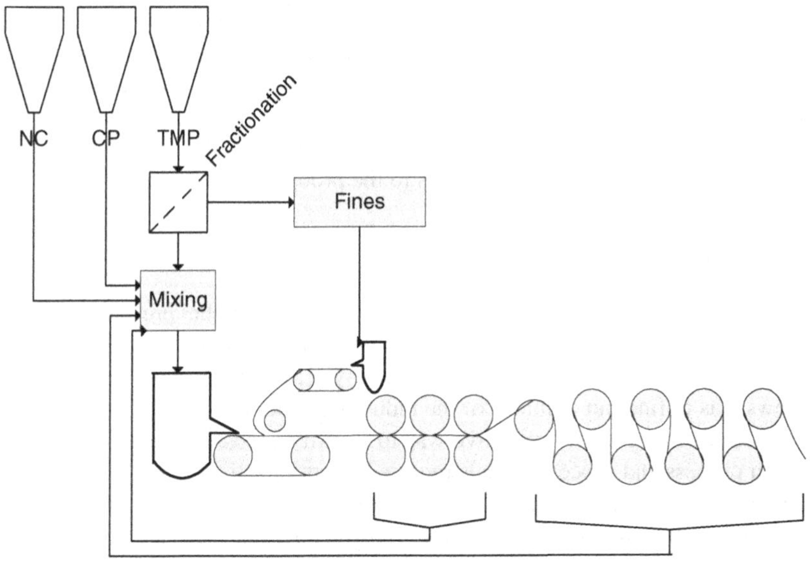

Fig. 5. Simplified Flow Chart of the Case Process

3 Assessment of the Model with a Case Study and Interviews

3.1 Case Study

The presented design method was studied on a example of a paper machine design project including pulp storages and mixing [21]. The paper machine is designed to have two headboxes, one for the base of the paper and one for the surface.(Fig. 5) The raw materials consisted of chemical and thermomechanical pulp. Also a small dose of nanocellulose was used to make the base of the paper stronger. The thermomechanical pulp was fractionated so that the largest fibres are used in making of the base of the paper in first headbox and the finest fibers were used in the surface layer to make the surface of the paper smoother. Nanocellulose was used to compensate the strength of the paper in base layer.

The scenario of this design project is applicable to a product design project or a conceptual design of a renewal delivery project. The optimizing design was used for minimizing the deviation of the strength and fast changes in the controls. In addition, there are constraints in flows and volumes of the towers.

When designing a new facility, and there is no previous measurement data of the target process, all assumptions made of the system are based on physical laws, statistics of another processes or expert knowledge. Part of the process can be modeled accurately based on physics. Such issues in this facility are e.g. the flows between process equipments. Some phenomena are more difficult to model, but there are models that are detailed enough for the optimization needs. For example the outflow of a tower can be modeled as a plug flow, ideal mixture or a combination of those. The accuracy of the model can be grown with more detailed physical models or by adjusting the model parameters according to the available statistical data.

The difficulty in paper making is that there are phenomena that are quite poorly understood and can only be modeled statistically. Such issues in this case are the relations between the process design and control and the quality properties of paper, e.g. the effect of changes in material flows to the probability of breaks.

3.2 Interviews

A set of interviews was performed in order to evaluate the presented model of collaborative design and model usage. The interviewees included experts from both paper and pulp industry and vendors of process systems. The primary objective of the interviews was to find out opinions of the industrial experts about the feasibility of the design process. A secondary objective was to identify necessary development targets for design process and process models required in optimization.

The general comment from the interviewees about the feasibility of the model of collaborative design was that it is feasible in principle, but there are some important reservations. Particularly, the applicability of optimization in a design process depends on the characteristics of the project. For small projects, where the possible benefits of optimization are smaller and time constraints also can be quite tight, the presented model of a design process is quite likely too heavy. However, for large greenfield projects the situation is different. The proper stage for optimization in such a project would be the pre-feasibility stage, i.e. before a contract between the vendor and a customer is made. The customers typically want to have an exact estimate of the costs of an investment before coming to one. Another possible situation where the presented model could be applied in a somewhat modified form is a product development project of a vendor.

The most important development target for the design process identified during the interviews was development of a more light weight model for utilizing optimization during delivery projects. Such a model would make it feasible to utilize optimization also in smaller projects. In order to do this, part of the optimization related work has to be moved into a phase preceding individual customer projects, e.g. into product development projects. The question how optimization work should be divided between product development and delivery projects is maybe the most important further development target for the model of collaborative optimization.

The general observation during the interviews concerning the modeling required for optimizing design was that it is exactly the challenging part of the whole approach. There are well-known and important phenomena in paper and pulp processes for which models usable for optimization do not exist. In addition to this, in design projects there are situations when there is not enough usable data even when applicable models would exist. However, on the other hand there are also some other design tasks for which models and data are available. As a conclusion, taking into account the currently availability of process models and data, optimization should be applied to selected parts of the design problem and combined with other design methods. Identifying the limits of optimization in the design of paper and pulp processes is on-going research and combination of optimization with other design methods an essential topic for further development.

There are a few different ways how the challenges concerning the model usage in optimizing design could be approached. First, the utilization of the existing models could be developed. Optimization studies during R&D projects could be used to update design knowledge, which is then utilized during customer projects, e.g. in a form of design rules. Another option is to develop new optimization methods and models for restricted targets during R&D projects and apply them during customer projects. A third option is to build a library of process models and more systematically utilize them for optimization in delivery projects. A fourth option is to develop new optimizing methods that require less effort on process modeling even at the expense of higher computational requirements.

4 Conclusions

We have presented a business process model for utilizing optimization in the design of continuous processes. The model is applicable to the conceptual design phase. The model includes organizational boundaries, roles, knowledge, data and a coarse workflow. The model has then evaluated with a case study from the paper industry and a set of expert interviews. The model should be considered as a start point when defining a design process for an enterprise. The process is intended to be fitted and specified for the needs of a particular organization.

This research has revealed that applying optimizing design for pulp and paper facilities not only requires development of optimization methods and tools but also changes in the business processes of design organizations and also customers. An enterprise offering collaborative optimizing design can't compete with traditional design enterprises, if the customer is not aware of the different approach with different time and budget requirements. A rough estimation of the time needed for an optimizing design by an experienced team to solve a similar task would be 3-4 months because of the data collection, iterations and computational time of the simulations. Therefore, the design organization has to be able to convince the customer that optimizing design will benefit the project.

Assessing the outcome of BLMOO method is difficult without conducting studies in real scale plant designs. It can though be stated that even very small improvements in conceptual design phase can lead to great benefits during the life cycle of the plant. Having an efficient business process in design helps the design organization to sell the concept of BLMOO to the customer. In our further research we will analyze the design process comparing it to the characteristics of design processes other than traditional plant design process.

One great challenge is the trustworthiness of the models. In order to convince the customer to invest in a separate optimization project or to allow longer and more expensive conceptual phase, the models have to match with experiential data. The optimizing design has the greatest potential, when the solution is outside the conventional solution area. Therefore the models should be proved to be valid also when extrapolated outside the area where the data for the model has been gathered. By performing sensitivity analysis, it is possible to estimate the influence of the accuracy of the model parameters. However, estimating the validity of models that are constructed using statistical data is more difficult. The conditions where the statistical data is gathered and the assumptions of the validity outside the normal operating area should be taken into account when assessing the trustworthiness of the models.

Acknowledgements. This research was supported by Forestcluster Ltd and its Effnet program. Other research partners in the research project were from Tampere University of Technology, University of Eastern Finland, University of Jyväskylä and VTT Technical Research Center of Finland.

References

1. Eichfelder, G.: Multiobjective bilevel optimization. Mathematical Programming 123, 419–449 (2010)
2. Dahlquist, E.: Use of modeling and simulation in pulp and paper industry. EU COST and Malardalen University Press, Sverige (2008)
3. Blanco, A., Dahlquist, E., Kappen, J., Manninen, J., Negro, C., Ritala, R.: Use of modelling and simulation in the pulp and paper industry. Mathematical and Computer Modelling of Dynamical Systems 15, 409–423 (2009)
4. Bogusch, R., Marquardt, W.: A formal representation of process model equations. Computers & Chemical Engineering 21, 1105–1115 (1997)
5. Kahrs, O., Marquardt, W.: The validity domain of hybrid models and its application in process optimization. Chemical Engineering and Processing: Process Intensification 46, 1054–1066 (2007)
6. Hämäläinen, J., Miettinen, K., Madetoja, E., Mäkelä, M.M., Tarvainen, P.: Multiobjective decision making for papermaking. CD-Rom Proceedings of MCDM (2004)
7. Pajula, E.: Studies on Computer Aided Process and Equipment Design in Process Industry. Helsinki University of Technology (2006)
8. Ropponen, A., Ritala, R., Pistikopoulos, E.N.: Optimization issues of the broke management system in papermaking. Computers and Chemical Engineering 35, 2510–2520 (2011)
9. Linnala: Dynamic simulation and optimization of an SC papermaking line - illustrated with case studies. Nordic Pulp and Paper Research Journal 25, 213–220 (2010)
10. Marquardt, W., Nagl, M.: Workflow and information centered support of design processes—the IMPROVE perspective. Computers & Chemical Engineering 29, 65–82 (2004)
11. Seuranen, T., Pajula, E., Hurme, M.: Applying CBR and Object Database Techniques in Chemical Process Design. In: Aha, D.W., Watson, I. (eds.) ICCBR 2001. LNCS (LNAI), vol. 2080, pp. 731–743. Springer, Heidelberg (2001)
12. Bañares-Alcántara, R., King, J.M.P.: Design support systems for process engineering — III. Design rationale as a requirement for effective support. Computers & Chemical Engineering 21, 263–276 (1996)
13. Theissen, M., Hai, R., Marquardt, W.: Design Process Modeling in Chemical Engineering. Journal of Computing and Information Science in Engineering 8, 011007 (2008)

14. Theißen, M., Hai, R., Marquardt, W.: A framework for work process modeling in the chemical industries. Computers & Chemical Engineering 35, 679–691 (2011)
15. Catledge, L.D., Potts, C.: Collaboration during conceptual design. IEEE Comput. Soc. Press, pp. 182–189 (1996)
16. Kettinger, W.J., Grover, V.: Special section: toward a theory of business process change management. J. Manage. Inf. Syst. 12, 9–30 (1995)
17. Schein, E.: Process Consultation Revisited: Building the Helping Relationship. Prentice Hall (1998)
18. Kettinger, W.J., Teng, J.T.C., Guha, S.: Business Process Change: A Study of Methodologies, Techniques, and Tools. MIS Quarterly 21, 55–80 (1997)
19. Davenport, T., James, S.: The new industrial engineering: information technology and business process redesign. Massachusetts Institute of Technology (1990)
20. Strömman, M., Seilonen, I., Peltola, J., Koskinen, K.: Design process model for optimizing design of continuous production processes. In: Proceedings of 1st International Conference on Simulation and Modeling Methodologies, Technologies and Applications, SIMULTECH 2011 , pp. 492-501 (2011)
21. Ropponen, A., Ritala, R.: Operational Optimization of Flow Management in Papermaking. Presented at the Papercon 2012, New Orleans, USA (2012)

Hydrodynamic Shape Optimization of Fishing Gear Trawl-Doors

Leifur Leifsson, Slawomir Koziel, and Eirikur Jonsson

Engineering Optimization & Modeling Center,
School of Science and Engineering, Reykjavik University,
Menntavegur 1, 101 Reykjavik, Iceland
{leifurth,koziel,eirikurjon07}@ru.is

Abstract. Rising fuel prices and inefficient fishing gear are hampering the fishing industry. Any improvements of the equipment that lead to reduced operating costs of the fishing vessels are highly desirable. This chapter describes an efficient optimization algorithm for the design of trawl-door shapes using accurate high-fidelity computational fluid dynamic models. Usage of the algorithm is demonstrated on the re-design of typical trawl-doors at high- and low-angle of attack.

Keywords: Trawl-doors, CFD, Design Optimization, Space Mapping, Surrogate-based Optimization, Variable-resolution Modeling, Simulation-driven Design.

1 Introduction

Efficient trawling is vital for the fishing industry. Rising fuel prices have an enormous impact on the fishing industry as a whole as the cost of fuel is a major part of the operation cost. Although trawling is normally performed at low speeds (less than 4 knots), most of the fuel is spent during the trawling operation (often over 80%). The reason is the high drag of the fishing gear assembly. Therefore, a careful study and redesign of the assembly may lead to a reduction in the fuel consumption and improve efficiency.

A typical fishing gear assembly for trawling, shown in Fig. 1(a), consists of a large net, a pair of trawl-doors to keep the net open, and a cable assembly extending from the trawl-doors to the boat and the net. The trawl-doors are a small part of the fishing gear, typically their span is 6-8 m and chord 2-3 m (see Fig. 1(b)), while the cables are a few hundred meter long and the net tens of meters). In spite of this, the trawl-doors may be responsible for up to 30% of the total drag [1]. Almost all the trawl-doors that have been developed over the years are fundamentally the same, i.e., highly cambered thin plates with leading edge (and some with trailing edge) elements. Minor design changes have been made to trawl-doors over the years, mainly because their designs are solely based on time consuming and expensive physical experiments.

Although, computational fluid dynamics (CFD) is widely used in design of a variety of engineering devices, such as aircraft, ships, and cars, very few applications are reported for trawl-doors in the literature [2]. Therefore, there is an opportunity to apply state-of-the-art CFD methods and optimization techniques to analyse and redesign the

M.S. Obaidat et al. (eds.), *Simulation and Modeling Methodologies, Technologies and Applications*,
Advances in Intelligent Systems and Computing 256,
DOI: 10.1007/978-3-319-03581-9_22,

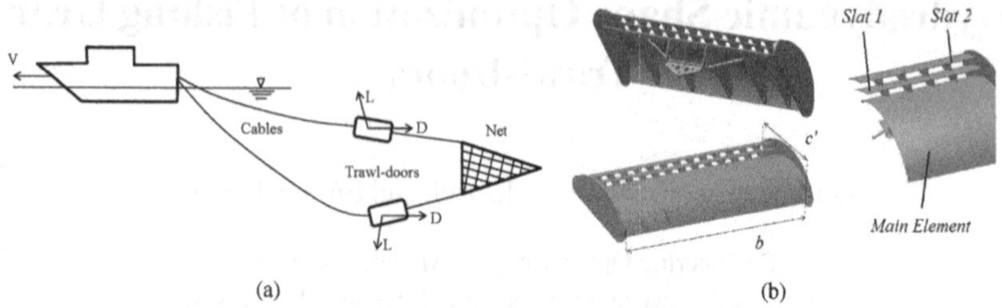

Fig. 1. A sketch of a fishing vessel (not drawn to scale); (a) the main parts of the fishing gear, (b) typical trawl-door shape

trawl-doors. The trawl-doors have a low aspect ratio and are operated at high angles of attack (up to 50 degrees). As a result, the flow is highly three-dimensional and transient. A full three-dimensional simulation is required to capture the flow physics accurately. However, such simulation can be time consuming, and if used directly within the optimization loop (requiring a large number of simulations), the overall time of the design process becomes prohibitive. An efficient design methodology is therefore essential for such design applications.

This chapter describes an accurate high-fidelity CFD model of trawl-doors and an efficient approach for their hydrodynamic design optimization [3,4]. The layout of the chapter is as follows. First, the CFD model is described and results of a validation study is given. Then, the details of the optimization algorithm are provided. The CFD model is then applied to the hydrodynamic analysis of a typical trawl-door shape. Finally, the CFD model and the optimization algorithm are applied to the hydrodynamic design of trawl-doors.

2 Computational Fluid Dynamic Modeling

This section describes a two-dimensional trawl-door CFD model. In particular, the geometry, the computational grid, and the flow solver are described. Results of a grid convergence study are presented, as well as validation of the CFD model.

2.1 Geometry

A simple chord-wise cross-sectional cut of a typical trawl-door (shown in Fig. 1(b)) is considered. A two-dimensional cut is shown in Fig. 2. There are three elements, the main element (ME), which is the largest element of the assembly, slat 1, the middle element, and slat 2, the element farthest from the ME. The trawl-door is normalized with the chord length of the ME (c).

2.2 Flow Equations

Trawl-doors are devices used in seawater, hence, we can safely assume that the flow is incompressible. Furthermore, we can assume that the flow is steady, viscous and with no body forces. The Reynolds-averaged Navier-Stokes equations are taken to be the governing equations with Menter's k-ω-SST turbulence model (see for example [5]).

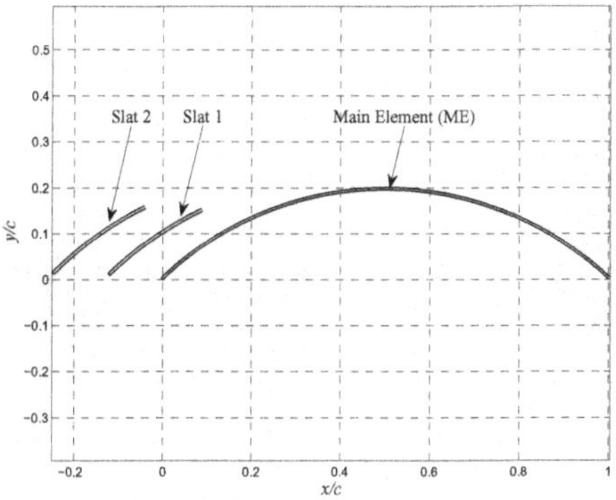

Fig. 2. Cross-section of a typical trawl-door with three elements, main element (ME), slat 1 and slat 2

2.3 Computational Mesh

The farfield is configured by a box-topology where the trawl-door geometry is placed in the center of the box. The main element leading edge (LE) is placed as the origin $(x/c, y/c) = (0, 0)$, with the farfield extending 100 main element chord lengths away from the origin. The grid is an unstructured triangular grid where the elements are clustered around the trawl-door geometry, growing in size as they move away from the origin. The maximum element size on the geometry is set to 0.1% of c. The maximum element size in domain away from the trawl-door is $10c$. In order to capture the viscous boundary layer well, a prismatic inflation layer is extruded from all surfaces. The inflation layer has a initial height of $5 \times 10^{-6}c$, growing with exponential growth ratio of 1.2 and extending 20 layers from the surface. The initial layer height is chosen so that $y^+ < 1$. In the wake region aft of the trawl-door, the grid is made denser by applying a density grid with an element size of 5% of c, extending $20c$ aft of the trawl-door geometry. The density mesh is configured in an adaptive manner so that it aligns with the flow direction. An example grid is shown in Fig. 3.

2.4 Numerical Solver

Numerical fluid flow simulations are performed using the computer code ANSYS FLUENT [6]. The flow solver is set to a coupled velocity-pressure-based formulation. The spatial discretization schemes are second order for all variables and the gradient information is found using the Green-Gauss node based method. Additionally, due to the difficult flow condition at high angle of attacks, the pseudo-transient option and high-order relaxation terms are used in order to get a stable converged solution [6]. The iterative solution is performed with relaxation factors to prevent a numerical oscillation of the solution that can lead to a no solution or errors. The residuals, which are the sum of the L^2 norm of all governing equations in each cell, are monitored and checked for convergence. The convergence criterion for the high-fidelity model is such that a solution is considered to be converged if the residuals have dropped by six orders of

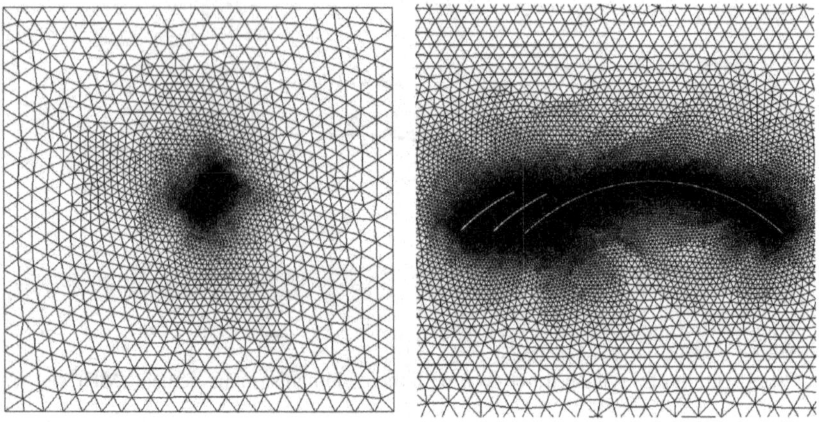

(a) (b)

Fig. 3. High-fidelity mesh for the angle of attack $\alpha = 50$ degrees; (a) the whole solution domain, and (b) a zoom in close to the trawl-door

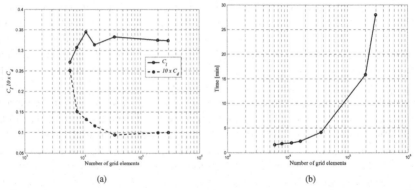

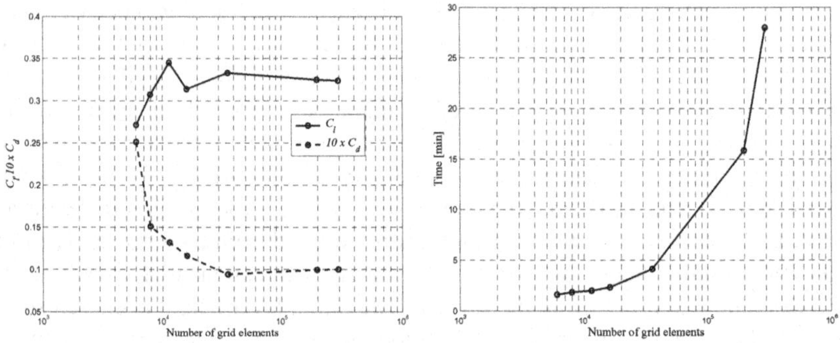

(a) (b)

Fig. 4. Grid convergence study using the NACA 0012 airfoil at $V_\infty = 2m/s$, $Re = 2 \times 10^6$ and angle of attack $\alpha = 3°$. a) Lift (C_l) and drag (C_d) coefficient versus number of grid elements, b) simulation time versus number of grid elements

magnitude, or the total number of iterations has reached 1000. Also, the lift and drag coefficients are monitored for convergence.

The working fluid used is water and the inlet boundary is a velocity-inlet with a freestream velocity $V_\infty = 2\ m/s$, (which is typical during trawling), split into its x and y components depending on the angle of attack α. The outlet boundary is a pressure-outlet. Reynolds number is $Re_c = 2 \times 10^{-6}$. The inlet flow is assumed to be calm, with turbulent intensity and viscosity ratio of 0.05% and 1, respectively.

2.5 Grid Convergence

A sufficiently fine enough mesh is found by carrying out a grid convergence study using the NACA 0012 airfoil at $V_\infty = 2\ m/s$, $Re = 2 \times 10^6$, and $\alpha = 3°$. The results are

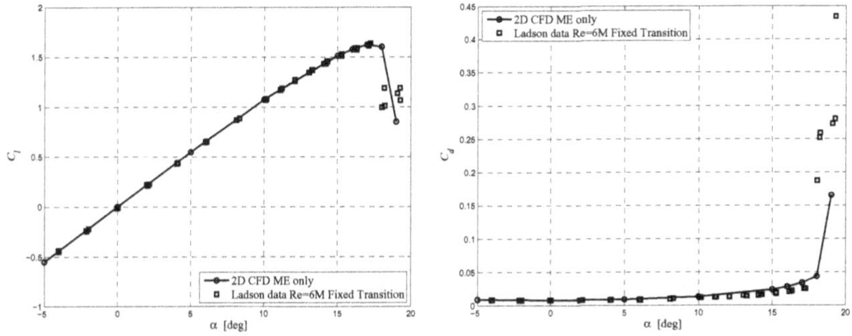

(a) (b)

Fig. 5. Results of the CFD model validation. The model is compared with experimental data [8] at $Re_c = 6 \times 10^6$. (a) Lift curve, and (b) drag curve.

shown in Fig. 4 and reveal that 197,620 grid elements are needed for convergence. The overall simulation time needed for one high-fidelity CFD simulation was around 16 minutes, executed on four Intel-i7-2600 processors in parallel (1000 solver iterations where required).

2.6 Model Validation

Due to lack of available two-dimensional experimental data for trawl-door shapes, other types of geometries are used to validate the high-fidelity CFD model. The NACA four-digit airfoils [7] have been studied extensively in the past and we consider the NACA 0012 airfoil as the validation case. The results are shown in Fig. 5. There is good agreement between the CFD model and the experimental data in terms of lift up to the stall region, and up to $\alpha \approx 10^\circ$ for drag.

3 Optimization with Space Mapping

The trawl-door design is carried out in a computationally efficient manner by exploiting the space mapping (SM) methodology [9,10]. Space mapping replaces the direct optimization of an expensive (high-fidelity or fine) airfoil model f obtained through high-fidelity CFD simulation, by an iterative updating and re-optimization of a cheaper surrogate model s. The key component of SM is a physics-based low-fidelity (or coarse) model c that embeds certain knowledge about the system under consideration and allows us to construct a reliable surrogate using a limited amount of high-fidelity model data. Here, the low-fidelity model is evaluated using the same CFD solver as the high-fidelity one, so that both models share the same knowledge of the trawl-door performance.

3.1 Optimization Problem

Simulation-driven design can be generally formulated as a nonlinear minimization problem[9]

$$\mathbf{x}^* = \arg\min_{\mathbf{x}} H\left(f(\mathbf{x})\right), \tag{1}$$

where $\mathbf{x}$ is a vector of design parameters, f the high-fidelity model to be minimized at $\mathbf{x}$ and H is the objective function. $\mathbf{x}^*$ is the optimum design vector. The high-fidelity model will represent the hydrodynamic forces, i.e., the lift and drag coefficients. The response will have to form

$$f(\mathbf{x}) = \left[C_{l.f}(\mathbf{x})\; C_{d.f}(\mathbf{x})\right]^T, \tag{2}$$

where $C_{l.f}$ and $C_{d.f}$ are the lift and drag coefficient, respectively, generated by the high-fidelity CFD model. We are interested in minimizing the drag for a given lift, so the objective function will take the form of

$$H\left(f(\mathbf{x})\right) = C_d, \tag{3}$$

with the design constraint written as

$$c\left(f(\mathbf{x})\right) = -C_{l.f}(\mathbf{x}) + C_{l.\min} \leq 0, \tag{4}$$

where $C_{l.\min}$ is the minimum required lift coefficient.

3.2 Space Mapping Basics

Starting from an initial design $\mathbf{x}^{(0)}$, a generic space mapping algorithm produces a sequence $\mathbf{x}^{(i)}, i = 0, 1 \ldots$ of approximate solution to (1) as [10]

$$\mathbf{x}^{(i+1)} = \arg\min_{\mathbf{x}} H\left(s^{(i)}(\mathbf{x})\right), \tag{5}$$

where

$$s^{(i)}(\mathbf{x}) = \left[C_{l.s}^{(i)}(\mathbf{x})\; C_{d.s}^{(i)}(\mathbf{x})\right]^T, \tag{6}$$

is the surrogate model at iteration i. As previously described, the high-fidelity CFD model f is accurate, but computationally expensive. Using Space Mapping, the surrogate s is a composition of the low-fidelity CFD model c and a simple linear transformations to correct the low-fidelity model response [9]. The corrected response is denoted as $s(\mathbf{x}, \mathbf{p})$ where $\mathbf{p}$ represent a set of model parameters and at iteration i the surrogate is

$$s^{(i)}(\mathbf{x}) = s(\mathbf{x}, \mathbf{p}). \tag{7}$$

The SM parameters $\mathbf{p}$ are determined through a parameter extraction (PE) process. In general, this process is a nonlinear optimization problem where the objective is to minimize the misalignment of surrogate response at some or all previous iteration high-fidelity model data points [9]. The PE optimization problem can be defined as [9]

$$\mathbf{p}^{(i)} = \arg\min_{\mathbf{p}} \sum_{k=0}^{i} w_{i,k} \| f(\mathbf{x}^{(k)}) - s(\mathbf{x}^{(k)}, \mathbf{p}) \|^2, \tag{8}$$

where $w_{i,k}$ are weight factors that control how much impact previous iterations affect the SM parameters.

3.3 Low-Fidelity CFD Model

The general underlying low-fidelity model c used for all cases is constructed in the same way as the high-fidelity model f, but with a coarser grid discretization and relaxed convergence criteria. Referring back to the grid study, carried out in Section 2.5, and inspecting Fig. 4, we select the coarse low-fidelity model. Based on time and accuracy, with respect to lift and drag, we select the grid parameters representing the fourth point from the right, giving 16,160 elements for the low-fidelity CFD model. The evaluation time of the low-fidelity model is 2.3 minutes on four Intel-i7-2600 processors in parallel. The solution has converged within 200 iterations. However, the maximum number of iterations for the low-fidelity model is set to three times that, or 700 iterations, due to the nature of problem and different geometries to be optimized. This reduces the overall simulation time to 1.6 minutes. The ratio of simulation times of the high- and low-fidelity model in this case is high/low = 16/1.6 = 10. This is based on the solver uses all 700 iterations in the low-fidelity model to obtain a solution.

3.4 Surrogate Model Construction

As mentioned above, the SM surrogate model s is a composition of the low-fidelity CFD model c and corrections or linear transformations where model parameters $\mathbf{p}$ are extracted using one of the PE processes described above. Parameter extraction and surrogate optimization create a certain overhead on the whole process which can be up to 80-90% of the computational cost. This is due to the fact that physics-based low-fidelity models are in general relatively expensive to evaluate.

To alleviate this problem, an output SM with both multiplicative and additive response correction is exploited here with the surrogate model parameters extracted analytically. We use the following formulation [10]

$$\begin{aligned} s^{(i)}(\mathbf{x}) &= \mathbf{A}^{(i)} \circ c(\mathbf{x}) + \mathbf{D}^{(i)} + \mathbf{q}^{(i)} \\ &= \left[a_l^{(i)} C_{l.c}(\mathbf{x}) + d_l^{(i)} + q_l^{(i)} \quad a_d^{(i)} C_{d.c}(\mathbf{x}) + d_d^{(i)} + q_d^{(i)} \right]^T . \end{aligned} \tag{9}$$

The parameters $\mathbf{A}^{(i)}$ and $\mathbf{D}^{(i)}$ are obtained by solving

$$\left[\mathbf{A}^{(i)}, \mathbf{D}^{(i)} \right] = \arg\min_{\mathbf{A},\mathbf{D}} \sum_{k=0}^{i} \| f\left(\mathbf{x}^{(k)} \right) - \mathbf{A} \circ c\left(\mathbf{x}^{(k)} \right) + \mathbf{D} \|^2, \tag{10}$$

where $w_{i,k} = 1$, i.e., all previous iteration points are used to improve globally the response of the low-fidelity model. The additive term $q^{(i)}$ is defined such it ensures a perfect match between the surrogate and the high-fidelity model at design $\mathbf{x}^{(i)}$, namely $f(\mathbf{x}^{(i)}) = s(\mathbf{x}^{(i)})$ or a zero-order consistency [11]. We can write the additive term as

$$q^{(i)} = f\left(\mathbf{x}^{(i)}\right) - \left[\mathbf{A}^{(i)} \circ c(\mathbf{x}^{(i)}) + \mathbf{D}^{(i)}\right]. \tag{11}$$

Since an analytical solution exists for $\mathbf{A}^{(i)}$, $\mathbf{D}^{(i)}$ and $\mathbf{q}^{(i)}$, there is no need to perform non-linear optimization to solve Eq. 8 to obtain the parameters. $\mathbf{A}^{(i)}$ and $\mathbf{D}^{(i)}$ can be obtained by solving [10]

$$\begin{bmatrix} a_l^{(i)} \\ d_l^{(i)} \end{bmatrix} = \left(\mathbf{C}_l^T \mathbf{C}_l\right)^{-1} \mathbf{C}_l^T \mathbf{F}_l \,, \tag{12}$$

$$\begin{bmatrix} a_d^{(i)} \\ d_d^{(i)} \end{bmatrix} = \left(\mathbf{C}_d^T \mathbf{C}_d\right)^{-1} \mathbf{C}_d^T \mathbf{F}_d \,, \tag{13}$$

where

$$\mathbf{C}_l = \begin{bmatrix} C_{l.c}(\mathbf{x}^{(0)}) & C_{l.c}(\mathbf{x}^{(1)}) & \ldots & C_{l.c}(\mathbf{x}^{(i)}) \\ 1 & 1 & \ldots & 1 \end{bmatrix}^T , \tag{14}$$

$$\mathbf{F}_l = \begin{bmatrix} C_{l.f}(\mathbf{x}^{(0)}) & C_{l.f}(\mathbf{x}^{(1)}) & \ldots & C_{l.f}(\mathbf{x}^{(i)}) \\ 1 & 1 & \ldots & 1 \end{bmatrix}^T , \tag{15}$$

$$\mathbf{C}_d = \begin{bmatrix} C_{d.c}(\mathbf{x}^{(0)}) & C_{d.c}(\mathbf{x}^{(1)}) & \ldots & C_{d.c}(\mathbf{x}^{(i)}) \\ 1 & 1 & \ldots & 1 \end{bmatrix}^T , \tag{16}$$

$$\mathbf{F}_d = \begin{bmatrix} C_{d.f}(\mathbf{x}^{(0)}) & C_{d.f}(\mathbf{x}^{(1)}) & \ldots & C_{d.f}(\mathbf{x}^{(i)}) \\ 1 & 1 & \ldots & 1 \end{bmatrix}^T , \tag{17}$$

which are the least-square optimal solutions to the linear regression problems

$$\mathbf{C}_l a_l^{(i)} + d_l^{(i)} = \mathbf{F}_l \,, \tag{18}$$

$$\mathbf{C}_d a_d^{(i)} + d_d^{(i)} = \mathbf{F}_d \,. \tag{19}$$

Note that $\mathbf{C}_l^T \mathbf{C}_l$ and $\mathbf{C}_d^T \mathbf{C}_d$ are non-singular for $i > 1$ and assuming that $\mathbf{x}^{(k)} \neq \mathbf{x}^{(i)}$ for $k \neq i$. For $i = 1$ only the multiplicative SM correction with $\mathbf{A}^{(i)}$ is used.

3.5 Optimization Algorithm

The optimization algorithm exploits the SM based surrogate and a trust-region convergence safeguard [12]. The trust-region parameter λ is updated after each iteration. The optimization algorithm is as follows

1. Set $i = 0$; Select λ, the trust region radius; Evaluate the high-fidelity model at the initial solution, $f(\mathbf{x}^{(0)})$;
2. Using data from the low-fidelity model c, and f at $\mathbf{x}^{(k)}, k = 0, 1, \ldots, i$, setup the SM surrogate $s^{(i)}$; Perform PE;
3. Optimize $s^{(i)}$ to obtain $x^{(i+1)}$;
4. Evaluate $f(\mathbf{x}^{(i+1)})$;
5. If $H(f(\mathbf{x}^{(i+1)})) < H(f(\mathbf{x}^{(i)}))$, accept $\mathbf{x}^{(i+1)}$; Otherwise set $\mathbf{x}^{(i+1)} = \mathbf{x}^{(i)}$;
6. Update λ;
7. Set $i = i + 1$;
8. If the termination condition is not satisfied, go to 2, else proceed;
9. End; Return $\mathbf{x}^{(i)}$ as the optimum solution.

The termination condition is set as $||\mathbf{x}^{(i)} - \mathbf{x}^{(i-1)}|| < 10^{-3}$.

4 Hydrodynamic Characteristics

The high-fidelity CFD model is applied to the analysis of a typical trawl-door (shown in Fig. 2). In particular, the lift and drag responses of the trawl-door are obtained by varying the angle of attack. A parametric study of the trawl-door is performed by varying the position of slats at a given angle of attack and observe the change in response. All the studies are performed at a free-stream velocity $V_\infty = 2\ m/s$ (3.9 knots) and Reynolds number $Re_c = 2 \times 10^6$.

4.1 Lift and Drag Curves

The performance the trawl-door is evaluated at a number of different angles of attack or from $\alpha = -5°$ to $\alpha = 60°$ with 5 degree increments. Three different configurations were studied, the main element only, the main element with one slat, and the main element with two slats. Figure 6 shows the lift and drag curves.

Inspecting Fig. 6(a) reveals that the flow remains attached for relatively low angles of attack when considering the main element only, and stall occurs close to $\alpha = 10°$. For the main element with one slat, the stall occurs at $\alpha = 20°$ and for the F11 design (the main element with two slats) the stall occurs at $\alpha = 25°$. Adding the slats, therefore, improves the performance by delaying the stall, as well as increasing the $C_{l.\max}$. This effect is expected as it is well known for multi-element high-lift devices on aircraft.

We can split the drag curve (Fig. 6(b)) into three regions, first where the stall has not occurred where $\alpha < 20°$, and after stalling $20° < \alpha < 35°$ and $\alpha > 35°$. As seen in part $\alpha < 20°$, prior to stall, the drag increases as more slats are added. As the angle of attack is increased beyond stall, the drag rises due to a massive flow separation (Fig. 7). Effectiveness of slats are evident here, the flow remains attached at a higher angle of attack and, thus, preventing separation and reducing drag. In the last region, $\alpha > 35°$, the drag increases rapidly due to massively separated flow on the upper surface (Fig. 7(a)).

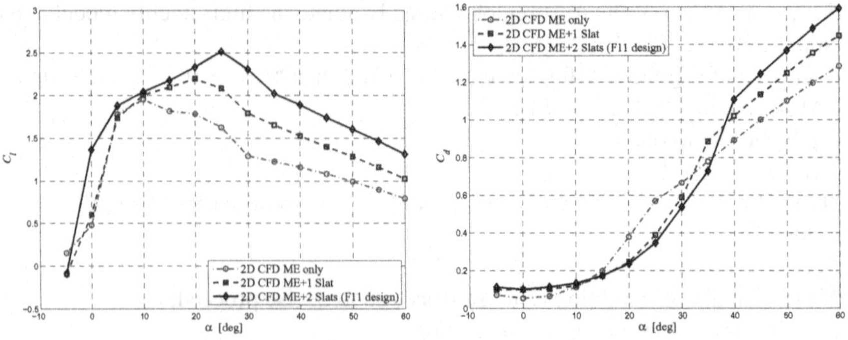

(a) (b)

Fig. 6. Section lift and drag curves of a typical trawl-door (shown in Fig. 2) at $V_\infty = 2\ m/s$ and $Re_c = 2 \times 10^6$

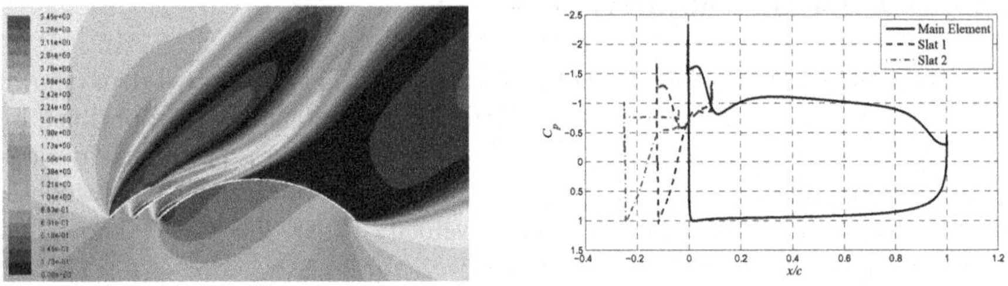

(a) (b)

Fig. 7. The trawl-door characteristics at $V_\infty = 2\ m/s$, $Re_c = 2 \times 10^6$, and angle of attack $\alpha = 30°$; (a) velocity contours, and (b) pressure distribution

4.2 Slat Position Sensitivity

The change in the performance of the conventional trawl-door by varying the position of the slats was investigated. The slats were translated in the horizontal and the vertical direction with the distance between the slats kept fixed (see Fig. 8(a)). The results of the CFD analysis at a fixed angle of attack $\alpha = 30°$ are shown in Fig. 8(b). At this angle of attack, the lift-to-drag ratio is 4.6.

A significant increase in the section lift-to-drag ratio can be obtained by translating the slats in the horizontal direction away from the main element. In particular, an approximately 10% increase in the lift-to-drag ratio is obtained when moving the slats in the negative horizontal direction by 8% of the main element chord. Translating the slat in the positive horizontal direction reduces the lift-to-drag ratio. A reduction in the lift-to-drag ratio is obtained by a positive and a negative vertical translation, aside from a slight rise for a small negative one.

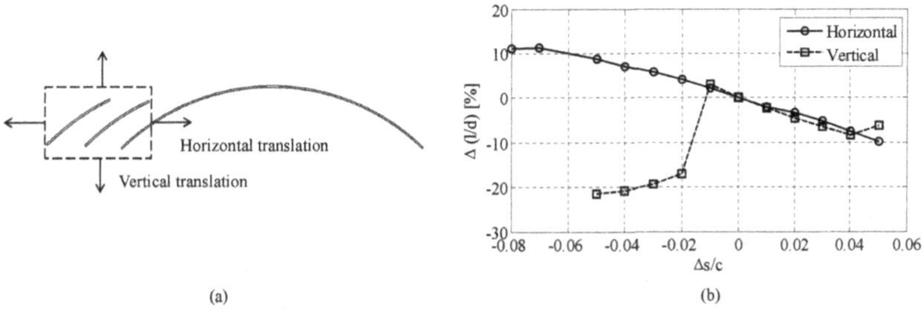

Fig. 8. Results of a parametric study of the slat position on a conventional trawl-door; (a) slat translation directions, and (b) percentage change in lift-to-drag ratio with slat translation ($\Delta s/c$) at $\alpha = 30°$.

5 Design Optimization

The objective of the design studies presented here is to optimize the position and orientation of the slat of the conventional trawl-door. To simplify the geometry for these preliminary studies, only one slat is used. The design problem is formulated so that the drag coefficient is minimized for a given lift coefficient constraint. Maximum overlaTwo design optimization case studies are presented. In one case, the angle of attack is fixed to 45 degrees, wich is a typical operating setting. In the other case, the angle of attack is set as a design variable, bounded by values close to zero degrees.

5.1 Problem Formulation

The objective is to minimize the drag coefficient C_d subject to a constraint on the lift coefficient $C_l \geq C_{l.\min}$. The design optimization is formulated as a nonlinear minimization problem

$$\mathbf{x}^* = \arg\min_{\mathbf{x}} C_d, \tag{20}$$

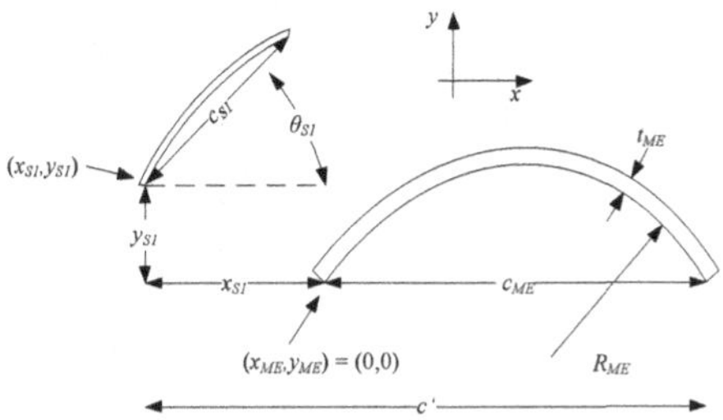

Fig. 9. Definition of the slat position and orientation relative to the leading-edge of the main element of the trawl-door

subject to

$$
\begin{aligned}
C_{l,f} &\geq C_{l,\min}, \\
Gap &\geq Gap_{\min}, \\
Overlap &\leq Overlap_{\max},
\end{aligned}
\tag{21}
$$

with the design variable vector $\mathbf{x} = [x_{S1}/c \quad y_{S1}/c \quad \theta_{S1}]^T$, where $(x_{S1}/c, y_{S1}/c)$ is the position of the leading-edge of the slat and θ_{S1} the orientation (these parameters are defined in Fig. 9). Additional constraints include a geometry validity check where the design is checked at every iteration such that the optimizer rejects designs if the slat and the main element cross, or violate a minimum gap, $Gap \geq Gap_{\min}$, or a maximum overlap, $Overlap \leq Overlap_{\max}$. The minimum gap between elements is defined as the minimum distance from any point on the main element to any point on the slat.p is defined as the distance x/c which the trailing edge of the slat overlaps the leading edge of the main element.

In both design cases, the free-stream velocity is fixed at $V_\infty = 2\ m/s$ and the Reynolds number is $Re_c = 2 \times 10^6$. The minimum lift coefficient is $C_{l,\min} = 1.5$, the minimum gap is $Gap_{\min} = 0.05$, and the maximum overlap is $Overlap_{\max} = 0.1$. The design variable bounds are $-0.3 \leq x_{S1}/c \leq 0.2$, $-0.3 \leq y_{S1}/c \leq 0.2$, and $20° \leq \theta_{S1} \leq 50°$. In Case 1, the angle of attack is fixed to $\alpha = 45°$, and in Case 2 it is set as a design variable with the bounds $2° \leq \alpha \leq 8°$.

5.2 Results

The numerical results of the optimization cases are shown in Table 1. The initial and optimized geometries and flow fields are shown in Fig. 10. In Case 1, the angle of attack is fixed to 45 degrees. From the initial design, the slat has moved 6.8% of c to the left (away from the main element) and 5.1% of c down, as well as adding about 16 degrees to the orientation. As a result, the lift coefficient increases by 9% and the drag is reduced by 23%. The lift-to-drag ratio increases from 1.24 to 1.76, or by 42%. Although, the

Table 1. Numerical results of the design optimization cases. The free-stream velocity is $V_\infty = 2\ m/s$ and the Reynolds number is $Re_c = 2 \times 10^6$.

	Case 1		Case 2	
Variable	**Initial**	**Optimized**	**Initial**	**Optimized**
x_{S1}/c	-0.1192	-0.1870	-0.1192	-0.2107
y_{S1}/c	0.0085	-0.0427	0.0085	-0.0100
θ_{S1} [deg]	33.9	49.6479	33.9	24.0113
α [deg]	45.0000	45.0000	30.0	2.0000
C_l	1.4592	1.5902	1.7925	1.4382
C_d	1.1782	0.9016	0.5875	0.0614
l/d	1.2385	1.7637	3.0511	23.4235

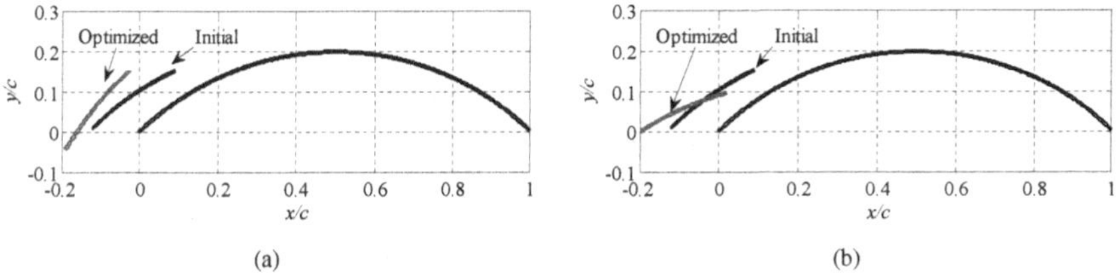

Fig. 10. Initial and optimized trawl-doors for (a) Case 1, and (b) Case 2. Velocity contours of optimized trawl-doors for (c) Case 1, and (d) Case 2.

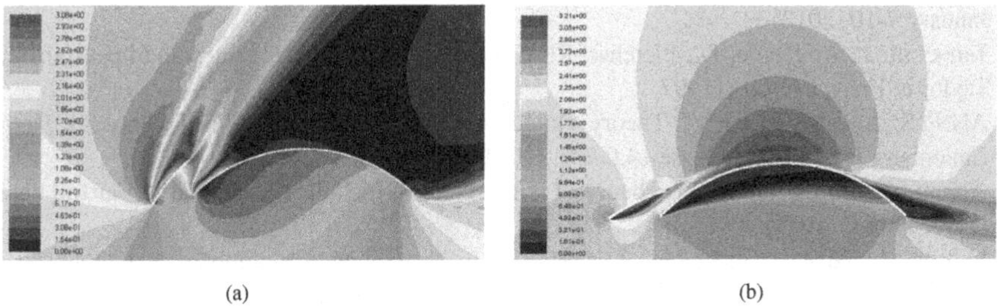

Fig. 11. Velocity contours of optimized trawl-doors for (a) Case 1, and (b) Case 2

efficiency has increased somewhat, there still is a massive flow separation over the main element (Fig. 11(a)).

The second design case considers the angle of attack as a design variable between 2 to 8 degrees. In the optimized design for Case 2, the angle of attack hits the lower bound of 2 degrees. The optimizer aligns the slat to the oncoming flow (reducing the orientation angle from 34° to 24°) and moves it slightly away from the main element (9.2% of c to the left and 2.9% down). The resulting design has a lift-to-drag ratio of 23.4. The optimized design has attached flow on the upper surface at the low angle of attack (see Fig. 11(b)). It should be noted here, that a control system is necessary in this case as the trawl-door is being operated in an unstable region, thus complicating the design significantly.

6 Conclusions

A two-dimensional computational fluid dynamic model of the flow past trawl-doors of fishing gear has been described. Analysis of typical trawl-door shapes show that their performance is poor. Design studies using an efficient optimization algorithm indicate that their performance can be significantly improved by minor redesigns, such as a careful repositioning of the slat elements. Introducing improved trawl-door designs through a CFD-driven optimization process will be crucial in achieving significant fuel consumption reduction of fishing vessels – which is critical for the fishing industry.

References

1. Garner, J.: Botnvarpan og bunadur hennar. Fiskifelag Islands (1967)
2. Haraldsson, H.O., Brynjolfsson, S., Jonsson, V.K.: Reiknileg hermun idustreymis umhverfis toghlera. Velabrogd (University of Iceland) 17, 31–35 (1996)
3. Jonsson, E., Leifsson, L., Koziel, S.: Trawl-door performance analysis and design optimization with cfd. In: 2nd Int. Conf. on Simulation and Modeling Methodologies, Technologies, and Applications (SIMULTECH), Rome, Italy, July 28-31 (2012)
4. Jonsson, E., Hermannsson, E., Juliusson, M., Koziel, S., Leifsson, L.: Computational fluid dynamic analysis and shape optimization of trawl-doors. In: 51st AIAA Aerospace Sciences Meeting including the New Horizons Forum and Aerospace Exposition, Grapevine, Texas, January 7-10 (2013)
5. Tannehill, J., Anderson, D., Pletcher, R.: Computational fluid mechanics and heat transfer. Taylor & Francis Group (1997)
6. ANSYS: ANSYS FLUENT Theory Guide. ANSYS, Southpointe 275 Technology Drive Canonburg PA 15317. Release 14.0 edn. (2011)
7. Abbott, I., Von Doenhoff, A.: Theory of wing sections: including a summary of airfoil data. Dover Pubns. (1959)
8. Ladson, C.: Effects of independent variation of mach and reynolds numbers on the low-speed aerodynamic characteristics of the naca 0012 airfoil section (1988)
9. Bandler, J., Cheng, Q., Dakroury, S., Mohamed, A., Bakr, M., Madsen, K., Sondergaard, J.: Space mapping: the state of the art. IEEE Transactions on Microwave Theory and Techniques 52, 337–361 (2004)
10. Koziel, S., Leifsson, L.: Knowledge-based airfoil shape optimization using space mapping. In: 30th AIAA Applied Aerodynamics Conference, AIAA 2012-3016, January 25-28 (2012)
11. Alexandrov, N., Lewis, R.: An overview of first-order model management for engineering optimization. Optimization and Engineering 2, 413–430 (2001)
12. Forrester, A., Keane, A.: Recent advances in surrogate-based optimization. Progress in Aerospace Sciences 45, 50–79 (2009)

Wing Aerodynamic Shape Optimization by Space Mapping

Leifur Leifsson, Slawomir Koziel, and Eirikur Jonsson

Engineering Optimization & Modeling Center,
School of Science and Engineering, Reykjavik University,
Menntavegur 1, 101 Reykjavik, Iceland
{leifurth,koziel,eirikurjon07}@ru.is

Abstract. This chapter describes an efficient aerodynamic design optimization methodology for wings in transonic flow. The approach replaces a computationally expensive high-fidelity computational fluid dynamic model (CFD) in an iterative optimization process with a corrected polynomial approximation model constructed by a cheap low-fidelity CFD model. The output space mapping technique is used to correct the approximation model to yield an accurate predictor of the high-fidelity one. The algorithm is applied to two transonic wing design problems.

Keywords: Transonic Wing Design, CFD, Surrogate-based Optimization, Variable-resolution Modeling, Space Mapping.

1 Introduction

The wing is the most important component of an aircraft, significantly affecting its overall performance. As the wing provides lift, it is at the same time the main source of drag, responsible for about two-thirds of the total drag of the aircraft [1]. Reducing this wing drag by a better design is often the primary objective of modern aircraft design. Nowadays, aerodynamic design using high-fidelity computational fluid dynamic (CFD) models is ubiquitous and plays an important role in aircraft development. Traditional design optimization techniques, such as gradient-based or population-based ones, involve a large number of simulations. Consequently, direct aerodynamic optimization with high-fidelity CFD models using traditional optimization techniques is impractical, even when using cheap adjoint sensitivities.

One of the overall objectives of surrogate-based optimization (SBO) [2, 3] is to reduce the number of evaluations of expensive simulations, thereby making the design process more efficient. This is achieved by an iterative correction-prediction process where a surrogate model (a computationally cheap representation of the high-fidelity one) is constructed and subsequently exploited to obtain approximate location of the high-fidelity model optimal design. The surrogate model can be constructed by approximating sampled high-fidelity model data using, e.g., polynomial approximation [2], radial basis functions [3, 4], kriging [5–8], neural networks [9, 10], or support vector regression [11] (response surface approximation surrogates) or by correcting/enhancing a physics-based low-fidelity model (physical surrogates) [12].

M.S. Obaidat et al. (eds.), *Simulation and Modeling Methodologies, Technologies and Applications*, Advances in Intelligent Systems and Computing 256,
DOI: 10.1007/978-3-319-03581-9_23,

Approximation surrogates usually require a significant number of high-fidelity model evaluations to ensure decent accuracy. Furthermore, the number of samples typically grows exponentially with the number of design variables. On the other hand, approximation surrogates can be a basis of efficient global optimization techniques [3]. Various techniques of updating the training data set (so-called infill criteria [3]) have been developed that aim at obtaining global modeling accuracy, locating globally optimal design, or the trade-offs between the two, particularly in the context of kriging interpolation [3].

Physics-based surrogate models are not as versatile as approximation ones because they rely on an underlying low-fidelity model (a simplified description of the system under consideration), typically problem specific. The physics-based models can be obtained by a number of ways, such as by neglecting certain second-order effects, using simplified equations, or, which is probably the most versatile approach, by exploiting the same CFD solver as used to evaluate the high-fidelity model but with coarser mesh and/or relaxed convergence criteria (so called variable-resolution modeling) [13]. The physics-based surrogate models contain knowledge about the system of interest. Due to this, a limited amount of high-fidelity model data is necessary to ensure a required accuracy of the surrogate. For the same reason, these physics-based models have good generalization capabilities.

There have been proposed several SBO algorithms using physics-based surrogates in the literature, including the approximation and model management optimization (AMMO) [14], space mapping (SM) [15, 16], manifold mapping (MM) [17], and, more recently, the shape-preserving response prediction (SPRP) [18]. All of these methods differ in a specific way of how the low-fidelity model is used to construct the surrogate. Space mapping is probably the most popular approach of this kind. It was originally developed for simulation-driven design in microwave engineering [15] however, it is currently becoming more and more popular in other areas of engineering and science (cf. Refs. [15, 16] and references therein). Despite its potential, space mapping has not become popular in aerodynamic shape optimization. The only work reported so far is by Robinson et al. [19], where the so-called corrected SM was applied, among other methods, to airfoil design, however no significant design speed up has been reported.

This chapter describes a space mapping algorithm for the aerodynamic shape optimization of wings in transonic flow [20, 21]. The chapter gives the details of the CFD model and the optimization algorithm. A couple of numerical examples involving constrained lift maximization and drag minimization are used to demonstrate the algorithm usage and effectiveness.

2 Computational Fluid Dynamic Model

The CFD model is described in this section. In particular, the details of the governing equations, grid generation, and flow solver are provided. Furthermore, results of a grid convergence study are shown.

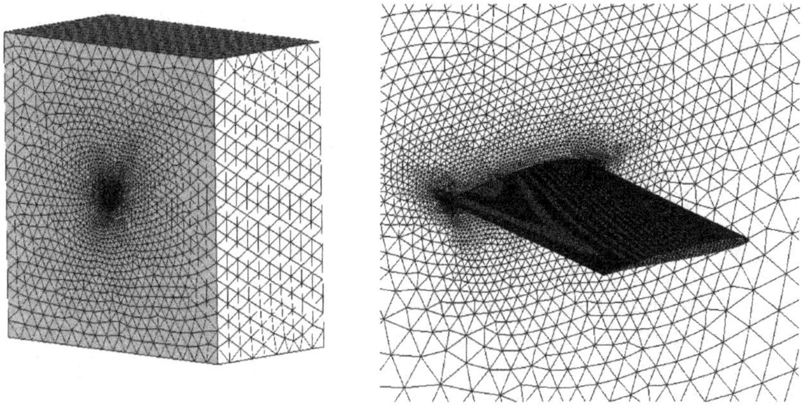

(a) (b)

Fig. 1. A view of of the computational grid, (a) the farfield grid, and (b) a close-up of the wing shell grid

2.1 Governing Equations

Commercial transport aircraft operate in the transonic flow regime where the flow is compressible. The flow is assumed to be steady, viscous, and without body forces, mass-diffusion, chemical reactions or external heat addition. The Reynolds-averaged Navier-Stokes equations are solved with the one equation Spalart-Allmaras turbulence model [22]. The air properties are modelled by the ideal gas law and the Sutherland law for dynamic viscosity.

2.2 Computational Grid

The farfield is configured in a box topology where the wing root airfoil is placed in the center of the symmetry plane with its leading edge placed at the origin $(x, y, z) = (0, 0, 0)$. The farfield extends 100 chord lengths in all directions from the wing, up-stream, above, below and aft of the wing where the maximum element size in the flow domain is 11 chord lengths.

An unstructured tri/tetra shell grid is created on all surfaces. The shell grid from the wing is then extruded into the volume where the volume is flooded with tri/tetra elements. The grid is made dense close to the wing where it then gradually grows in size as moving away from the wing surfaces. To capture the viscous boundary layer an inflation layer or a prism layer is created on the wing surfaces as well. In the stream-wise direction, the number of elements on the wing is set to 100 on both upper and lower surface.

The bi-geometric bunching law with a growth ratio of 1.2 is employed in the stream-wise direction over the wing to obtain a more dense element distribution at the leading edge and the trailing edge. This is done in order to capture the high pressure gradient at the leading edge and the separation at the trailing edge. The minimum element size of the wing in the stream-wise direction is set to $0.1\%c$, and it is located at the leading and trailing edge. In the span-wise direction elements are distributed uniformly and number of elements set to 100 over the semi-span. A prism layer is used to capture the viscous boundary layer. This layer consists of a number of structured elements that grow in size normal to the wing surface into the domain volume. The inflation layer has a initial height of $5 \times 10^{-6}c$ where it is grown 20 layers into the volume using a exponential growth law with ratio of 1.2. The initial layer height is chosen so that $y^+ < 1$ at all nodes on the wing. The resulting grid is shown in Fig. 1.

2.3 Flow Solver

An implicit density-based solver is applied using the Roe-FDS flux type. The numerical fluid flow simulations are performed using the computer code ANSYS FLUENT [23]. The spatial discretization schemes are set to second order for all variables, and the gradient information is found using the Green-Gauss node based method. The residuals, which are the sum of the L^2 norm of all governing equations in each cell, are monitored and checked for convergence. The convergence criterion for the high-fidelity model is such that a solution is considered to be converged if the residuals have dropped by six orders of magnitude, or the total number of iterations has reached 1000. To reflect the compressible nature of this problem, two types of boundaries are used. The pressure-farfield is applied to the boundary on all surfaces, except where the wing penetrates the symmetry boundary.

2.4 Grid Convergence

A grid convergence study is conducted using the ONERA M6 wing [24]. The flow past the ONERA M6 wing is simulated at various grid resolutions at $Re_{\infty,c_{mac}} = 11.72 \times 10^6$, $M_\infty = 0.8395$ and angle of attack $\alpha = 3.06°$, where c_{mac} is the mean aerodynamic chord length. The flow conditions are selected to match experimental flow conditions of an ONERA M6 wing experiment 2308 conducted by Schmitt, V. and F. Charpin [25].

The grid convergence study, shown in Fig. 2(a), revealed that 1,576,413 cells are needed for convergence in lift. The drag, however, can still be improved as evident from Fig. 2(a), where convergence has not been reached due to limitations in the computational resources. We proceed, however, with this grid as the high-fidelity model grid. The overall simulation time needed for one high-fidelity CFD simulation was around 223 minutes, as shown in Fig. 2(b), executed on four Intel-i7-2600 processors in parallel. This execution time is based on 1000 solver iterations, where the solver terminated due to the maximum number of iterations limit.

3 Optimization with Space Mapping

The wing design is carried out in a computationally efficient manner by exploiting the space mapping (SM) methodology [15]. Space mapping replaces the direct optimization of an expensive (high-fidelity or fine) airfoil model f obtained through high-fidelity CFD simulation, by iterative updating and re-optimization of a cheaper surrogate model

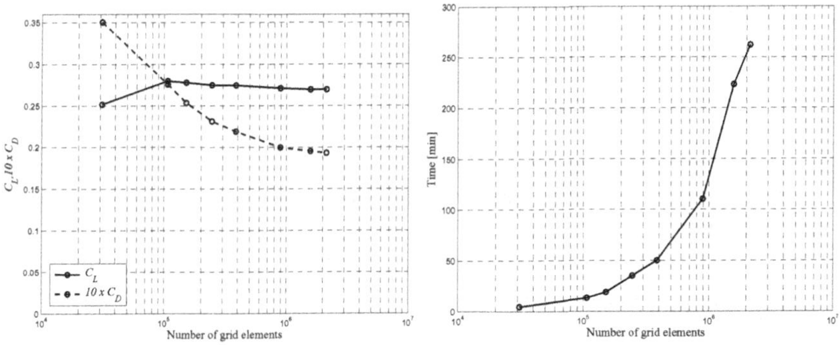

Fig. 2. Grid convergence study using the ONERA M6 wing at $Re_{\infty,c_{mac}} = 11.72 \times 10^6$, $M_\infty = 0.8395$ and angle of attack $\alpha = 3.06°$. (a) Lift (C_L) and drag (C_D) coefficients versus the number of grid elements, and (b) simulation time versus the number of grid elements.

s. The key component of SM is the physics-based low-fidelity (or coarse) model c that embeds certain knowledge about the system under consideration and allows us to construct a reliable surrogate using a limited amount of high-fidelity model data.

3.1 Optimization Problem

Simulation-driven design can be generally formulated as a nonlinear minimization problem

$$\mathbf{x}^* = \arg\min_{\mathbf{x}} H\left(f(\mathbf{x})\right), \tag{1}$$

where $\mathbf{x}$ is a vector of design parameters, f the high-fidelity model to be minimized at $\mathbf{x}$ and H is the objective function. $\mathbf{x}^*$ is the optimum design vector. The high-fidelity model will represent aerodynamic forces, such as the lift and drag coefficients, as well as other scalar responses such as cross-sectional area A of the wing at locations of interest. The cross-sectional area can be of a vector form $\mathbf{A}$ if one requires multiple area cross-sectional constraints at various locations on the wing, e.g., the wing root and the wing tip. The response will have to form

$$f(\mathbf{x}) = \left[C_{L.f}(\mathbf{x}) \quad C_{D.f}(\mathbf{x}) \quad A_f(\mathbf{x})\right]^T, \tag{2}$$

where $C_{L.f}$ and $C_{D.f}$ are the lift and drag coefficient for a three-dimensional wing, respectively, generated by the high-fidelity model. We are interested in the maximizing lift case, so the objective function will take the form of

$$H\left(f(\mathbf{x})\right) = -C_L, \tag{3}$$

the design constraints denoted as

$$C\left(f(\mathbf{x})\right) = \left[c_1\left(f(\mathbf{x})\right), \; \ldots \; , c_k\left(f(\mathbf{x})\right)\right]^T. \tag{4}$$

Maximizing lift will yield two nonlinear design constraints for drag and area,

$$c_1(f(\mathbf{x})) = C_{D.f}(\mathbf{x}) - C_{D.\max} \leq 0, \tag{5}$$

$$c_2(f(\mathbf{x})) = -A_f(\mathbf{x}) + A_{\min} \leq 0. \tag{6}$$

where $C_{D.\max}$ and $A_{\min}$ are the maximum allowable drag and minimum allowable cross-sectional area, respectively. The contrained drag minimization case is formulated in a similar fashion.

3.2 Space Mapping Basics

Starting from an initial design $\mathbf{x}^{(0)}$, a generic space mapping algorithm produces a sequence $\mathbf{x}^{(i)}, i = 0, 1 \ldots$ of approximate solutions to Eq. (1) as

$$\mathbf{x}^{(i+1)} = \arg\min_{\mathbf{x}} H\left(s^{(i)}(\mathbf{x})\right), \tag{7}$$

where

$$s^{(i)}(\mathbf{x}) = \left[C_{L.s}^{(i)}(\mathbf{x}) \quad C_{D.s}^{(i)}(\mathbf{x}) \quad A_s(\mathbf{x})^{(i)}\right]^T, \tag{8}$$

is the surrogate model at iteration i. As previously described, the accurate high-fidelity CFD model f is accurate but computationally expensive. Using space mapping, the surrogate s is a composition of the low-fidelity CFD model c and a simple linear transformation to correct the low-fidelity model response [15]. The corrected response is denoted as $s(\mathbf{x}, \mathbf{p})$, where $\mathbf{p}$ represents a set of model parameters and at iteration i the surrogate is

$$s^{(i)}(\mathbf{x}) = s(\mathbf{x}, \mathbf{p}). \tag{9}$$

The SM parameters $\mathbf{p}$ are determined through a parameter extraction (PE) process. In general, this process is a nonlinear optimization problem where the objective is to minimize the misalignment of surrogate response at some or all previous iteration high-fidelity model data points [15]. The PE optimization problem can be defined as

$$\mathbf{p}^{(i)} = \arg\min_{\mathbf{p}} \sum_{k=0}^{i} w_{i,k} \| f(\mathbf{x}^{(k)}) - s(\mathbf{x}^{(k)}, \mathbf{p}) \|^2, \tag{10}$$

where $w_{i,k}$ are weight factors that control how much impact previous iterations affect the SM parameters. Popular choices are

$$w_{i,k} = 1 \quad \forall i, k, \tag{11}$$

and

$$w_{i,k} = \begin{cases} 1 & k = i \\ 0 & \text{otherwise} \end{cases}. \tag{12}$$

In the first case, all previous SM iterations influence the parameters; in the second case, the parameters depend only on the most recent SM iteration.

3.3 Low-Fidelity CFD Model

The low-fidelity model c is constructed in the same way as the high-fidelity model f, but with a coarser grid discretization and with a relaxed convergence criteria - so-called variable-resolution modeling. Referring back to the grid study made in Section 2.4 and inspecting Fig. 2(a), a coarse low-fidelity model is selected. Based on time and accuracy with respect of lift and drag, the grid parameters that represent the second point from left are selected, giving 107,054 mesh elements. The time taken to evaluate the low-fidelity model is 13.2 minutes on four Intel-i7-2600 processors in parallel. The flow solver converged within 500 iterations. The maximum number of iterations for the low-fidelity model is therefore set to 500 iterations. This reduces the overall simulation time to 6.6 minutes. The ratio of simulation times of the high- and low- fidelity model in this case is high/low = $223/6.6 \approx 34$. This is based on the solver uses all 500 iterations in the low-fidelity model to obtain a solution.

The low-fidelity CFD model c turns out to be noisy. In order to alleviate the problem, a second order polynomial approximation model is constructed [5] using $n_c = 50$ training points sampled using latin hypercube sampling (LHS) [3] using the low-fidelity CFD model. The polynomial approximation model is defined as

$$\bar{c}(\mathbf{x}) = c_0 + \mathbf{c}_1^T \mathbf{x} + \mathbf{x}^T \mathbf{c}_2 \mathbf{x}, \tag{13}$$

where $\mathbf{c}_1 = [c_{1.1} \quad c_{1.2} \quad c_{1.3}]^T$ and $\mathbf{c}_2 = [c_{2.ij}]_{i,j=1,2,3}$. The coefficients c_0, $\mathbf{c}_1$, $\mathbf{c}_2$ are found by solving a linear regression problem

$$\bar{c}(\mathbf{x}^k) = c(\mathbf{x}^k), \tag{14}$$

where $k = 1, \ldots, n_c$. The resulting second order polynomial model $\bar{c}$ has nice analytical properties, such as smoothness and convexity.

3.4 Surrogate Model Construction

As mentioned above, the SM surrogate model s is a composition of the low-fidelity CFD model c and corrections or linear transformations where the model parameters $\mathbf{p}$ are extracted using one of the PE processes described above. The parameter extraction and the surrogate optimization create a certain overhead on the whole process and this overhead can be up to 80-90% of the computational cost. This is due to the fact that the physics-based low-fidelity models are in general relatively expensive to evaluate compared to the functional-based ones. Despite this, SM may be beneficial [26].

To alleviate this problem, the output SM with both multiplicative and additive response correction is exploited here with the surrogate model parameters extracted analytically. We use the following formulation

$$\begin{aligned} s^{(i)}(\mathbf{x}) &= \mathbf{A}^{(i)} \circ \bar{c}(\mathbf{x}) + \mathbf{D}^{(i)} + \mathbf{q}^{(i)} \\ &= \left[a_L^{(i)} C_{L.c}(\mathbf{x}) + d_L^{(i)} + q_L^{(i)} \quad a_D^{(i)} C_{D.c}(\mathbf{x}) + d_D^{(i)} + q_D^{(i)} \quad A_c(\mathbf{x}) \right]^T, \end{aligned} \tag{15}$$

where $\circ$ is a component-wise multiplication. No mapping is needed for the area $A_c(\mathbf{x})$ where, $A_c(\mathbf{x}) = A_f(\mathbf{x}) \quad \forall \mathbf{x}$ since low- and high-fidelity model represent the same geometry. Parameters $\mathbf{A}^{(i)}$ and $\mathbf{D}^{(i)}$ are obtained using

$$\left[\mathbf{A}^{(i)},\mathbf{D}^{(i)}\right] = \arg\min_{\mathbf{A},\mathbf{D}} \sum_{k=0}^{i} \| f\left(\mathbf{x}^{(k)}\right) - \mathbf{A}\circ\bar{c}\left(\mathbf{x}^{(k)}\right) + \mathbf{D}\|^2, \tag{16}$$

where $w_{i,k} = 1$, i.e., all the previous iteration points are used to improve globally the response of the low-fidelity model. The additive term $q^{(i)}$ is defined such that is ensures a perfect match between the surrogate and the high-fidelity model at design $\mathbf{x}^{(i)}$, namely $f\left(\mathbf{x}^{(i)}\right) = s\left(\mathbf{x}^{(i)}\right)$ or a zero-order consistency [14]. The additive term can be written as

$$q^{(i)} = f\left(\mathbf{x}^{(i)}\right) - \left[\mathbf{A}^{(i)}\circ\bar{c}(\mathbf{x}^{(i)}) + \mathbf{D}^{(i)}\right]. \tag{17}$$

Since an analytical solution exists for $\mathbf{A}^{(i)}, \mathbf{D}^{(i)}$ and $\mathbf{q}^{(i)}$ there is no need to perform non-linear optimization to solve Eq. (10) to obtain the parameters. $\mathbf{A}^{(i)}$ and $\mathbf{D}^{(i)}$ can be obtained by solving

$$\begin{bmatrix} a_L^{(i)} \\ d_L^{(i)} \end{bmatrix} = \left(\mathbf{C}_L^T\mathbf{C}_L\right)^{-1}\mathbf{C}_L^T\mathbf{F}_L\,, \tag{18}$$

$$\begin{bmatrix} a_D^{(i)} \\ d_D^{(i)} \end{bmatrix} = \left(\mathbf{C}_D^T\mathbf{C}_D\right)^{-1}\mathbf{C}_D^T\mathbf{F}_D\,, \tag{19}$$

where

$$\mathbf{C}_L = \begin{bmatrix} C_{L.c}(\mathbf{x}^{(0)}) & C_{L.c}(\mathbf{x}^{(1)}) & \ldots & C_{L.c}(\mathbf{x}^{(i)}) \\ 1 & 1 & \ldots & 1 \end{bmatrix}^T, \tag{20}$$

$$\mathbf{F}_L = \begin{bmatrix} C_{L.f}(\mathbf{x}^{(0)}) & C_{L.f}(\mathbf{x}^{(1)}) & \ldots & C_{L.f}(\mathbf{x}^{(i)}) \\ 1 & 1 & \ldots & 1 \end{bmatrix}^T, \tag{21}$$

$$\mathbf{C}_D = \begin{bmatrix} C_{D.c}(\mathbf{x}^{(0)}) & C_{D.c}(\mathbf{x}^{(1)}) & \ldots & C_{D.c}(\mathbf{x}^{(i)}) \\ 1 & 1 & \ldots & 1 \end{bmatrix}^T, \tag{22}$$

$$\mathbf{F}_D = \begin{bmatrix} C_{D.f}(\mathbf{x}^{(0)}) & C_{D.f}(\mathbf{x}^{(1)}) & \ldots & C_{D.f}(\mathbf{x}^{(i)}) \\ 1 & 1 & \ldots & 1 \end{bmatrix}^T, \tag{23}$$

which are the least-square optimal solutions to the linear regression problems

$$\mathbf{C}_L a_L^{(i)} + d_L^{(i)} = \mathbf{F}_L\,, \tag{24}$$

$$\mathbf{C}_D a_D^{(i)} + d_D^{(i)} = \mathbf{F}_D\,. \tag{25}$$

Note that $\mathbf{C}_L^T\mathbf{C}_L$ and $\mathbf{C}_D^T\mathbf{C}_D$ are non-singular for $i > 1$ and assuming that $\mathbf{x}^{(k)} \neq \mathbf{x}^{(i)}$ for $k \neq i$. For $i = 1$, only the multiplicative SM correction with $\mathbf{A}^{(i)}$ is used.

3.5 Optimization Algorithm

The optimization algorithm exploits the SM-based surrogate and a trust-region convergence safeguard [3]. The trust-region parameter λ is updated after each iteration. The optimization algorithm is as follows

1. Set $i = 0$; Select λ, the trust region radius; Evaluate the high-fidelity model at the initial solution, $f(\mathbf{x}^{(0)})$;
2. Using data from the low-fidelity model $\bar{c}$, and f at $\mathbf{x}^{(k)}, k = 0, 1, \ldots, i$, setup the SM surrogate $s^{(i)}$; Perform PE;
3. Optimize $s^{(i)}$ to obtain $x^{(i+1)}$;
4. Evaluate $f(\mathbf{x}^{(i+1)})$;
5. If $H(f(\mathbf{x}^{(i+1)})) < H(f(\mathbf{x}^{(i)}))$, accept $\mathbf{x}^{(i+1)}$; Otherwise set $\mathbf{x}^{(i+1)} = \mathbf{x}^{(i)}$;
6. Update λ;
7. Set $i = i + 1$;
8. If the termination condition is not satisfied, go to 2, else proceed;
9. End; Return $\mathbf{x}^{(i)}$ as the optimum solution.

The termination condition is set as $\|\mathbf{x}^{(i)} - \mathbf{x}^{(i-1)}\| < 10^{-3}$.

4 Numerical Examples

The optimization algorithm is demonstrated on two wing design cases at transonic conditions: lift maximization of a rectangular unswept wing, and a drag minimization of a swept wing. The surrogate model is optimized using the pattern-search algorithm [27].

4.1 Case 1: Unswept Wing

The wing is unswept and untwisted and is constructed by two NACA four-digit airfoils, located at the root and tip. The root airfoil shape is fixed to NACA 2412. The tip airfoil is to be designed. The initial design for the wing tip is chosen at random within the design space at the start of the optimization run. The normalized semi-wingspan is set as twice the wing chord length c, or $(b/2) = 2c$. All other wing parameters are kept fixed. The design vector can be written as $\mathbf{x} = [m\ p\ t/c]^T$, where the variables represent the wing tip NACA four-digit airfoil parameters.

The objective is to maximize the wing lift coefficient $C_{L.f}$, subject to constraints on the drag coefficient $C_{D.f} \leq C_{D.\max} = 0.03$ and the wing tip normalized cross-sectional area $A \geq A_{\min} = 0.01$. The bounds on the design variables are $0.02 \leq m \leq 0.03$, $0.7 \leq p \leq 0.9$ and $0.06 \leq t/c \leq 0.08$. The free-stream Mach number is $M_\infty = 0.8395$ and the angle of attack is fixed at $\alpha = 3.06°$. Two optimization runs were performed, denoted as Run 1 and Run 2. The numerical results are given in Table 1, and the initial and optimized airfoil cross-sections are shown in Fig. 3(a) and Fig. 3(b), respectively.

In Run 1, the lift is increased by +10% and the drag is pushed above its constraint at $C_{D,\max} = 0.03$, where the optimized drag coefficient is $C_D = 0.0311$. The drag constraint is violated slightly, or by +4%, which is within the 5% constraint tolerance band. The lift-to-drag ratio is decreased by -14%. The algorithm requires less than 10 equivalent high-fidelity model evaluations (Total Cost in Table 1), where 50 low-fidelity model evaluations (N_c) are used to create the approximation model and 8 high-fidelity model evaluations (N_f) the design iterations. It is evident that the optimized wing tip airfoil is thicker as the normalized cross-sectional area is increased by +26%, and the

Table 1. Numerical results of Case 1

	Run 1		Run 2	
Variable	Initial	Optimized	Initial	Optimized
m	0.0200	0.0200	0.0259	0.0232
p	0.7000	0.8725	0.8531	0.8550
t/c	0.0628	0.0793	0.0750	0.0600
C_L	0.2759	0.3047	0.3426	0.3388
C_D	0.0241	0.0311	0.0344	0.0307
C_L/C_D	11.4481	9.7974	9.9593	11.0358
A	0.0422	0.0534	0.0505	0.0404
N_c	-	50	-	50
N_f	-	8	-	7
Total Cost	-	< 10	-	< 9

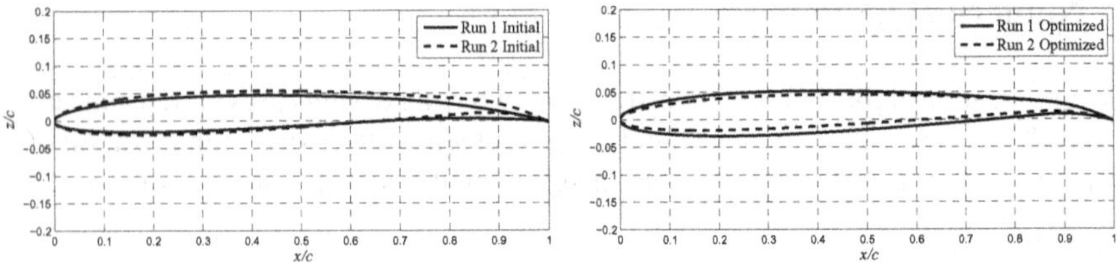

(a) Initial (b) Optimized

Fig. 3. Case 1 (a) initial and (b) optimized designs

increased drag can be related to the increment in area. No change is in the camber m, but the location of the maximum camber p has moved slightly aft.

The initial design for Run 2 violates the drag constraint. The proposed method is, however, able to push the drag to its constraint limit where the optimized drag coefficient is slightly violated (by +2%). While the drag is decreased by -11%, the lift is maintained and only drops by -1%. As a result, the lift-to-drag ratio is increased by +11%. The algorithm requires less than 9 high-fidelity model evaluations (50 low-fidelity model evaluations used to create the approximation model and 7 high-fidelity model evaluations). The optimized wing tip airfoil is thinner than the initial design (the normalized cross-sectional area is reduced by -20%). Little changes are made to the camber m and the maximum camber location p. Comparing runs 1 and 2, we note that although starting from different initial designs the optimized designs show similarities in two of three design variables, the maximum camber m and maximum camber location p. The third design variable, the airfoil thickness t/c, differs by approximately 2%.

Pressure coefficient surface contour plots of the initial and optimized designs for Run 1 are shown in Fig. 4. The shock on the mid wing has been moved aft on both the upper and the lower surfaces. Also, a second shock as formed near the tip on the upper surface. This causes the drag rise, as well as an increase in lift since the pressure distribution has opened up (as can be seen from Fig. 5).

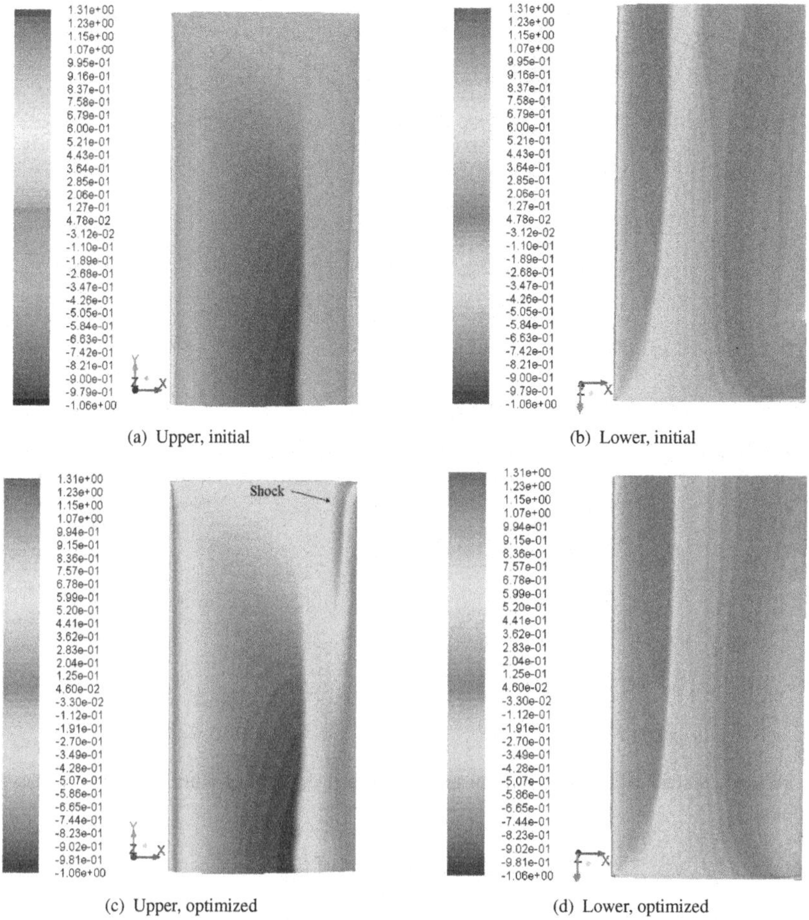

Fig. 4. Planform pressure coefficient contour plots for Run 1 of Case 1

4.2 Case 2: Swept Wing

An untwisted and swept wing is considered. The wing planform is fixed and set the same as the ONERA M6 wing [24], i.e., the wing planform properties are: semi-span $b/2 = 1.1963\ m$, root chord $c_r = 0.8059\ m$, taper ratio $\lambda = 0.562$, and leading-edge sweep $\Lambda = 30°$. The root section profile is fixed to a fit of the ONERA D airfoil section: NACA four-digit airfoil with $m = 0$, $p = 0$, and $t/c = 0.1019$. The tip section is to be designed and is parameterized by an NACA profile with the same design variables as in Case 1.

The objective is to minimize the wing drag coefficient $C_{D.f}$, subject to constraints on the lift coefficient $C_{L.f} \geq C_{L.\min} = 0.0265$, and the wing tip normalized cross-sectional area $A \geq A_{\min} = 0.05$. The bounds on the design variables are $0.018 \leq m \leq 0.03$, $0.6 \leq p \leq 0.8$ and $0.07 \leq t/c \leq 0.11$. The operating condition is the same as in Case 1.

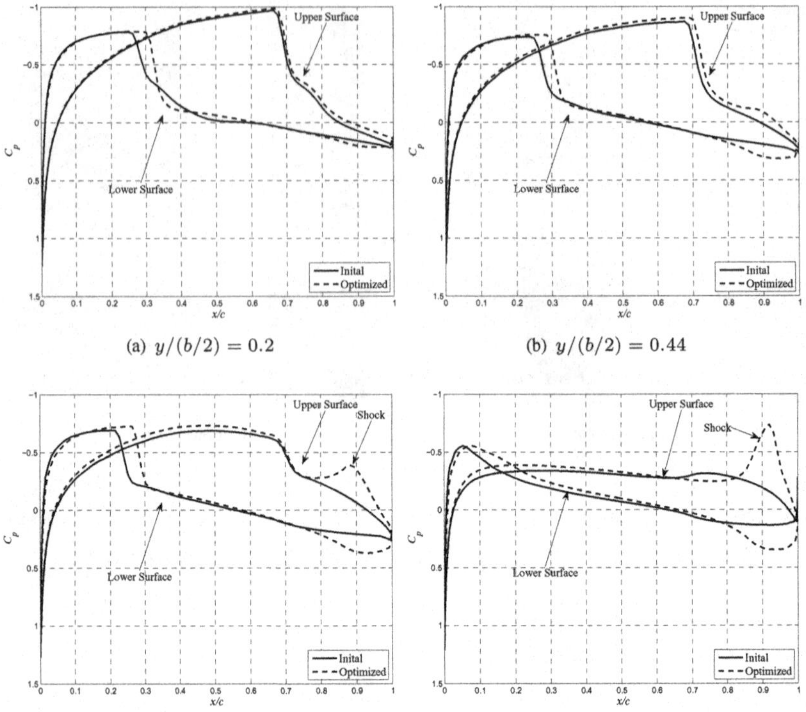

(a) $y/(b/2) = 0.2$ (b) $y/(b/2) = 0.44$

(c) $y/(b/2) = 0.65$ (d) $y/(b/2) = 0.95$

Fig. 5. Pressure distributions of the initial and optimized designs of Run 1 of Case 1 at a few spanstations

Table 2. Numerical results of Case 2

Variable	Initial	Optimized
m	0.0200	0.0180
p	0.8000	0.4701
t/c	0.1019	0.0742
C_L	0.3610	0.3202
C_D	0.0283	0.0227
C_L/C_D	12.7562	14.1057
A	0.0686	0.0500
N_c	-	71
N_f	-	3
Total Cost	-	< 5

Numerical results of the optimization are given in Table 2. The optimizer reduces the aifoil thickness (from 10.2% to 7.4%) to reduce the drag by 56 counts. The maximum camber (m) hits the lower bound and the location of maximum camber (p) is moved to approximately mid chord. As a result, the cross-sectional area constraint is active. However, the lift constraint is not active.

The optimization algorithm requires 71 low-fidelity model evaluations (N_c), and 3 high-fidelity ones (N_f), yielding a total computational cost of less than 5 equivalent high-fidelity model evaluations. The number of low-fidelity evaluations is due to: (1) 43 evaluations to obtain a starting point for the space mapping algorithm, and (2) a total of 28 samples to construct the local quadratic response surface.

5 Conclusions

A robust and efficient aerodynamic design optimization methodology for wings using high-fidelity CFD models has been described. The approach exploits a cheap surrogate model to obtain an approximate optimum design of an expensive high-fidelity computational fluid dynamic (CFD) model. The surrogate model is constructed using a corrected second-order polynomial approximation model derived from low-fidelity CFD model data. The correction is performed using the output space mapping technique. The space mapping correction is applied both to the objectives and the constraints, ensuring zero-order consistency and a perfect alignment between the surrogate and the high-fidelity model. The approach performs well and optimized designs are obtained using only a few high-fidelity model evaluations.

Acknowledgements. This work was funded by RANNIS, the Icelandic Research Fund for Graduate Students, grant ID: 110395-0061.

References

1. Raymer, D.: Aircraft design: a conceptual approach. American Institute of Aeronautics and Astronautics (2006)
2. Queipo, N., Haftka, R., Shyy, W., Goel, T., Vaidyanathan, R., Kevin Tucker, P.: Surrogate-based analysis and optimization. Progress in Aerospace Sciences 41, 1–28 (2005)
3. Forrester, A., Keane, A.: Recent advances in surrogate-based optimization. Progress in Aerospace Sciences 45, 50–79 (2009)
4. Wild, S., Regis, R., Shoemaker, C.: Orbit: Optimization by radial basis function interpolation in trust-regions. SIAM Journal on Scientific Computing 30, 3197–3219 (2008)
5. Koziel, S., Ciaurri, D.E., Leifsson, L.: Surrogate-based methods. In: Koziel, S., Yang, X.-S. (eds.) COMA. SCI, vol. 356, pp. 33–59. Springer, Heidelberg (2011)
6. Simpson, T., Poplinski, J., Koch, P., Allen, J.: Metamodels for computer-based engineering design: survey and recommendations. Engineering with Computers 17, 129–150 (2001)
7. Journel, A., Huijbregts, C.: Mining geostatistics. Academic Press, Number London [ua] (1978)
8. O'Hagan, A., Kingman, J.: Curve fitting and optimal design for prediction. Journal of the Royal Statistical Society. Series B (Methodological), 1–42 (1978)
9. Haikin, S.: Neural Networks: A Comprehensive Foundation. Prentice Hall (1998)
10. Minsky, M., Papert, S.: Perceptrons: An introduction to computational geometry. The MIT Press, Cambridge (1969)
11. Smola, A., Schölkopf, B.: A tutorial on support vector regression. Statistics and Computing 14, 199–222 (2004)

12. Søndergaard, J.: Optimization using surrogate models by the space mapping technique. PhD thesis, PhD. Thesis, Technical University of Denmark, Informatics and mathematical modelling (2003)
13. Leifsson, L., Koziel, S.: Variable-fidelity aerodynamic shape optimization. In: Koziel, S., Yang, X.-S. (eds.) COMA. SCI, vol. 356, pp. 179–210. Springer, Heidelberg (2011)
14. Alexandrov, N., Lewis, R.: An overview of first-order model management for engineering optimization. Optimization and Engineering 2, 413–430 (2001)
15. Bandler, J., Cheng, Q., Dakroury, S., Mohamed, A., Bakr, M., Madsen, K., Sondergaard, J.: Space mapping: the state of the art. IEEE Transactions on Microwave Theory and Techniques 52, 337–361 (2004)
16. Koziel, S., Cheng, Q., Bandler, J.: Space mapping. IEEE Microwave Magazine 9, 105–122 (2008)
17. Echeverria, D., Hemker, P.: Space mapping and defect correction. Computational Methods in Applied Mathematics 5, 107–136 (2005)
18. Koziel, S., Leifsson, L.: Airfoil shape optimization using variable-fidelity modeling and shape-preserving response prediction. In: Yang, X.-S., Koziel, S. (eds.) Computational Optimization and Applications. SCI, vol. 359, pp. 99–124. Springer, Heidelberg (2011)
19. Robinson, T., Willcox, K., Eldred, M., Haimes, R.: Multifidelity optimization for variable-complexity design. In: Proceedings of the 11th AIAA/ISSMO Multidisciplinary Analysis and Optimization Conference, Portsmouth, VA (2006)
20. Jonsson, E., Leifsson, L., Koziel, S.: Transonic wing optimization by variable-resolution modeling and space mapping. In: 2nd Int. Conf. on Simulation and Modeling Methodologies, Technologies, and Applications (SIMULTECH), Rome, Italy, July 28-31 (2012)
21. Jonsson, E., Leifsson, L., Koziel, S.: Aerodynamic optimization of wings by space mapping. In: 51st AIAA Aerospace Sciences Meeting Including the New Horizons Forum and Aerospace Exposition, Grapevine, Texas, January 7-10 (2013)
22. Tannehill, J., Anderson, D., Pletcher, R.: Computational fluid mechanics and heat transfer. Taylor & Francis Group (1997)
23. ANSYS: ANSYS FLUENT Theory Guide. ANSYS, Southpointe 275 Thecnology Drive Canonburg PA 15317. Release 13.0 edn. (2010)
24. NASA: Onera-m6-wing validation case (2008), `http://www.grc.nasa.gov/WWW/wind/valid/m6wing/m6wing.html`
25. Schmitt, V., Charpin, F.: Pressure distributions on the onera-m6-wing at transonic mach numbers. In: Experimental Data Base for Computer Program Assessment Report of the Fluid Dynamics Panel Working Group 04, AGARD AR 138 (May 1979)
26. Zhu, J., Bandler, J., Nikolova, N., Koziel, S.: Antenna optimization through space mapping. IEEE Transactions on Antennas and Propagation 55, 651–658 (2007)
27. Koziel, S.: Multi-fidelity multi-grid design optimization of planar microwave structures with sonnet. International Review of Progress in Applied Computational Electromagnetics, 719–724 (2010)

Efficient Design Optimization of Microwave Structures Using Adjoint Sensitivity

Slawomir Koziel, Leifur Leifsson, and Stanislav Ogurtsov

Engineering Optimization & Modeling Center, School of Science and Engineering, Reykjavik University, Menntavegur 1, Reykjavik, IS-101, Iceland
{koziel,leifurth,stanislav}@ru.is

Abstract. An important step of the microwave design process is the adjustment of geometry and material parameters of the structure under consideration to make it meet given performance requirements. Nowadays, it is typically conducted using full-wave electromagnetic (EM) simulations. Because accurate high-fidelity simulations are computationally expensive, automation of this process is quite challenging. In particular, the use of conventional numerical optimization algorithms may be prohibitive as these methods normally require a large number of objective function evaluations (and, consequently, EM simulations) to converge. The adjoint sensitivity technique that recently become available in commercial EM simulation software packages can be utilized to speed up the EM-driven design optimization process either by utilizing the sensitivity information in conventional gradient-based algorithms or by combining it with surrogate-based approaches. Here, several recent methods and algorithms for microwave design optimization using adjoint sensitivity are reviewed. We discuss advantages and disadvantages of these techniques and illustrate them through numerical examples.

Keywords: Computer-Aided Design (CAD), Simulation-driven Design, Microwave Design Optimization, Electromagnetic Simulation, Adjoint Sensitivity.

1 Introduction

Modern microwave engineering heavily relies on electromagnetic (EM) simulation. Simulations with discrete EM solvers are used for design verification as well as in the design process itself, e.g., for adjusting dimensions and/or material parameters of the structure under design. As a matter of fact accurate EM simulations of realistic microwave components are both CPU intensive and memory demanding. Speeding up computations is possible with parallelization (OpenMP, MPI, and GPU) and distributed computing. Nevertheless, the bottleneck of the EM-simulation-based optimization techniques remains to be the large number of simulations required by conventional optimization algorithms. Another issue is the present of numerical noise in simulation-based objective functions. Because of that noise local search methods often fail to find the optimal design and even fail to get a substantial improvement. Many commercial EM simulation packages have implemented basic design automation methods (mostly conventional gradient-based and derivative-free approaches such as Quasi-Newton or

M.S. Obaidat et al. (eds.), *Simulation and Modeling Methodologies, Technologies and Applications*, Advances in Intelligent Systems and Computing 256,
DOI: 10.1007/978-3-319-03581-9_24,

Nelder-Mead algorithms, population-based algorithms such as genetic algorithms), yet it is a common practice to each for a satisfactory design with tedious and time-consuming parameter sweeps (sweeping one parameter at a time) involving numerous simulations guided by the engineer.

Efficient simulation-driven design processes can be realized through surrogate based optimization (SBO). The most successful SBO techniques in microwave engineering include space mapping (SM) [1]-[3], simulation-based tuning [4], [5], manifold mapping (MM) [6], as well as shape-preserving response prediction [7]. These SBO techniques can be extremely efficient, nevertheless, they are not straightforward to automate and use them as reliable "push-button"-like approaches working for various microwave problems. SBO techniques typically require some experience [2]. Most of them are not globally convergent so that whether a satisfactory design is obtained or not may depend on a proper implementation as well as on certain knowledge particularly while constructing the surrogate model.

Another approach to improve efficiency of the simulation-driven design process is an utilization of adjoint sensitivity that allows obtaining derivative information of the system with little or no extra computational cost [8]-[12]. Until recently, adjoint sensitivities were not commercially available, which means that they were not available for most designers. Situation changed a few years ago when adjoint sensitivities were implemented for instance in CST Microwave Studio [13].

In this paper, we review several recent techniques that exploit adjoint sensitivity in order to speed up the EM-simulation-driven microwave design process. These techniques include gradient-based search methods embedded in trust region framework, as well as surrogate-based methods, specifically space mapping [2] and manifold mapping [6], enhanced by adjoint sensitivity in order to improve their convergence properties and reduce the computational cost of surrogate model optimization step. The efficiency of the presented approaches is demonstrated using several microwave design cases. Comparison with other optimization techniques, including Matlab's *fminimax* [14] and a Quasi-Newton type of algorithm [15] is also provided.

2 Optimization Using Trust Regions and Adjoint Sensitivities

In this section, we discuss gradient-based optimization of microwave structures accelerated through adjoint sensitivity. Our considerations are illustrated using the waveguide filter example.

2.1 Microwave Design Problem Formulation

The microwave design task can be formulated as a nonlinear minimization problem

$$\boldsymbol{x}_f^* \in \arg\min_{\boldsymbol{x}} U\left(\boldsymbol{R}_f(\boldsymbol{x})\right) \tag{1}$$

where $\boldsymbol{R}_f \in R^m$ denotes the response vector of a high-fidelity (or fine) model of the microwave structure under design evaluated through expensive high-fidelity EM simulation; $\boldsymbol{x} \in R^n$ is a vector of designable variables. Typically, these are geometry and/or material parameters. The response $\boldsymbol{R}_f(\boldsymbol{x})$ might be, e.g., the modulus of the transmission coefficient $|S_{21}|$ evaluated at m different frequencies. In some cases, $\boldsymbol{R}_f$ may consists of several vectors representing, e.g., filter reflection and transmission coefficients, or an antenna reflection coefficient, antenna gain, etc. U is a given scalar merit function, e.g., a norm, or a minimax function with upper and lower specifications. U is formulated so that a better design corresponds to a smaller value of U. $\boldsymbol{x}_f^*$ is the optimal design to be determined.

Direct solution of (1) using conventional algorithm may be prohibitive because it usually requires a large number of fine model evaluations, each of which is computationally expensive by itself. For many microwave structures, the evaluation time may be as long as a few hours.

2.2 Optimization Using Adjoint Sensitivity and Trust Regions

The algorithm proposed in [16] uses the 1st-order model (the surrogate) $\boldsymbol{S}(\boldsymbol{x})$ of the high-fidelity model $\boldsymbol{R}_f$. $\boldsymbol{S}(i)(\boldsymbol{x})$ is nothing else but a linear function, a first-order Taylor expansion of $\boldsymbol{R}f$ at $\boldsymbol{x}(i)$ of the form:

$$\boldsymbol{S}^{(i)}(\boldsymbol{x},\boldsymbol{x}^{(i)},\boldsymbol{R}_f(\boldsymbol{x}^{(i)}),J_{\boldsymbol{R}_f}(\boldsymbol{x}^{(i)}))=\boldsymbol{R}_f(\boldsymbol{x}^{(i)})+J_{\boldsymbol{R}_f}(\boldsymbol{x}^{(i)})\cdot(\boldsymbol{x}-\boldsymbol{x}^{(i)}) \tag{2}$$

$J_{Rf}(\boldsymbol{x})$ is an estimated Jacobian of $\boldsymbol{R}_f$ at $\boldsymbol{x}$, $J_{Rf}(\boldsymbol{x}) = [\partial \boldsymbol{R}_{fi}/\partial x_j]_{i=1,\ldots,m;\, j=1,\ldots,n}$, obtained using adjoint sensitivity (if available) or finite differentiation $\partial \boldsymbol{R}_{fi}/\partial x_j \cong [\boldsymbol{R}_{fi}([x_1 \ldots x_j+d_j \ldots x_n]^T) - \boldsymbol{R}_{fi}(\boldsymbol{x})]/d_j$ for all the other parameters.

The optimization algorithm framework is the following (r_0 is the initial trust region radius)

1. $i = 0$; $r = r_0$;
2. Optimize a linear model: $\boldsymbol{x}_{tmp} = \text{argmin}\{\|\boldsymbol{x} - \boldsymbol{x}^{(i)}\| \leq r: \boldsymbol{S}^{(i)}(\boldsymbol{x},\boldsymbol{x}^{(i)},\boldsymbol{R}_f(\boldsymbol{x}^{(i)}),J_F(\boldsymbol{x}^{(i)}))\}$;
3. Calculate gain ratio: $\rho = [U(\boldsymbol{R}_f(\boldsymbol{x}^{(i)}) - U(\boldsymbol{R}_f(\boldsymbol{x}_{tmp}))]/[U(\boldsymbol{R}_f(\boldsymbol{x}^{(i)}) - U(\boldsymbol{S}(\boldsymbol{x}_{tmp}))]$;
4. If $U(\boldsymbol{R}_f(x_{tmp})) < U(\boldsymbol{R}_f(x^{(i)}))$ then $\boldsymbol{x}^{(i+1)} = \boldsymbol{x}_{tmp}$; $i = i + 1$;
5. Update r: $\rho < r_{decr}$ then $r = r/m_{decr}$; else if $\rho > r_{incr}$ then $r = r \cdot m_{incr}$;
6. If termination condition is not satisfied, go to step 2; else, END.

Here, r_{decr} and r_{incr} denote threshold values for decreasing or increasing the trust region radius by the corresponding factors m_{decr} and m_{incr}. The algorithm is terminated if either of the following conditions is satisfied: $r < \varepsilon_r$, $n_F < n_{Fmax}$, $\|\boldsymbol{x}^{(i)} - \boldsymbol{x}^{(i-1)}\| < \varepsilon_x$, or $\|U(\boldsymbol{R}_f(\boldsymbol{x}^{(i)})) - U(\boldsymbol{R}_f(\boldsymbol{x}^{(i-1)}))\| < \varepsilon_F$, where ε_r, ε_x, ε_F, n_{Fmax}, are user defined parameters, whereas n_F is the number of high-fidelity model evaluations.

The response Jacobian is recalculated after each successful iteration (i.e., when $U(\boldsymbol{R}_f(\boldsymbol{x}_{tmp})) < U(\boldsymbol{R}_f(\boldsymbol{x}^{(i)}))$) for those variables where adjoint sensitivity is available. The finite-difference sensitivity is not recalculated as long as the new iteration is successful in order to reduce the number of high fidelity function evaluations.

The above algorithm is a local-search method. Assuming that the exact sensitivity of $\boldsymbol{R}_f$ at $\boldsymbol{x}^{(i)}$ is used to define the first-order model $\boldsymbol{S}^{(i)}$, $\boldsymbol{S}^{(i)}$ satisfies both zero- and first-order consistency conditions with the high-fidelity model $\boldsymbol{R}_f$, i.e., $\boldsymbol{S}^{(i)}(\boldsymbol{x}^{(i)}) = \boldsymbol{R}_f(\boldsymbol{x}^{(i)})$ and $J_S(\boldsymbol{x}^{(i)}) = J_{Rf}(\boldsymbol{x}^{(i)})$. This is sufficient for the global convergence of the algorithm at least to a local optimum of the high-fidelity model provided that $\boldsymbol{R}_f$ is sufficiently

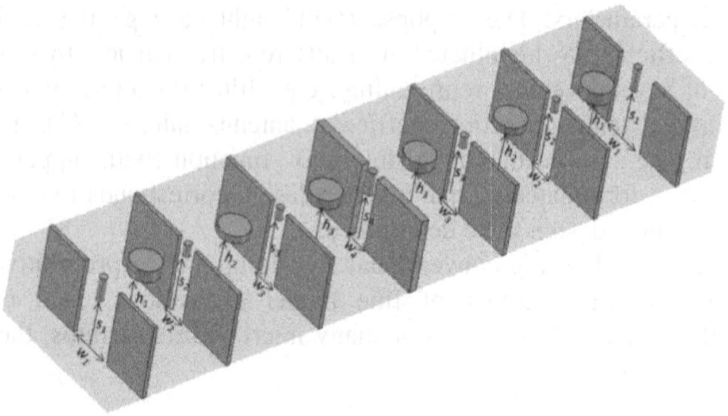

Fig. 1. Geometry of a waveguide bandpass filter

smooth [17]. In practice, the high-fidelity model is noisy; nevertheless, the performance of the algorithm is quite remarkable [16].

Responses obtained using EM solvers are inherently noisy (except, perhaps, when the mesh topology is fixed). The major reason is that the mesh topology is a discontinuous function of the design variables (or, more general, of geometry of the structure under considerations). The minor reason is that the evaluation process itself is noisy (e.g., due to finite tolerances used to terminate the EM simulation). This poses some problems for the optimization process. In particular, finite differentiation with small increments (e.g., 10^{-8}) will not work: the value of the derivative obtained this way will be completely unreliable, regardless of the model discretization density. The reason is that the change of the response due to the small perturbation of any given design variable will be, most likely, much smaller than the amplitude of the numerical noise. Therefore, for noisy functions, better and more consistent gradient estimation can be obtained using larger finite differentiation step sizes. Based on the above considerations, our algorithm uses relatively large steps for finite differentiation, typically, 10^{-3} or even larger (depending on the absolute values of the design variables).

2.3 Design Example: A Waveguide Bandpass Filter

Consider the waveguide filter shown in Fig. 1. Design variables are $[h_1\ h_2\ h_3\ s_1\ s_2\ s_3\ s_4\ w_1\ w_2\ w_3\ w_4]^T$. Design specifications are $|S_{11}| \leq -20$ dB for 667.5 MHz $\leq \omega \leq$ 675 MHz. The initial design is $\boldsymbol{x}^{(0)}$ = [163.5 172 165.3 160.5 160.5 160.5 130.5 –60.5 –29.5 –28.5 –27.5]^T (minimax specification error +19.2 dB). Optimization results are shown in Table 1. Figure 2 shows the filter responses at the initial design and at the optimized design found by the algorithm of Section 2.2. In this case, the algorithm performs substantially better than the methods used for comparison both with respect to the computational cost of the design and the quality of the final design. It should be noted that the computational complexity of our algorithm using finite-differences derivatives is comparable to that of Matlab's *fminimax*, even though the latter exploits adjoint sensitivity.

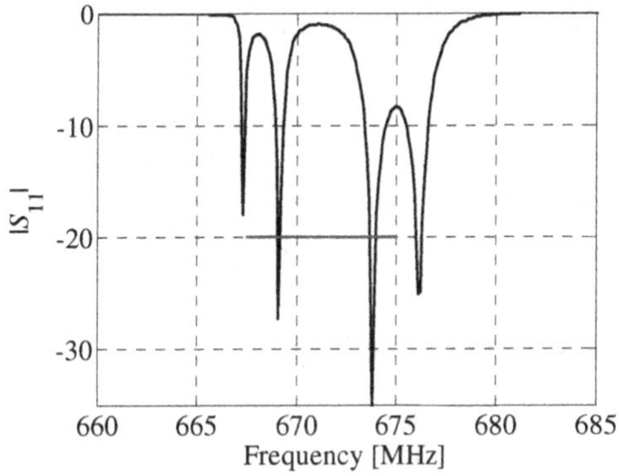

(a)

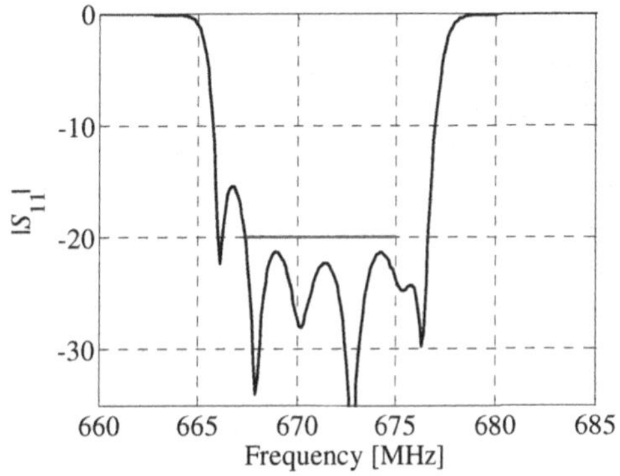

(b)

Fig. 2. Waveguide bandpass filter: (a) responses at the initial design $\boldsymbol{x}^{(0)}$; (b) responses at the optimized design found by the proposed algorithm using mixed adjoint and finite-difference sensitivities

Table 1. Optimization results for the waveguide bandpass filter

Optimization Algorithm		Final Specification Error	Number of Function Evaluations
Quasi-Newton optimizer		+5.3 dB	1454
Matlab's *fminimax*		+1.2 dB	88
This work (Algorithm of Section 2)	Adjoint sensitivity	–2.2 dB	16
	Mixed adjoint / finite-difference sensitivity*	–1.3 dB	46
	Finite-difference sensitivity	–0.4 dB	107

* Adjoint sensitivity for the first seven variables, finite-differences for the remaining four variables.

3 Optimization Using Adjoint Sensitivity and Surrogate Models

In this section we consider various ways to utilize adjoint sensitivity to speed up as well as improve the robustness of microwave design optimization exploiting surrogate models. We discuss optimization exploiting first order Taylor-expansion models as well as techniques based on space mapping and manifold mapping.

3.1 Surrogate-Based Optimization Basics

A generic surrogate-based optimization (SBO) algorithm [18], [19] generates a sequence of approximate solutions to (1), $\boldsymbol{x}^{(i)}$, as follows

$$\boldsymbol{x}^{(i+1)} = \arg\min_{\boldsymbol{x}} U\left(\boldsymbol{R}_s^{(i)}(\boldsymbol{x})\right) \tag{3}$$

where $\boldsymbol{R}_s^{(i)}$ is the surrogate model at iteration i. Here, $\boldsymbol{x}^{(0)}$ is the initial design. $\boldsymbol{R}_s^{(i)}$ is assumed to be a computationally cheap yet reliable representation of $\boldsymbol{R}_f$, particularly in the neighborhood of $\boldsymbol{x}^{(i)}$. Under these assumptions, the algorithm (3) is likely to produce a sequence of designs that quickly approach $\boldsymbol{x}_f^*$. Usually, $\boldsymbol{R}_f$ is only evaluated once per iteration (at every new design $\boldsymbol{x}^{(i+1)}$) for verification purposes and to obtain the data necessary to update (to correct) the surrogate model. Because of the low computational cost of the surrogate model, its optimization cost can usually be neglected. Thus the total optimization cost is determined by the evaluation of $\boldsymbol{R}_f$. The key point here is that the number of calls of $\boldsymbol{R}_f$ for a well performing surrogate-based algorithm is substantially smaller than for conventional optimization methods.

3.2 Robustness of Surrogate-Based Optimization: Trust Region Methods

Robustness of the surrogate-based optimization process (3) depends on the quality of the surrogate model $\boldsymbol{R}_s^{(i)}$. In general, in order to ensure convergence of the algorithm (3) to at least local optimum of the high-fidelity model, the first-order consistency conditions should to be met [20], i.e., one has to have $\boldsymbol{R}_s^{(i)}(\boldsymbol{x}^{(i)}) = \boldsymbol{R}_f(\boldsymbol{x}^{(i)})$ and $\boldsymbol{J}_{\boldsymbol{R}_s^{(i)}}(\boldsymbol{x}^{(i)}) = \boldsymbol{J}_{\boldsymbol{R}_f}(\boldsymbol{x}^{(i)})$, where $\boldsymbol{J}$ stands for the Jacobian of the respective model. Also, the process (3) has to be embedded in the trust-region (TR) framework [21], i.e.,

$$\boldsymbol{x}^{(i+1)} = \arg\min_{\boldsymbol{x}:\|\boldsymbol{x}-\boldsymbol{x}^{(i)}\|\le\delta^{(i)}} U(\boldsymbol{R}_s^{(i)}(\boldsymbol{x})) \tag{4}$$

where the TR radius $\delta^{(i)}$ is updated using classical rules [21]. In general, the SBO algorithm (4) can be successfully utilized without satisfying the aforementioned conditions [1], [2]. However, in these cases, the quality of the underlying low-fidelity model may be critical for performance (including the algorithm convergence) [22]. Also the optimum design may not be located an accurately.

Availability of cheap adjoint sensitivity [8], [13] makes it possible to satisfy consistency conditions in an easy way without excessive computations. Options exploiting this possibility are discussed in the next section.

3.3 SBO with First-Order Taylor Model and Trust Regions

The simplest way of exploiting adjoint sensitivity for antenna optimization is to use the following surrogate model for the SBO scheme (4):

$$\boldsymbol{R}_s^{(i)}(\boldsymbol{x}) = \boldsymbol{R}_f(\boldsymbol{x}^{(i)}) + \boldsymbol{J}_{\boldsymbol{R}_f}(\boldsymbol{x}^{(i)})\cdot(\boldsymbol{x}-\boldsymbol{x}^{(i)}) \tag{5}$$

where $\boldsymbol{J}_{Rf}$ is the Jacobian of $\boldsymbol{R}_f$ obtained using the adjoint sensitivity technique. The key point of the algorithm is to find a new design $\boldsymbol{x}^{(i)}$ and the updating process for the search radius $\delta^{(i)}$. Here, instead of the standard rules, we use the following strategy ($\boldsymbol{x}^{(i-1)}$ and $\delta^{(i-1)}$ are the previous design and the search radius, respectively):

1. For $\delta_k = k\cdot\delta^{(i-1)}$, $k = 0, 1, 2$, solve: $\boldsymbol{x}^k = \arg\min_{\boldsymbol{x}:\|\boldsymbol{x}-\boldsymbol{x}^{(i)}\|\le\delta_k} U(\boldsymbol{R}_s^{(i)}(\boldsymbol{x}))$. Note that $\boldsymbol{x}^0 = \boldsymbol{x}^{(i-1)}$.

   The values of δ_k and $U_k = U(\boldsymbol{R}_s^{(i)}(\boldsymbol{x}^k))$ are interpolated with a 2nd-order polynomial to find δ^* that gives the smallest (estimated) value of the specification error (δ^* is limited to $3\cdot\delta^{(i-1)}$). Set $\delta^{(i)} = \delta^*$.
2. Find a new design $\boldsymbol{x}^{(i)}$ by solving (4) with the current $\delta^{(i)}$.
3. Calculate the gain ratio $\rho = [U(\boldsymbol{R}_f(\boldsymbol{x}^{(i)})) - U_0]/\ [U(\boldsymbol{R}_s^{(i)}(\boldsymbol{x}^{(i)})) - U_0]$; If $\rho < 0.25$ then $\delta^{(i)} = \delta^{(i)}/3$; else if $\rho > 0.75$ then $\delta^{(i)} = 2\cdot\delta^{(i)}$;
4. If $\rho < 0$ go to 2;
5. Return $\boldsymbol{x}^{(i)}$ and $\delta^{(i)}$;

The trial points $\boldsymbol{x}^k$ are used to find the best value of the search radius, which is further updated based on the gain ratio ρ which is the actual versus the expected objective function improvement. If the new design is worse than the previous one, the search radius is reduced to find $\boldsymbol{x}^{(i)}$ again, which eventually will bring the improvement of U because the models $\boldsymbol{R}_s^{(i)}$ and $\boldsymbol{R}_f$ are first-order consistent [20]. This precaution is necessary because the procedure in Step 1 only gives an estimation of the search radius.

As an example, consider a wideband hybrid antenna [23] shown in Fig. 3, which is a quarter-wavelength monopole loaded by a dielectric ring resonator. The design goal is to have $|S_{11}| \le -20$ dB for 8-to-13 GHz. The design variables are $\boldsymbol{x} = [h_1\ h_2\ r_1\ r_2\ g]^T$. The initial design is $\boldsymbol{x}^{(0)} = [2.5\ 9.4\ 2.3\ 3.0\ 0.5]^T$ mm. Other parameters are fixed. The final design with the proposed algorithm is $\boldsymbol{x}^{(0)} = [3.94\ 10.01\ 2.23\ 3.68\ 0.0]^T$ mm. Table 2 and Fig. 4 compare the design cost and quality of the final design found by the algorithm described above and Matlab's *fminimax*. It can be observed that our algorithm yields better design at significantly smaller computational cost (75 percent design time reduction).

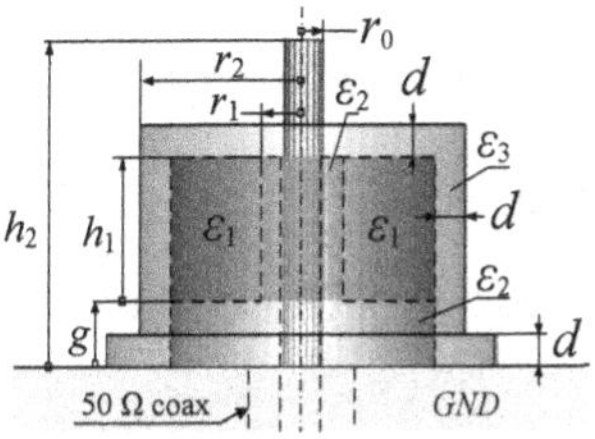

Fig. 3. Wideband hybrid antenna: geometry

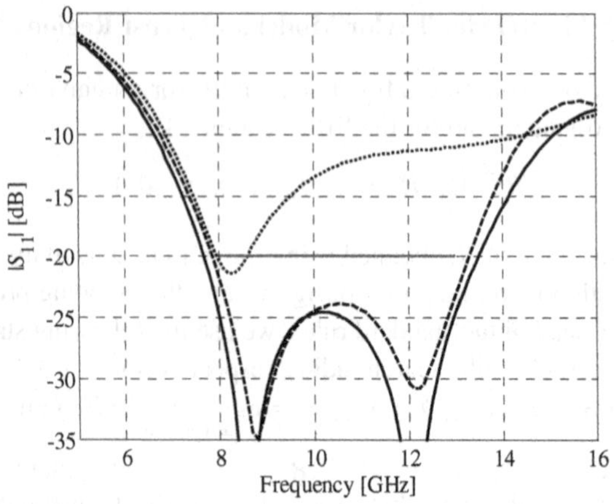

Fig. 4. Wideband hybrid antenna: reflection response at the initial design (· · ·), at the final design by Matlab's *fminimax* (- - -), and by the proposed algorithm (—)

Table 2. Wideband Hybrid Antenna: Design Results

Algorithm	max$\|S_{11}\|$ for 8 to 13 GHz at Final Design	Design Cost (Number of EM Analyses)
Matlab's *fminimax*	–22.6 dB	98
This work	–24.6 dB	24

3.4 Optimization Using Space Mapping and Manifold Mapping

Construction of the surrogate model can also be based on the underlying low-fidelity model $\boldsymbol{R}_c$ obtained from coarse-discretization simulation data. Two methods considered here that use this approach are space mapping (SM) [2] and manifold mapping (MM) [6]. Usually, some knowledge about the component under design is embedded in the low-fidelity model allows us to reduce the number of high-fidelity model evaluations necessary to find an optimum design.

The SM surrogate is constructed using input and output SM [1] of the form:

$$\boldsymbol{R}_s^{(i)}(\boldsymbol{x}) = \boldsymbol{R}_c(\boldsymbol{x} + \boldsymbol{c}^{(i)}) + \boldsymbol{d}^{(i)} + \boldsymbol{E}^{(i)}(\boldsymbol{x} - \boldsymbol{x}^{(i)}) \tag{6}$$

Here, only the input SM vector $\boldsymbol{c}^{(i)}$ is obtained through the nonlinear parameter extraction process

$$\boldsymbol{c}^{(i)} = \arg\min_{\boldsymbol{c}} \| \boldsymbol{R}_f(\boldsymbol{x}^{(i)}) - \boldsymbol{R}_c(\boldsymbol{x}^{(i)} + \boldsymbol{c}) \| \tag{7}$$

Output SM parameters are calculated as

$$\boldsymbol{d}^{(i)} = \boldsymbol{R}_f(\boldsymbol{x}^{(i)}) - \boldsymbol{R}_c(\boldsymbol{x}^{(i)} + \boldsymbol{c}^{(i)}) \tag{8}$$

and

$$\boldsymbol{E}^{(i)} = \boldsymbol{J}_{\boldsymbol{R}_f}(\boldsymbol{x}^{(i)}) - \boldsymbol{J}_{\boldsymbol{R}_c}(\boldsymbol{x}^{(i)} + \boldsymbol{c}^{(i)}) \tag{9}$$

Formulation (6)-(9) ensures zero- and first-order consistency [20] between the surrogate and the fine model.

The manifold mapping (MM) surrogate model is defined as [6]

$$\boldsymbol{R}_s^{(i)}(\boldsymbol{x}) = \boldsymbol{R}_f(\boldsymbol{x}^{(i)}) + \boldsymbol{S}^{(i)}\left(\boldsymbol{R}_c(\boldsymbol{x}) - \boldsymbol{R}_c(\boldsymbol{x}^{(i)})\right) \quad (10)$$

where $\boldsymbol{S}^{(i)}$ is the $m \times m$ correction matrix defined as

$$\boldsymbol{S}^{(i)} = \boldsymbol{J}_{\boldsymbol{R}_f}(\boldsymbol{x}^{(i)}) \cdot \boldsymbol{J}_{\boldsymbol{R}_c}(\boldsymbol{x}^{(i)})^{\dagger} \quad (11)$$

The pseudoinverse, denoted by †, is defined as

$$\boldsymbol{J}_{\boldsymbol{R}_c}{}^{\dagger} = \boldsymbol{V}_{\boldsymbol{J}_{\boldsymbol{R}_c}} \boldsymbol{\Sigma}_{\boldsymbol{J}_{\boldsymbol{R}_c}}^{\dagger} \boldsymbol{U}_{\boldsymbol{J}_{\boldsymbol{R}_c}}^T \quad (12)$$

where $\boldsymbol{U}_{JRc}$, Σ_{JRc}, and $\boldsymbol{V}_{JRc}$ are the factors in the singular value decomposition of $\boldsymbol{J}_{\boldsymbol{R}_c}$. The matrix $\boldsymbol{\Sigma}_{JRc}{}^{\dagger}$ is the result of inverting the nonzero entries in $\boldsymbol{\Sigma}_{JRc}$, leaving the zeroes invariant [6]. Using the sensitivity data as in (12) ensures that the surrogate model (10) is first-order consistent with the fine model. In our implementation, the coarse model is preconditioned using input space mapping of the form (7) in order to improve its initial alignment with the fine model.

Both the parameter extraction (7) and surrogate model optimization processes (4) are implemented by exploiting adjoint sensitivity data of the low-fidelity model, which allows for further cost savings. More details of the implementations can be found in [24].

As an illustration of operation and performance of the SM and MM algorithms, consider an ultra-wideband (UWB) antenna shown in Fig. 5. The antenna and its models include: a microstrip monopole, housing, edge mount SMA connector, section of the feeding coax. The design variables are $\boldsymbol{x} = [l_1\ l_2\ l_3\ w_1]^T$. Simulation time of the low-fidelity model $\boldsymbol{R}_c$ (156,000 mesh cells) is 1 min, and that of the high-fidelity model $\boldsymbol{R}_f$ (1,992,060 mesh cells) is 40 min (both at the initial design). Both models are simulated with the transient solver of CST Microwave Studio [13]. The design specifications for reflection are $|S_{11}| \le -12$ dB for 3.1 GHz to 10.6 GHz. The initial design is $\boldsymbol{x}^{init} = [20\ 2\ 0\ 25]^T$ mm.

The antenna was optimized using the SBO algorithm (4) with both the SM and MM surrogate models. Fig. 6(a) shows the responses of $\boldsymbol{R}_f$ and $\boldsymbol{R}_c$ at $\boldsymbol{x}^{init}$. Fig. 6(b) shows the response of the high-fidelity model at the final design $\boldsymbol{x}^{(2)} = [20.22\ 2.43\ 0.128\ 19.48]^T$ ($|S_{11}| \le -12.5$ dB for 3.1 to 10.6 GHz) obtained after only two SBO iterations with the MM surrogate, i.e. only 4 evaluations of the high-fidelity model (Table 3). The number of function evaluations is larger than the number of MM iterations because some designs have been rejected by the TR mechanism. The algorithm using the SM surrogate required three iterations and the final design is $\boldsymbol{x}^{(3)} = [20.29\ 2.27\ 0.058\ 19.63]^T$ ($|S_{11}| \le -12.8$ dB for 3.1 to 10.6 GHz) obtained after three SM iterations. The total optimization cost (Table 4) is equivalent to around 6 evaluations of the fine model. Figure 7 shows the evolution of the specification algorithm for the manifold mapping algorithm.

As another example, consider a third-order Chebyshev bandpass filter [25] shown in Fig. 8. The design variables are $\boldsymbol{x} = [L_1\ L_2\ S_1\ S_2]^T$ mm. Other parameters are: $W_1 = W_2 = 0.4$ mm. Both fine (396,550 mesh cells, evaluation time 45 min) and coarse (82,350 mesh cells, evaluation time 1 min) models are evaluated by the CST MWS transient solver [13].

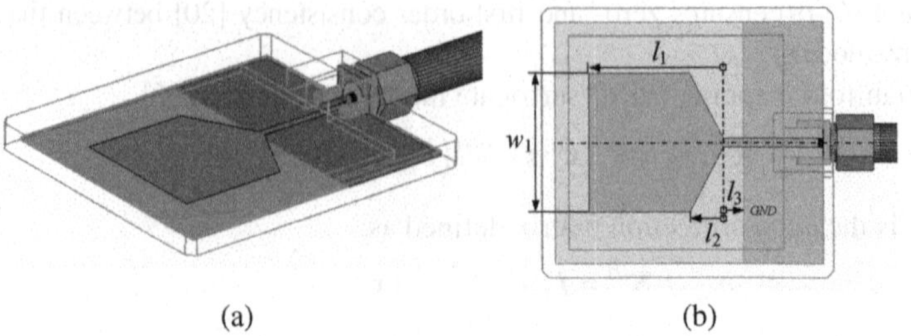

Fig. 5. UWB antenna: (a) 3D view; (b) top view. The housing is shown transparent.

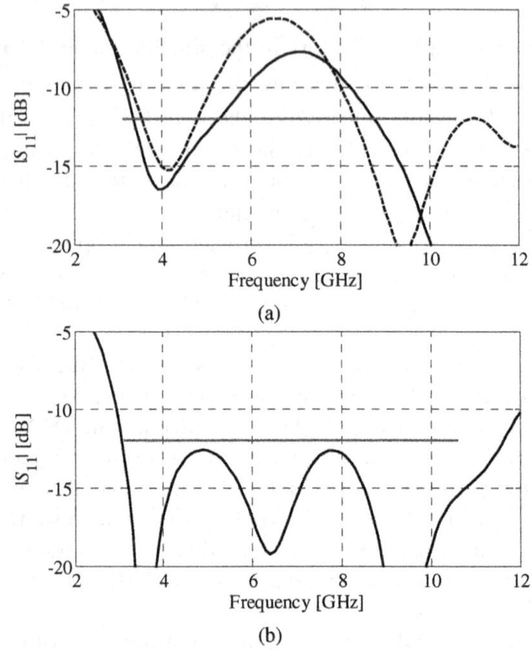

Fig. 6. UWB antenna optimized using the manifold mapping algorithm: (a) responses of $\boldsymbol{R}_f$ (—) and $\boldsymbol{R}_c$ (- - -) at the initial design $\boldsymbol{x}^{init}$; (b) response of $\boldsymbol{R}_f$ (—) at the final design

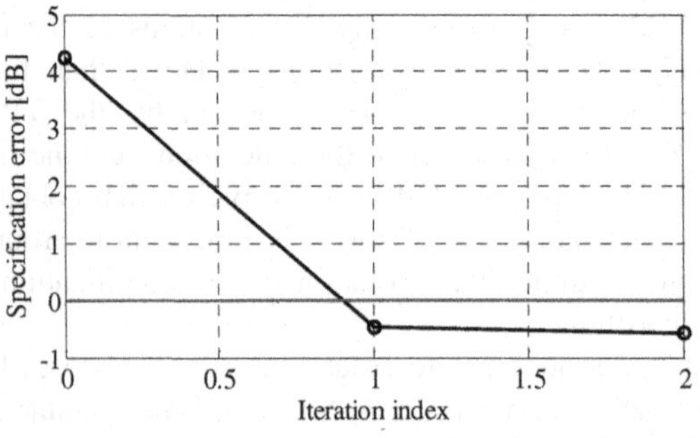

Fig. 7. UWB antenna: Minimax specification error versus manifold mapping algorithm iteration index

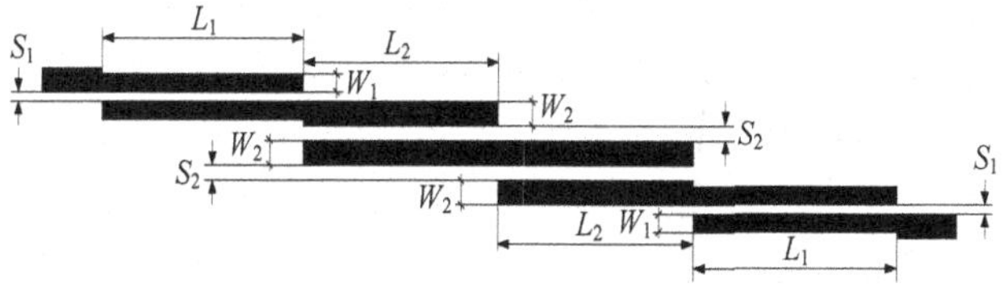

Fig. 8. Third-order Chebyshev bandpass filter: geometry

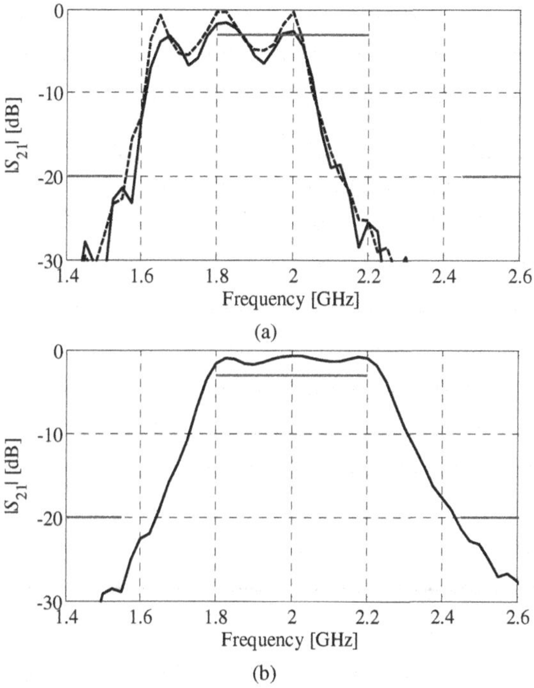

Fig. 9. Third-order Chebyshev filter: (a) responses of $\boldsymbol{R}_f$ (—) and $\boldsymbol{R}_c$ (- - -) at the initial design $\boldsymbol{x}^{init}$; (b) response of $\boldsymbol{R}_f$ (—) at the final design

The design specifications are $|S_{21}| \geq -3$ dB for 1.8 GHz $\leq \omega \leq$ 2.2 GHz, and $|S_{21}| \leq -$20 dB for 1.0 GHz $\leq \omega \leq$ 1.55GHz and 2.45 GHz $\leq \omega \leq$ 3.0 GHz. The initial design is $\boldsymbol{x}^{init} = [16\ 16\ 1\ 1]^T$ mm.

The filter was optimized using the SM algorithm. Optimization results are shown in Fig. 9 and Table 5. The final design $\boldsymbol{x}^{(5)} = [14.58\ 14.57\ 0.93\ 0.56]^T$ is obtained after five SM iterations. As before, the optimization cost is very low. Also, thanks to sensitivity information as well as trust region, the algorithm improves the specification error at each iteration, see Fig. 10. Such an algorithm behavior is not necessarily the case for conventional space mapping [1].

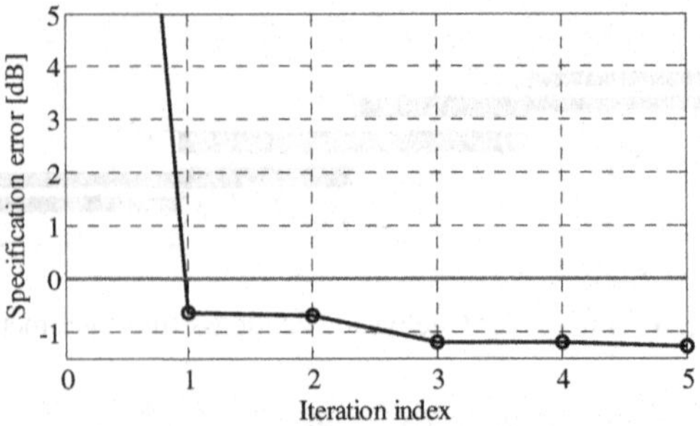

Fig. 10. Third-order Chebyshev filter: minimax specification error versus SM iteration index

Table 3. UWB Antenna: Optimization Results Using Manifold Mapping

Algorithm Component	Number of Model Evaluations*	CPU Time	
		Absolute	Relative to $\boldsymbol{R}_f$
Evaluation of $\boldsymbol{R}_c$	31	31 min	0.8
Evaluation of $\boldsymbol{R}_f$	4	120 min	4.0
Total cost*	N/A	151 min	**4.8**

* Includes $\boldsymbol{R}_f$ evaluation at the initial design.

Table 4. UWB Antenna: Optimization Results Using Space Mapping

Algorithm Component	Number of Model Evaluations*	CPU Time	
		Absolute	Relative to $\boldsymbol{R}_f$
Evaluation of $\boldsymbol{R}_c$	45	45 min	1.1
Evaluation of $\boldsymbol{R}_f$	5	200 min	5.0
Total cost*	N/A	205 min	**6.1**

* Includes $\boldsymbol{R}_f$ evaluation at the initial design.

Table 5. Third-order Chebyshev Filter: Optimization Results Using Space Mapping

Algorithm Component	Number of Model Evaluations*	CPU Time	
		Absolute	Relative to $\boldsymbol{R}_f$
Evaluation of $\boldsymbol{R}_c$	67	67 min	1.5
Evaluation of $\boldsymbol{R}_f$	6	270 min	6.0
Total cost*	N/A	337 min	**7.5**

* Includes ***Rf*** evaluation at the initial design.

4 Conclusions

A review of recently developed microwave design optimization techniques exploiting adjoint sensitivity has been presented. It was demonstrated that by exploiting cheap derivative information, the EM-simulation-driven design process can be performed efficiently and in a robust way. Adjoint sensitivity can also be used to improve performance of the surrogate-based optimization algorithm as illustrated on the example of space mapping and manifold mapping techniques.

References

1. Bandler, J.W., Cheng, Q.S., Dakroury, S.A., Mohamed, A.S., Bakr, M.H., Madsen, K., Søndergaard, J.: Space mapping: the state of the art. IEEE Trans. Microwave Theory Tech. 52(1), 337 (2004)
2. Koziel, S., Cheng, Q.S., Bandler, J.W.: Space mapping. IEEE Microwave Magazine 9(6), 105–122 (2008)
3. Amari, S., LeDrew, C., Menzel, W.: Space-mapping optimization of planar coupled-resonator microwave filters. IEEE Trans. Microwave Theory Tech. 54(5), 2153–2159 (2006)
4. Swanson, D., Macchiarella, G.: Microwave filter design by synthesis and optimization. IEEE Microwave Magazine 8(2), 55–69 (2007)
5. Rautio, J.C.: Perfectly calibrated internal ports in EM analysis of planar circuits. In: IEEE MTT-S Int. Microwave Symp. Dig., Atlanta, GA, pp. 1373–1376 (2008)
6. Echeverria, D., Hemker, P.W.: Space mapping and defect correction. CMAM The International Mathematical Journal Computational Methods in Applied Mathematics 5(2), 107–136 (2005)
7. Koziel, S.: Shape-preserving response prediction for microwave design optimization. IEEE Trans. Microwave Theory and Tech. 58(11), 2829–2837 (2010)
8. Nair, D., Webb, J.P.: Optimization of microwave devices using 3-D finite elements and the design sensitivity of the frequency response. IEEE Trans. Magn. 39(3), 1325–1328 (2003)
9. El Sabbagh, M.A., Bakr, M.H., Nikolova, N.K.: Sensitivity analysis of the scattering parameters of microwave filters using the adjoint network method. Int. J. RF and Microwave Computer-Aided Eng. 16(6), 596–606 (2006)
10. Kiziltas, G., Psychoudakis, D., Volakis, J.L., Kikuchi, N.: Topology design optimization of dielectric substrates for bandwidth improvement of a patch antenna. IEEE Trans. Antennas Prop. 51(10), 2732–2743 (2003)
11. Uchida, N., Nishiwaki, S., Izui, K., Yoshimura, M., Nomura, T., Sato, K.: Simultaneous shape and topology optimization for the design of patch antennas. In: European Conf. Antennas Prop., pp. 103–107 (2009)
12. Bakr, M.H., Ghassemi, M., Sangary, N.: Bandwidth enhancement of narrow band antennas exploiting adjoint-based geometry evolution. In: IEEE Int. Symp. Antennas Prop., pp. 2909–2911 (2011)
13. CST Microwave Studio, 2012. CST AG, Bad Nauheimer Str. 19, D-64289 Darmstadt, Germany (2012)
14. Matlab™, Version 7.6, The MathWorks, Inc., 3 Apple Hill Drive, Natick, MA 01760-2098 (2008)
15. Nocedal, J., Wright, S.J.: Numerical Optimization. Springer Series in Operations Research. Springer (2000)
16. Koziel, S., Mosler, F., Reitzinger, S., Thoma, P.: Robust microwave design optimization using adjoint sensitivity and trust regions. Int. J. RF and Microwave CAE 22(1), 10–19 (2012)
17. Alexandrov, N.M., Dennis, J.E., Lewis, R.M., Torczon, V.: A trust region framework for managing use of approximation models in optimization. Struct. Multidisciplinary Optim. 15(1), 16–23 (1998)
18. Koziel, S., Yang, X.-S. (eds.): Computational Optimization, Methods and Algorithms. SCI, vol. 356. Springer, Heidelberg (2011)
19. Forrester, A.I.J., Keane, A.J.: Recent advances in surrogate-based optimization, Prog. Aerospace Sciences 45(1-3), 50–79 (2009)

20. Alexandrov, N.M., Lewis, R.M.: An overview of first-order model management for engineering optimization. Optimization Eng. 2(4), 413–430 (2001)
21. Conn, A.R., Gould, N.I.M., Toint, P.L.: Trust Region Methods. MPS-SIAM Series on Optimization (2000)
22. Koziel, S., Bandler, J.W., Madsen, K.: Quality assessment of coarse models and surrogates for space mapping optimization. Optimization and Engineering 9(4), 375–391 (2008)
23. Petosa, A.: Dielectric Resonator Antenna Handbook. Artech House (2007)
24. Koziel, S., Ogurtsov, S., Bandler, J.W., Cheng, Q.S.: Robust space mapping optimization exploiting EM-based models with adjoint sensitivity. In: IEEE MTT-S Int. Microwave Symp. Dig (2012)
25. Kuo, J.T., Chen, S.P., Jiang, M.: Parallel-coupled microstrip filters with over-coupled end stages for suppression of spurious responses. IEEE Microwave and Wireless Comp. Lett. 13(10), 440–442 (2003)

Author Index